Gas–Solid Reactions

GAS–SOLID REACTIONS

Julian Szekely

DEPARTMENT OF MATERIALS SCIENCE AND ENGINEERING
MASSACHUSETTS INSTITUTE OF TECHNOLOGY
CAMBRIDGE, MASSACHUSETTS

James W. Evans

DEPARTMENT OF MATERIALS SCIENCE AND ENGINEERING
UNIVERSITY OF CALIFORNIA
BERKELEY, CALIFORNIA

Hong Yong Sohn

DEPARTMENT OF MINING, METALLURGICAL, AND FUELS ENGINEERING
UNIVERSITY OF UTAH
SALT LAKE CITY, UTAH

ACADEMIC PRESS   New York   San Francisco   London   1976

A Subsidiary of Harcourt Brace Jovanovich, Publishers

ACADEMIC PRESS, INC.
111 Fifth Avenue, New York, New York 10003

United Kingdom Edition published by
ACADEMIC PRESS, INC. (LONDON) LTD.
24/28 Oval Road, London NW1

Library of Congress Cataloging in Publication Data

Szekely, Julian,   Date
    Gas-solid reactions.

    Bibliography: p.
    1.   Chemical reaction, Conditions and laws of.
2.   Gases.   3.   Solids.   I.   Evans, James William,
    Date   joint author.   II.   Sohn, Hong Yong, joint
author.   III.   Title.
QD501.S948   541'.39   75-30476
ISBN 0–12–680850–3

登 記 證：內 版 台 業 字 第 1423 號
發 行 人：楊　　　明　　　德
　　　　　台北市襄陽路 13～2 號
發 行 所：高 立 圖 書 有 限 公 司
　　　　　台北市襄陽路 13～2 號
電　話：3 6 1 5 3 3 0 號
郵政劃撥帳戶 105614 號
印 刷 所：東 雅 印 製 廠 有 限 公 司
　　　　　台北市西藏路52-530號
中華民國72年　 月　 日

TO THE MEMORY OF
Joseph A. Bergantz

Contents

Chapter 4   Reactions of Porous Solids

Chapter 5   Reactions between Solids Proceeding through Gaseous Intermediates

Chapter 6   Experimental Techniques for the Study of Gas–Solid Reactions

Preface

In recent years considerable advances have been made in our understanding of gas–solid reaction systems. These advances are due in part to the development of more sophisticated mathematical models in which account is taken of such structural effects as pore size, grain size, and pore diffusion. Another important contributory factor has been the use of more sophisticated experimental techniques such as electron microscopy, X-ray diffraction, and porosimetry, which together with pore diffusion measurements provide information on the key structural parameters and make possible the critical assessment of this new generation of models.

These new developments were motivated to a great extent by the societal and economic importance of gas–solid reaction systems due to their relevance for a broad range of processing operations, including iron oxide reduction, the combustion of solid fuels, the desulfurization of the fuel gases, and the incineration of solid wastes.

The purpose of this monograph is to present in an integrated form a description of gas–solid reaction systems, where full account is taken of these new developments and where structural models of single particle systems, experimental techniques, the interpretation of measurements, the design of gas–solids contacting systems, and practical applications are treated in a unified manner.

The actual approach to be developed here is based on methodology similar to that employed in chemical reaction engineering in the interpretation of rate data and the design of process systems in heterogeneous catalysis. More specifically, through the use of this methodology, the individual components of the overall reaction sequence are studied and examined in isolation and the description of the system is then synthesized from these components. This approach provides greatly improved insight and at the same time allows a much broader generalization of the results than is possible through the use of empirical models.

While there is a close parallel between heterogeneous catalytic reaction systems and gas–solid reactions, the latter systems are rather more complicated because of the direct *participation* of the solid in the overall reaction. As the solid is consumed or undergoes chemical change, its structure changes continuously, making the system inherently transient. It follows that the analysis of gas–solid reactions involves an additional dimension, that of time, which is not necessarily needed in the study of heterogeneous gas–solid reactions. The inherently unsteady nature of gas–solid reaction systems introduces a number of complicating factors which render the tackling of these problems a definitely nonroutine task requiring originality.

It is noted here that while the discussion in this text is devoted to gas–solid reactions, with little modification the treatment developed here should be applicable to liquid–solid reaction systems.

The material presented here could form part of a one-semester graduate level course to be given to students either in metallurgy (materials engineering) or in chemical engineering. It is hoped, moreover, that the book will appeal to the growing number of practicing engineers engaged in process research, development, and design in the many fields where gas–solid reactions are of importance.

Acknowledgments

As is perhaps natural, in the preparation of this manuscript we have leaned heavily on our own experience and published work in the fields of gas–solid reaction kinetics and reaction engineering. It is fitting, however, to make a specific acknowledgment here of the important contributions made to this field by Professors J. M. Smith, E. E. Petersen, C. Y. Wen, R. L. Pigford, W. O. Philbrook, G. Bitsianes, and Drs. E. T. Turkdogan, R. W. Bartlett, and N. J. Themelis, in addition to the numerous citations that will appear in the text.

Thanks are also due to graduate students, past and present, including Dr. C. I. Lin, Dr. S. Song, Mr. Mark Propster, Mr. C.-H. Koo, and Mr. A. Hastaoglu, who have contributed materially through their work and through helpful comments on the manuscript.

We would also like to thank Mrs. Lucille Delmar and Mrs. Christine Jerome for the careful typing and retyping of the manuscript and Mrs. Gloria Pelatowski for the preparation of the drawings.

Chapter 1 | Introduction

Gas–solid reactions play a major role in the technology of most industrialized nations. Gas–solid reactions encompass a very broad field including the extraction of metals from their ores (iron oxide reduction, the roasting of sulfide ores, etc.), the combustion of solid fuels, coal gasification, and the incineration of solid refuse, to cite but a few of the representative examples. Typical equipment used for effecting gas–solid reactions is also rather diverse, as seen from Figs. 1.1–1.3.

Figure 1.1 is a schematic line diagram of a typical iron blast furnace installation where iron ore is being reduced and molten, hot metal is the final product. A typical iron blast furnace may be some 10 m in diameter and up to 100 m high and is fed with ore, pellets, or sinter, which may range in size from 1–10 cm.

Figure 1.2 shows the schematic layout of a modern moving-grate incinerator for the treatment of approximately 400 tons per day of refuse. The refuse supplied to such systems usually does not undergo any special preparation and may thus range in size from about 1–2 meters to a few centimeters.

Figure 1.3 is a schematic diagram of a fluidized bed arrangement for the roasting of copper sulfides. The solid reactants supplied to such a system are carefully sized and may be of the order of 0.01–0.2 cm.

Some typical gas–solid reactions are

1. $C(s) + O_2(g) \rightarrow CO_2(g)$; $\Delta H° = -94{,}052$ kcal/g-mole C.
2. $FeO(s) + H_2(g) \rightarrow Fe(s) + H_2O(g)$; $\Delta H° = +6{,}502$ kcal/g-mole FeO.
3. $FeS_2(s) + 11O_2(g) \rightarrow 2Fe_2O_3(s) + 8SO_2(g)$; $\Delta H° = -197{,}650$ kcal/g-mole FeS_2.
4. $C(s) + CO_2(g) \rightarrow 2CO(g)$; $\Delta H° = +41{,}220$ kcal/g-mole C.

5. $FeO(s) + C(s) \rightarrow Fe(s) + CO(g)$; $\Delta H° = +37,884$ kcal/g-mole FeO.
   A. $FeO(s) + CO(g) \rightarrow Fe(s) + CO_2(g)$; $\Delta H° = -3,336$ kcal/g-mole
   FeO.
   B. $CO_2(g) + C(s) \rightarrow 2CO(g)$; $\Delta H° = +41,220$ kcal/g-mole C.

Of these reactions the oxidation (combustion) of carbon and the oxidation (roasting) of pyrite are both highly exothermic, while the reduction of iron oxide with hydrogen is slightly endothermic. Both the reaction of carbon dioxide with carbon and the reaction of ferrous oxide with carbon are strongly endothermic. It is perhaps of interest to make some comments on the characteristics of the five reactions listed.

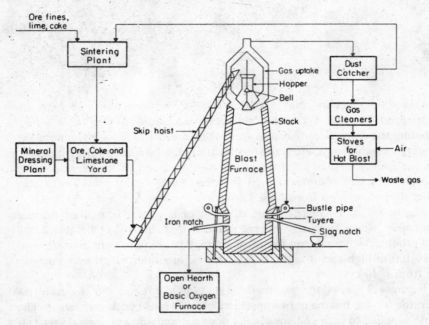

FIG. 1.1. Schematic diagram of an iron blast furnace and associated plant.

Systems 1 and 4 involve reaction between a gas and a solid where there is no solid reaction product, so that upon completion of the reaction the solid phase will disappear.† Some of these reactions are also termed " gasification."

In systems 2 and 3, the reaction between a gas and a solid results in both solid and gaseous reaction products.

† In the combustion or gasification of naturally occurring coal, this assertion would not be true, strictly speaking, because there would be an ash or a residue formed, although this would represent a rather small fraction of the total material reacted.

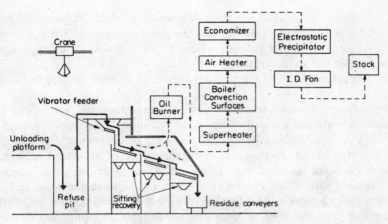

FIG. 1.2. Schematic diagram of a modern refuse incinerator.

Note that at first sight system 5 does not appear to be a gas–solid reaction but rather a reaction between two solid reactants; however, as seen through the subsequent equations, the overall reaction scheme does proceed through gaseous intermediates. In fact, the system may be regarded as being composed of two coupled gas–solid reactions.

A common feature of all gas–solid reaction systems is that the overall process may involve several intermediate steps. Typically, these intermediate steps involve the following:

(1) Gaseous diffusion (mass transfer) of reactants and products from the bulk of the gas phase to the internal surface of the reacting solid particle.

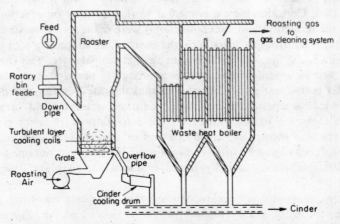

FIG. 1.3. Schematic diagram of a fluidized bed roaster for sulfide ores.

(2) Diffusion of gaseous reactants or gaseous products through the pores of a solid reaction product or through the pores of a partially reacted solid.

(3) Adsorption of the gaseous reactants on and desorption of reaction products from the solid surfaces.

(4) The actual chemical reaction between the adsorbed gas and the solid.

In studying gas–solid reactions then, we are concerned with these four phenomena (external mass transfer, pore diffusion, adsorption/desorption, and chemical reaction) as well as with several other phenomena that may affect the progress of reaction and the performance of industrial equipment in which gas–solid reactions are carried out. These other phenomena include heat transfer (both within the reacting solid and between the solid and surrounding gas), changes in the solid structure (such as sintering) that accompany the reaction, and the flow of gases and solids through equipment in which gas–solid reactions are taking place. The following chapters address themselves to these phenomena and to the way in which they affect the reaction of individual pellets, the design of experiments to measure reaction rates, and the performance of industrial gas–solid reactors. Although the discussion in this book will mainly be for gas–solid reactions, the treatment can easily be extended to most liquid–solid reaction systems.

Until the late 1960s, workers in the field of gas–solid reactions usually regarded the solid reactant as a dense continuum, i.e., devoid of structure. In certain cases (e.g., the reduction of nonporous lumps of iron ore) this approach was justified. This viewpoint was certainly convenient since the mathematical representation of such a reaction system is relatively simple. This period was also marked by much contention over which of the four intermediate steps listed above was "rate controlling" for a particular gas-solid reaction. Considerable experimental work was done on reactions such as iron oxide reduction and sweeping claims were made as to one or another step being rate controlling, usually on the basis of the shape of a plot of solid conversion against time or reaction rate against temperature. The somewhat oversimplified view represented by this approach is perhaps underscored if we consider that investigators have found that the rate "constants" deduced from their measurements for essentially identical systems could vary by as much as a factor of 10,000; moreover, the activation energies for given reactions were also shown to vary by factors of 3 or 4. This state of affairs perhaps explains why very little of the kinetic information obtained by the earlier investigators has actually found practical use in plant design and operation.

In recent years it has been recognized that reactions involving porous solids (e.g., the reduction of iron ore pellets or sinter and the adsorption of sulfur dioxide by limestone) are at least as important as reactions involving dense solids, and that an approach that regarded such a porous reacting solid as a dense continuum is inappropriate. More precisely, it is now accepted that such structural parameters as the porosity, specific surface area, and pore size distribution of the solid reactant will markedly affect the rate of reaction. Mathematical models have been developed to describe the reaction of porous solids and have been shown to fit experimental data in many cases. In this book, we are concerned mainly with this new "structural" viewpoint.

A further finding was the fact that often it is not realistic to talk about a single rate-controlling step because there may be numerous factors that have almost equal effects on the overall rate. Moreover, the relative importance of these factors can change in the course of the reaction, or if the temperature, pellet size, etc., are changed.

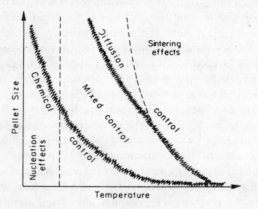

FIG. 1.4. Representation of how temperature and pellet size determine the rate-controlling step and the significance of nucleation and sintering effects in gas–solid reactions.

Figure 1.4 is a schematic representation of how two factors—temperature and the size of the reacting pellet or lump of solid—may determine the rate-controlling step. Other choices of factors, e.g., extent of reaction of the solid and pellet porosity, might have been made but Fig. 1.4 serves to summarize our present knowledge of the reaction between a gas and a solid. Under conditions where transfer of gas molecules to reaction sites within the solid is facile (e.g., small pellet size), it is reasonable to expect the reaction to be controlled by the chemical step. Under conditions where the chemical step is facile (e.g., high temperature) we may expect diffusional control. Between the two extremes is a region of mixed control.

Figure 1.4 also points to two complications encountered in gas–solid reactions. Many gas–solid reactions (e.g., 2, 3, and 5 above) entail the nucleation of a second solid phase. Particularly at lower temperatures, this nucleation process may be slow enough to take a significant fraction of the total time of reaction of the solid.

Even at temperatures considerably below their melting points most porous solids undergo sintering. Hence a reaction between a gas and a porous solid, or between a gas and a dense solid but yielding a porous solid product, may be affected by sintering if carried out at elevated temperatures. The phenomena of sintering and nucleation are not well understood and consequently our knowledge of the kinetics of gas–solid reactions outside the region delineated by the dashed lines in Fig. 1.4 is limited at present.

Fortunately, in many gas–solid reactions, nucleation and sintering effects may be ignored. This is not to imply that under these conditions analysis and use of kinetic data are straightforward. Due account must be taken of the interaction between mass transfer of gaseous reactants (and products) to (and from) the reaction sites within the solid and of the dependence of this transfer process, and the chemical steps at the reaction sites, on the structure of the solid. The aim of this volume is to present research carried out by the authors and by others that has led to an alternative approach to gas–solid reactions in which a more realistic view is taken of the nature of the type of solid encountered frequently in reactions of industrial importance.

In many ways this more modern approach to gas–solid reactions has borrowed heavily from developments in heterogeneous catalysis, where investigators have made extensive studies of coupled pore-diffusion and adsorption phenomena and of the characterization of porous solids. However, gas–solid reactions are rather more complex than heterogeneous catalytic reactions, because for gas–solid reactions the solid structure may change continuously as the solid reactant is consumed. The actual manifestation of these structural changes may include the progressive growth of a solid product layer, the enlargement of pores due to the reaction, the sintering of the partially reacted specimens, etc.

Our objective in the presentation of this material is to attempt to integrate these recent developments in the more fundamental understanding of gas–solid reaction systems, both with experimental data and with the design of industrial gas–solid reaction systems.

More specifically, in Chapter 2 we shall review the basic components of gas–solid reactions, that is, the information available from the individual chemical or physical steps (such as pore diffusion and adsorption phenomena, that actually provide the building blocks for composite, real gas–solid reaction systems.

In Chapter 3 we shall describe the reactions of individual nonporous solid particles. In Chapter 4 we shall present a treatment of the reaction of single porous particles. Chapter 5 will be devoted to the discussion of solid–solid reactions proceeding through gaseous intermediates, an example of which was given earlier. Chapter 6 will be concerned with a discussion of experimental approaches to the study of gas–solid reaction systems; more specifically, we shall discuss how experimental studies may be planned in a rational manner in the light of the theoretical developments described in the earlier chapters.

Chapter 7 will present a discussion of how information on single-particle behavior may be used for the design of multiparticle, large-scale assemblies, viz., packed- and fluidized-bed or moving-bed reaction systems.

Finally, in Chapter 8 we shall give a description of some specific gas–solid reaction systems together with some statistical indices indicating the economic importance of the systems and the processes based upon them.

A successful approach to the study of gas–solid reaction systems and the design of processes based on specific reactions requires a carefully chosen balance between analysis and measurements, and their integration. It has been our aim throughout the presentation of the material in the text to seek this balance.

| **The Elements of Gas–Solid Reaction Systems Involving Single Particles**

2.1 Introduction

The smallest representative unit of a gas–solid reaction system is the interaction of a single particle with a moving gas stream. Since the study of single-particle systems is quite convenient and, at least in principle, the results may be generalized to the more complex multiparticle assemblies, much of this book will be devoted to the discussion of single-particle systems.

Let us consider a gas–solid reaction of the type

$$A(g) + bB(s) = cC(g) + dD(s) \qquad (2.1.1)$$

where b, c, and d are the stoichiometric coefficients. Typical examples of practical systems that may be represented by this general scheme would include

$$Fe_2O_3 + 3CO = 2Fe + 3CO_2$$
$$NiO + H_2 = Ni + H_2O$$
$$2CuS + 3O_2 = 2CuO + 2SO_2, \quad \text{etc.}$$

A schematic representation of such a reacting particle is given in Fig. 2.1, where it is seen that the overall reaction process may involve the following individual steps:

REACTION–DIFFUSION

(1) Gas phase mass transfer of the gaseous reactant from the bulk of the gas stream to the external surface of the solid particle.

(2)   (i) Diffusion of the gaseous reactant through the pores of the solid matrix, which could consist of a mixture of solid reactants and products.

(ii)   Adsorption of the gaseous reactant on the surface of the solid matrix.

(iii)   Chemical reaction at the surface of the solid matrix.

(iv)   Desorption of the gaseous product from the surface of the solid matrix.

(v)   Diffusion of gaseous reaction product through the pores of the solid matrix.

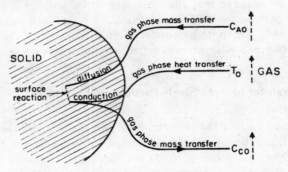

FIG. 2.1. Schematic representation of the endothermic reaction of a single solid pellet with a gas.

Steps (i)–(v), while necessarily sequential as far as the reaction is concerned, may take place simultaneously in a diffuse spatial domain.

(3)   Gas phase mass transfer of the gaseous product from the external surface of the solid to the bulk of the gas stream.

HEAT TRANSFER

For exothermic or endothermic reactions the diffusional and reaction steps will also be accompanied by

(1)   convective (and possibly radiative) heat transfer between the gas stream and the surface of the solid particle, and

(2)   conduction heat transfer within the solid reactant–product matrix.

STRUCTURAL CHANGES

The reaction and heat transfer process could lead to structural changes, such as sintering or changes in the pore structure, which in turn could have a marked effect on the overall reaction rate.

Ideally, in the proper representation of gas–solid reactions, all these effects should be taken into consideration. However, in most of the work that has been done up to the present, simplifying assumptions were introduced in the development of the mathematical models through which many of these intermediate steps were either neglected or "lumped together." In some instances these simplifying assumptions were justified, perhaps *a posteriori*. In other cases, insufficient information was the main reason for using relatively simple models.

In the following we shall present a brief discussion of the elementary steps that constitute the building blocks of each gas–solid reaction system. The reader will observe that the treatment to be given will be necessarily quite uneven because some of these elementary steps, such as gas phase heat and mass transfer, are quite well understood, while some of the other effects, such as sintering and adsorption kinetics, are much less well understood.

2.2 Mass Transfer between Single Particles and a Moving Gas Stream

As discussed earlier, the rate at which gaseous reactants are transferred from the bulk of the gas to the surface of the solid or the rate at which gaseous products are removed from the surface of the solid reactant may play an important role in determining the overall rate of reaction.

This mass transfer step has been extensively studied, primarily through the examination of simple physical transfer processes, such as vaporization or drying. Thus, the mass transfer component of the overall reaction sequence is perhaps the best understood. While it is possible to calculate the rate of mass transfer between a moving gas stream and a solid surface by the simultaneous solution of the appropriate fluid flow and diffusion equations [1, Chapter 17], here we shall adopt a more empirical approach through the use of mass transfer coefficients, although some comments will be made about the way these two approaches may be regarded as complementary.

MASS TRANSFER COEFFICIENTS

Let us consider the system sketched in Fig. 2.2, which is really an enlargement of a section of Fig. 2.1, where material is being transferred from a solid surface into a moving gas stream.

Let A denote the (gaseous) transferred species, the concentration of which is then designated by C_{As} and C_{A0} at the solid surface and in the bulk of the gas stream, respectively. Then the rate at which mass is being transferred from the solid into the fluid per unit solid surface area is given by

$$N_A = h_D(C_{As} - C_{A0}) \qquad (2.2.1)$$

where h_D is the mass transfer coefficient, which is defined by this equation. The concentrations may be expressed as either moles per unit volume or

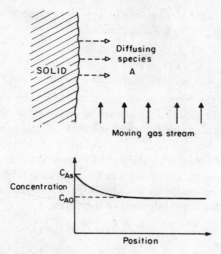

FIG. 2.2. The diffusion of species A from a solid surface into a moving gas stream.

mass per unit volume and then the flux of the transferred species, N_A, is given in the corresponding units (moles per unit area per unit time, or mass per unit area per unit time); thus h_D has the dimensions of length/time.

Some authors have expressed the concentrations in terms of partial pressures or mole fractions, leading to mass transfer coefficients different from the above.

Equation (2.2.1) implies that the molar flux is proportional to the concentration difference, which is a reasonable approximation for many gas–solid reaction systems. A somewhat more accurate mass transfer equation that has been proposed by Bird *et al.* [1, Chapter 21] takes the form

$$N_A = h_D(C_{As} - C_{A0}) + X_{As}(N_A + N_C) \qquad (2.2.2)$$

where N_C is the mass flux of second diffusing species C, and X_{As} the mole fraction of component A at the solid surface. The first term of the r.h.s. of Eq. (2.2.2) represents transfer due to the diffusion of component A, while the second term corresponds to bulk flow due to diffusion. We note that the bulk flow term vanishes in case of equimolar counterdiffusion, i.e., when $N_A = -N_C$. Thus, bulk flow effects will not play a role in many oxide reduction processes such as

$$FeO + H_2 = Fe + H_2O, \qquad NiO + CO = Ni + CO_2, \qquad \text{etc.}$$

However, there are many gas–solid reactions where $N_A \neq -N_C$ and here the full form of Eq. (2.2.2) has to be used, especially when X_{As}, the mole fraction of the gaseous reactant, is not small. Typical examples where equimolar counterdiffusion may not be assumed include

$$CuCl_2 + H_2 = Cu + 2HCl$$
$$C + CO_2 = 2CO$$
$$3Cu_2O + CH_4 = 6Cu + CO + 2H_2O, \quad \text{etc.}$$

As noted earlier, in the majority of practical cases, the *mass transfer coefficient* h_D appearing in Eq. (2.2.2) has to be obtained from empirical correlations rather than from the rigorous solution of the convective diffusion equations.

It may be shown either with the aid of dimensional analysis [2, Chapter 16; 3; 4] or by expressing the mass flux to a surface (obtained through the solution of the convective diffusion equation) in a dimensionless form that these empirical correlations should have the form

$$N_{Sh} = f(N_{Re}, N_{Sc}) \tag{2.2.3}$$

for forced convection,

$$N_{Sh} = f(N_{Gr}, N_{Sc}) \tag{2.2.4}$$

for natural convection, and

$$N_{Sh} = f(N_{Gr}, N_{Re}, N_{Sc}) \tag{2.2.5}$$

for systems where both forced and natural convection might be significant, where $N_{Sh} = h_D L/D$† is the Sherwood number; $N_{Re} = UL/v$ the Reynolds number; and $N_{Sc} = v/D$ the Schmidt number. $N_{Gr} = gL^3 \beta_D \Delta X_A/v^2$ is the Grashof number for mass transfer; L is a characteristic dimension, i.e., the diameter of a spherical particle, or the equivalent diameter of a nonspherical particle, etc.; v is the kinematic viscosity; D is the binary diffusion coefficient; U is the linear velocity of the gas stream flowing past the particle (measured outside the boundary layer surrounding the particle); g is the acceleration due to gravity; ΔX_A is a characteristic concentration difference, and

$$\beta_D = \left| \frac{1}{\rho} \left(\frac{\partial \rho}{\partial X_A} \right)_T \right|$$

where ρ is the density and T is the temperature.

Frequently the mass transfer coefficient is expressed as a Chilton–Colburn j_D factor, which is defined as

$$j_D = N_{Sh}/N_{Re} N_{Sc}^{1/3}$$

---

† It is noted that in the correlations cited in this chapter and also in Chapter 7, the characteristic length for the solid particles is taken as the particle diameter. In contrast, in Chapters 3 and 4, the characteristic length in the Sherwood number will be taken as the particle radius to preserve conformity with previous work. These latter Sherwood numbers will be denoted by N'_{Sh}.

Because of the industrial importance of operations involving mass transfer from drops or particles (e.g., spray drying, combustion of liquid fuels), this topic has received much attention. An early outstanding study was that of Ranz and Marshall [5], whose experimental results on the evaporation of drops were correlated by the equation

$$N_{Sh} = 2.0 + 0.6 N_{Re}^{1/2} N_{Sc}^{1/3} \qquad (2.2.6)$$

Experiments were confined to the Reynolds number range 0–200.

Frössling [6] had earlier presented an almost identical equation and showed theoretically that the Sherwood number should take on a value of 2 in a stagnant fluid; moreover, there were theoretical grounds for expecting the square root dependence on the Reynolds number that he had observed experimentally. Frössling's experiments concerned the evaporation of nitrobenzene, aniline, water, and napthalene into a hot air stream, within the Reynolds number range $2 < N_{Re} < 1300$.

Much of the earlier work on this topic was reviewed in a paper by Rowe and Claxton [7], who measured heat and mass transfer to air and water from spheres ranging from 1/2 to 1-1/2 in. in diameter and in the range $20 < N_{Re} < 2000$. One of their conclusions is worth quoting:

Heat or mass transfer data from a sphere to a fluid for any particular system can be described by

$$N_{Sh} = A + B N_{Sc}^{m} N_{Re}^{n} \qquad (2.2.7)$$

or

$$N_{Nu} = A + B N_{Pr}^{m} N_{Re}^{n} \qquad \text{(for heat transfer†)} \qquad (2.2.8)$$

but there is considerable latitude to vary the terms A, B, or n provided the other two constants are chosen appropriately. Present day experimental techniques are unlikely to provide the precision needed to resolve this and little importance should be attached to differences between reported values, in particular of the index n.

In the light of these arguments Rowe and Claxton proposed the following correlations:

$$N_{Sh} = 2.0 + 0.69 N_{Re}^{1/2} N_{Sc}^{1/3} \qquad \text{for air} \qquad (2.2.9)$$

$$N_{Sh} = 2.0 + 0.79 N_{Re}^{1/2} N_{Sc}^{1/3} \qquad \text{for water} \qquad (2.2.10)$$

(with analogous expressions for heat transfer). Rowe and Claxton suggested, furthermore, that their data and those of other investigators would be cor-

† The quantities N_{Nu} and N_{Pr} will be defined subsequently in the section dealing with heat transfer.

related better by using an expression where the exponent of the Reynolds number was allowed to increase with increasing values of N_{Re}.

The work of Steinberger and Treybal [8] is of interest, because they have also made allowance for the *effect of natural convection*, due to concentration differences. The following correlations were proposed:

$$N_{Sh} = N_{Sh_0} + 0.347(N_{Re} N_{Sc}^{1/2})^{0.62} \qquad (2.2.11)$$

where

$$N_{Sh_0} = 2.0 + 0.569(N_{Gr} N_{Sc})^{0.25} \qquad \text{for} \quad (N_{Gr} N_{Sc}) < 10^8 \quad (2.2.12)$$

and

$$N_{Sh_0} = 2.0 + 0.0254(N_{Gr} N_{Sc})^{1/3} N_{Sc}^{0.244} \qquad \text{for} \quad (N_{Gr} N_{Sc}) > 10^8 \quad (2.2.13)$$

While the precision of three significant figures appearing in Eqs. (2.2.11)–(2.2.13) is not warranted, these correlations were found to show a good fit with Steinberger and Treybal's own measurements ($10 < N_{Re} < 16{,}920$, $987 < N_{Sc} < 69{,}680$ and $5130 < N_{Gr} < 125{,}200$) as well as with those of numerous other workers. The effect of natural convection can become quite significant at low Reynolds number, which may cause the Sherwood number to be much larger than 2 in experiments with fluids that appear to be stagnant.

Evnochides and Thodos [9] evaporated water and nitrobenzene into air from celite spheres and obtained the following correlation for the mass transfer coefficient:

$$j_D = 0.33/N_{Re}^{0.4} \qquad (2.2.14)$$

where $j_D = (h_D L/D)N_{Re}^{-1} N_{Sc}^{-1/3}$ in the Reynolds number range 1500–13,000. This equation is not grossly inconsistent with those given above.

Friedlander [10] has demonstrated on the basis of theoretical considerations that at low Reynolds numbers and in the absence of natural convection, the Sherwood number is given by the following:

$$N_{Sh} = 0.991 N_{Pe}^{1/3} \qquad \text{for} \quad N_{Pe} > 10 \qquad (2.2.15)$$

where $N_{Pe} = LU/D$ is the Peclet number for mass transfer.

The subject of mass transfer from solid spheres has recently been examined by Keey and Glen [11], who maintained that the exponent of the Reynolds number in an equation of the Frössling type should increase with increasing Reynolds number due to the changing importance of wake transfer with Reynolds number and due to the onset of turbulence in the boundary layer. Keey and Glen recommended a different correlation for each Reynolds number region.

Hughmark [12] has provided correlations for heat and mass transfer from rigid spheres based on the data of numerous workers. His equations encompass all Schmidt numbers and $1 \leq N_{Re} \leq 10{,}000$.

Boundary layer theory has now developed to such an extent that predictions can be made of local (heat and) mass transfer coefficients for rigid spheres from the front pole of the sphere up to the point of flow separation. Lee and Barrow [13] were able to show good agreement between local Sherwood numbers measured on napthalene spheres evaporating in air and those calculated from boundary layer theory for the frontal position of a sphere. Integrating their theoretical local coefficient over the front half of the sphere and adding a correlation from experimental data for the back half of the sphere gave the following expression for the overall coefficient:

$$N_{Sh} = \underbrace{0.51 N_{Re}^{1/2} N_{Sc}^{1/3}}_{\text{front half}} + \underbrace{0.022 N_{Re}^{0.78} N_{Sc}^{1/3}}_{\text{back half}}, \qquad 200 \leq N_{Re} \leq 2.0 \times 10^5 \quad (2.2.16)$$

This equation (which showed good agreement with the experimental results of other workers) clearly indicates the increased importance of the wake region at the back of the sphere at high Reynolds numbers.

The effect of turbulence, generated by bodies upstream from the particle under study, on heat and mass transfer rates has been studied by few investigators. A recent significant paper is that of Lavender and Pei [14], who showed by experiments using electrically heated spheres in a wind tunnel containing "turbulence generators" (such as wire mesh screens) that turbulence within the fluid enhanced heat transfer. Furthermore, the data could be successfully correlated using a turbulent Reynolds number [N_{ReT} (in addition to the usual Reynolds and Schmidt numbers]:

$$N_{ReT} = \sqrt{u^2} \cdot (L\rho/\mu) \qquad (2.2.17)$$

where u is the fluctuating component of the fluid velocity. The correlations are

$$N_{Nu} = 2.0 + 0.629 N_{Re}^{0.5} N_{ReT}^{0.035} \qquad (N_{ReT} < 1000)$$

and

$$N_{Nu} = 2.0 + 0.145 N_{Re}^{0.5} N_{ReT}^{0.250} \qquad (N_{ReT} > 1000)$$

for air ($N_{Pr} = 0.674$). Analogous correlations can be assumed to apply for mass transfer.

The reader may find the number of correlations presented here for the mass transfer coefficients somewhat confusing. It is to be stressed that the expressions proposed by Frössling, Rantz and Marshall, and Rowe and Claxton give virtually identical results for the range of Reynolds and Schmidt numbers usually encountered in gas–solid reaction systems of practical interest. *Indeed the "rate expressions" for gas–solid mass transfer constitute an area where most investigators appear to be in quite good agreement.*

Note that if the Grashof number is high compared with the Reynolds number, i.e., low gas velocities or near stagnant conditions, then the equations of Steinberger and Treybal should prove more satisfactory. In the case of appreciable free stream turbulence from bodies upstream of the particle, attention should be directed to the investigation of Lavender and Pei.

USE OF THE MASS TRANSFER CORRELATIONS

In using the previously cited correlations information is required on the property values of the system. In general, data should be available on the characteristic dimension, L, of the specimen and the linear gas velocity U; furthermore, the density of the gas may be readily calculated for known temperature and pressure from the equation of state. The principal problem is then to provide reasonably accurate values for the diffusivities and the viscosity of the gas stream.

Gas phase diffusivities

A detailed discussion of gas phase diffusion is beyond the scope of this monograph. For a rigorous statement of diffusion problems (including multicomponent diffusion) the reader is referred to the texts by Bird *et al.* [1] and Hirschfelder *et al.* [15].

EXPERIMENTAL DIFFUSIVITIES

Molecular diffusivities can be determined experimentally in a number of ways. The rates at which systems with concentration gradients approach steady state can be measured or the fluxes and concentrations of each species in systems where a steady (or quasi-steady) state is maintained by some means can be determined. Some of these techniques are described in Chapter 6. A survey of gaseous diffusion coefficients has been carried out recently by Marrero [16]. Some typical diffusion coefficients are given in Table 2.1.

TABLE 2.1

Gas pair	Temperature (°K)	Diffusivity (cm²/sec)
N_2–CO	316	0.24
H_2–CH_4	316	0.81
H_2–O_2	316	0.89
H_2O–O_2	450	0.59
CO–CO_2	315	0.18
	473	0.38
H_2O–N_2	327	0.31
CO–air	447	0.43
CO_2–air	501	0.43

While in many instances one cannot find experimental data on the diffusion coefficients, it is fortunate that gas phase diffusion coefficients may be predicted quite accurately.

ESTIMATION OF DIFFUSIVITIES

For binary gas mixtures at low pressures, e.g., below 10 atm, the diffusion coefficient D_{AB} is inversely proportional to the pressure, increases with increasing temperature, and is almost independent of composition for a given gas pair. Slattery and Bird [17] have proposed the following expression for D_{AB} at low pressures:

$$\frac{p\tilde{D}_{AB}}{(p_{CA}\,p_{CB})^{1/3}(T_{CA}\,T_{CB})^{5/12}(1/M_A + 1/M_B)^{1/2}} = a\left(\frac{T}{\sqrt{T_{CA}\,T_{CB}}}\right)^b \quad (2.2.18)$$

where p is the pressure (atm), D_{AB} is in cm^2/sec, T is the temperature (°K), M the molecular weight, and the subscript C indicates critical conditions. Experimental data yield the following values for the constants a and b:

Nonpolar gas pairs: $a = 2.745 \times 10^{-4}$, $b = 1.823$
H$_2$O with a nonpolar gas: $a = 3.640 \times 10^{-4}$, $b = 2.334$

For nonpolar gases the prediction on the Chapman–Enskog kinetic theory is usually preferred [15]:

$$D_{AB} = 0.0018583\,\frac{\sqrt{T^3(1/M_A + 1/M_B)}}{p\sigma_{AB}\Omega_{AB}} \quad (2.2.19)$$

where D_{AB}, T, and p are in the same units as for Eq. (2.2.18), σ_{AB} is a constant in the Lennard-Jones 12-6 potential function, and Ω_{AB} is the collision integral (a function of KT/ε_{AB}, tabulated in Bird et al. [1] and Hirschfelder et al. [15]). The values of σ_{AB} and Ω_{AB} may be calculated from

$$\sigma_{AB} = \tfrac{1}{2}(\sigma_A + \sigma_B) \quad (2.2.20)$$

$$\varepsilon_{AB} = \sqrt{\varepsilon_A\,\varepsilon_B} \quad (2.2.21)$$

where σ and ε are tabulated for a number of gases [15, 18] and may be calculated from viscosity data on the pure gas [15]. Alternatively σ and ε can be estimated from the properties of the pure component at the critical point (indicated by subscript C), liquid at the normal boiling point (b), or solid at the melting point (m), i.e.,

$$\sigma = 0.841\tilde{V}_C^{1/3} \quad \text{or} \quad 2.44(T_C/p_C)^{1/3} \quad (2.2.22)$$

$$\varepsilon/\kappa = 0.77T_C \quad (2.2.23)$$

$$\sigma = 1.166\tilde{V}_{b,\,\text{liq}}^{1/3}; \quad \varepsilon/\kappa = 1.15T_b \quad (2.2.24)$$

$$\sigma = 1.222\tilde{V}_{m,\,\text{sol}}^{1/3}; \quad \varepsilon/\kappa = 1.92T_m \quad (2.2.25)$$

where $\tilde{V}$ is the molar volume (cm^3/g-mole) and ε/κ and T_C are in $°K$; κ is Boltzmann's constant.

These procedures for estimating binary diffusion coefficients are thought to be unsatisfactory for gas mixtures containing polar molecules, such as water vapor and HCl. A description of alternative techniques may be found in the literature [19–24].

Example 2.2.1   Calculate the diffusivity of a hydrogen (A)–water vapor (B) mixture at 352°K and 1 atm pressure, using:

(i)   the equation proposed by Slattery and Bird (2.2.18),
(ii) the Chapman–Enskog equation (2.2.19).

Compare these predictions with the measurement of Schwertz and Brow [125], viz., $D_{AB} = 1.19$ cm^2/sec at 352°K.

SOLUTION   (1)   *Using the Slattery–Bird equation*   The critical point data are

	H_2	H_2O
Critical temperature ($°K$):	33.2	647.3
Critical pressure (atm):	12.8	218.5

$$(T_{CA} T_{CB})^{5/12} = 64.2, \qquad (T_{CA} T_{CB})^{1/2} = 146.8, \qquad (P_{CA} P_{CB})^{1/3} = 14.1$$

$$M_A = 2.0, \qquad M_B = 18.0, \qquad (1/M_A + 1/M_B)^{1/2} = 0.745$$

At 352°K

$$[T/(T_{CA} T_{CB})^{1/2}]^b = (2.4)^{2.334} = 7.75$$

Thus

$$D_{AB} = \frac{a}{p} \left(\frac{T}{(T_{CA} T_{CB})^{1/2}} \right)^b \left(\frac{1}{M_A} + \frac{1}{M_B} \right)^{1/2} (T_{CA} T_{CB})^{5/12} (P_{CA} P_{CB})^{1/3}$$

$$= 3.640 \times 10^{-4} \times 7.75 \times 0.745 \times 64.2 \times 14.1 = 1.90 \text{ cm}^2/\text{sec}$$

(2)   *Using the Chapman–Enskog equation*   Svehla [18] gives the following Lennard-Jones constants for the gases:

	H_2	H_2O
σ (A):	2.827	2.641
ε/κ ($°K$):	59.7	809.1

These values were all obtained from viscosity measurements and can be considered reliable. If reliable values had not been available, we would have estimated σ and ε/κ from boiling point data:

$$\sigma_{AB} = \tfrac{1}{2}(\sigma_A + \sigma_B) = 2.734 \text{ A}$$

$$\varepsilon_{AB}/\kappa = [(\varepsilon_A/\kappa)(\varepsilon_B/\kappa)]^{1/2} = (484 \times 10^2)^{1/2} = 220°K$$

At 352°K,

$$T/\varepsilon_{AB}/\kappa = 352/220 = 1.60$$

At $T/(\varepsilon_{AB}/\kappa) = 1.60$, $\Omega_{AB} = 1.167$ (from Hirschfelder *et al.* [15]),

$$D_{AB} = 1.8583 \times 10^{-3} \times \sqrt{\frac{T^3(1/M_A + 1/M_B)}{p\sigma_{AB}^2\Omega_{AB}}}$$

$$= 1.8583 \times 10^{-3} \times \sqrt{\frac{0.352^3 \times 10^9 \times (1/2 + 1/18)}{1 \times 2.734^2 \times 1.167}} = 1.05 \text{ cm}^2/\text{sec}$$

It is seen that the predictions based on the Slattery–Bird and the Chapman–Enskog equations differ by nearly a factor of two and that neither of these provides particularly good agreement with the experimental measurements. This substantial discrepancy between measurements and predictions is due to the polar nature of the hydrogen–water vapor mixture. For nonpolar gases such as CO_2, N_2, and CO, the predictions would have been much closer to the measured values [1].

VISCOSITY OF GASES

At pressures below about 10 atm the viscosity of a gas is independent of pressure and increases with increasing temperature, although the temperature dependence of viscosity is much less marked than that of the diffusion coefficient. A good collection of experimentally measured values of the viscosity has been compiled by Svehla [18]; some typical values are presented in Table 2.2.

TABLE 2.2

Typical Values of Gas Phase Viscosities

Gas	Temperature (°C)	Viscosity (g/cm sec)
Hydrogen	0	8.4×10^{-5}
	229	12.6×10^{-5}
	490	16.7×10^{-5}
	825	21.4×10^{-5}
Air	0	1.71×10^{-4}
	409	3.41×10^{-4}
	810	4.42×10^{-4}
	1134	5.21×10^{-4}

PREDICTION OF GAS PHASE VISCOSITIES

The techniques available for predicting gas phase viscosities are quite accurate and parallel very closely the methods described previously for estimating diffusion coefficients. A readable discussion of these, including the recommended procedures for estimating the viscosity of gas mixtures is available in the text by Bird *et al.* [1].

RELATIONSHIP BETWEEN MASS TRANSFER AND THE CONVERSION OF SOLIDS

In recapitulating the preceding discussion, for a given geometry and reactant gas composition, if we know the surface concentration of the reactant and product gas we can calculate the net flux of gaseous reactant to the solid surface.

Then if we consider that all the gaseous reactant supplied to the solid surface is used up to react with the solid (the quasi-steady state assumption), then for the reaction

$$A + bB = cC + dD \tag{2.1.1}$$

we can write

$$\begin{pmatrix} \text{molar flux of} \\ \text{species A} \end{pmatrix} \times \begin{pmatrix} \text{external surface} \\ \text{area of solid} \end{pmatrix} \times \begin{pmatrix} \text{stoichiometric} \\ \text{coefficient} \end{pmatrix} = \begin{pmatrix} \text{molar rate of} \\ \text{reaction of B} \end{pmatrix}$$

$$\tag{2.2.26}$$

Thus we have

$$N_A = (V_p/bA_p)(1 - \varepsilon)\rho_s \, dX/dt = h_D(C_{As} - C_{A0}) \tag{2.2.27}$$

where V_p and A_p denote the volume and surface area of the solid pellet, respectively, ρ_s is the molar density of the pore-free solid reactant,† ε the porosity of the pellet, and X the fraction of the solid reacted. In the general case, the surface concentration of the reactant, C_{As}, will also depend on factors other than mass transfer, such as pore diffusion and chemical kinetics, as will be discussed subsequently.

However, when the overall rate is mass transfer controlled, then equilibrium will exist at the solid surface and for

irreversible reactions $\qquad\qquad C_{AS} = 0$

and $\tag{2.2.28}$

reversible reactions $\qquad\quad C_{AS} = C_{AS}|_E = C_{CS}/K_E$

---

† Here and in subsequent chapters, if the solid reactant is not pure but contains an appreciable quantity of inerts, then ρ_s is no longer the molar density of pore-free solid but rather the moles of pure solid reactant per unit volume of pore-free solid.

where $C_{AS}|_E$ is the equilibrium concentration of A, which corresponds to the gas composition where the reactant and product solids may coexist in equilibrium. Some comments will be made subsequently on the calculation of the equilibrium concentrations.

For mass-transfer-controlled reactions, Eq. (2.2.27), a similar equation for the gaseous product C, and Eq. (2.2.28) yield after some algebra and integration

$$X = \frac{bh_D(C_{A0} - (C_{C0}/K_E))}{(1-\varepsilon)\rho_s} \left(\frac{A_p}{V_p}\right) \frac{K_E}{1+K_E} t \qquad (2.2.29)$$

For an irreversible reaction, of course, $K_E \to \infty$. Equation (2.2.29) shows that for mass-transfer-controlled systems the extent of reaction is a linear function of time, a linear function of the driving force (i.e., C_{A0} in this instance), and also linearly dependent on the quantity (A_p/V_p).

Equation (2.2.29) thus provides a convenient means for testing whether a given reaction system is indeed mass transfer controlled. Many gas–solid reaction systems are mass transfer controlled at high temperatures and for moderate pellet sizes. Typical examples include

$$CuO + H_2 = Cu + H_2O \quad [25]$$
$$NiO + Cl_2 = NiCl_2 + \tfrac{1}{2}O_2 \quad [26] \quad \text{(above 1500°K)}$$
$$Fe_2O_3 + 3Cl_2 = 2FeCl_3 + \tfrac{3}{2}O_2$$

Example 2.2.2   A CuO sphere, 1 cm in diameter, is reacted with hydrogen at 673°K and at 1 atm pressure. If the linear velocity of the hydrogen is 105 cm/sec and the porosity of the pellet is 0.3, calculate:

(i)   the mass transfer coefficient, and
(ii) the time required for the complete reaction of the pellet.

ADDITIONAL DATA D_{A-B}, the binary diffusion coefficient of hydrogen and water vapor, may be calculated with the aid of the Chapman–Enskog equation as 3.42 cm²/sec at 673°K. The viscosity of hydrogen at this temperature is 1.53×10^{-4} g/cm sec. The molar density of the pellet is 4.46×10^{-2} g-mole/cm³. The reaction may be taken as irreversible and the gas may be considered to consist almost entirely of hydrogen.

SOLUTION   The density of hydrogen at 673°K and 1 atm is given by

$$PM/RT = 1 \times 2/82.0 \times 673 = 3.62 \times 10^{-5} \text{ g/cm}^{-3}$$

The Reynolds number is

$$N_{Re} = UL\rho/\mu = 105 \times 1 \times 3.62 \times 10^{-5}/1.53 \times 10^{-4} = 25$$

The Schmidt number is

$$N_{Sc} = \mu/\rho D = \nu/D = 1.53 \times 10^{-4}/3.42 \times 3.62 \times 10^{-5} \simeq 1.24$$

Thus using the Rantz–Marshall correlation [Eq. (2.2.6)] we have

$$N_{Sh} = 2.0 + 0.6N_{Re}^{1/2}N_{Sc}^{1/3} = 5.25 \qquad (2.2.6)$$

and then $h_D = 5.25 \times 3.42 = 18$ cm/sec. The time required for the complete reaction of the specimen may now be calculated from Eq. (2.2.29) by setting $X = 1$. Thus we have

$$C_{A0} = 1.81 \times 10^{-5} \text{ g-mole/cm}^3, \qquad A_p/V_p = 3$$

so that $t = (1 - 0.3) \times 4.46 \times 10^{-2}/3 \times 1.81 \times 10^{-5} \times 18 \simeq 32$ sec† which is quite rapid. The preceding example was concerned with a system where the reaction was irreversible and equimolar counterdiffusion was taking place. Let us conclude this section with a simple example to illustrate how these considerations may be generalized to systems where neither of these constraints is met.

Example 2.2.3   Let us develop an expression for the rate of reaction of a $CuCl_2$ pellet with hydrogen in the region of mass transfer control.

SOLUTION   The reaction is

$$CuCl_2 + H_2 = Cu + 2HCl$$

which is not equimolar counterdiffusion. We now proceed by calculating the mass transfer coefficient, e.g., from Eq. (2.2.6). Then the flux of A, i.e., N_A, is given by Eq. (2.2.2):

$$N_A = h_D(C_{As} - C_{A0}) + X_{As}(N_A + N_C) \qquad (2.2.2)$$

On noting that for the present case

$$N_C = -2N_A$$

$\quad$ (molar flux of HCl)   (molar flux of H_2)

we have

$$N_A = [h_D/(1 + X_{As})](C_{A0} - C_{As})$$

We may now substitute this value of N_A into Eq. (2.2.27) and proceed as in the previous case. It is to be noted that X_{As} and C_{As} will have to be calculated by using the appropriate equilibrium relationships. We stress, moreover, that this procedure would be valid for the case of mass transfer control.

---

† We note that the purpose of this example was to illustrate how conversion may be calculated for the case of mass transfer control. In fact purely mass transfer control will seldom occur unless the pellets are small, because in the later stages of reaction pore diffusion could offer a significant resistance.

2.3   Diffusion of the Gaseous Reactants and Products through the Pores of a Solid Matrix

Diffusion through a porous solid matrix or simply pore diffusion may play an important role in gas–solid reactions. When the reactant solid is porous, diffusion through the pore space is necessary for the reactant gas to gain access to the solid surface; in a similar manner, the removal of the gaseous products will also involve this process. However, pore diffusion may also be an important component in the reaction of nonporous solids, when the solid product layer formed is itself porous, because then the supply of gaseous reactant and the removal of gaseous products have to be accomplished by diffusion through this porous product layer.

In this section we shall consider the diffusion of a binary gas mixture in a porous medium, in the absence of chemical reaction, because the understanding of this process is an important component of the more complex "real life" situations where pore diffusion occurs simultaneously with chemical reaction. These problems will be discussed in Chapters 3 and 4.

Pore diffusion is inherently much more complex than diffusion in liquids or gases and, as a consequence, is much less well understood. Figure 2.3 is a schematic representation of gaseous diffusion in a porous medium.

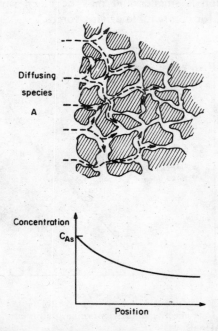

FIG. 2.3. The diffusion of species A into a porous solid.

Some of the complicating factors are immediately apparent:

(1)   The volume occupied by the solids (denoted by the shaded areas) is not available for diffusive transfer.

(2)   The actual diffusion path will not follow a straight line (as in one-dimensional diffusion problems) but will be quite tortuous, and the extent of this tortuosity will necessarily depend on the pore structure of the solids.

Other complicating factors may be introduced by

(3)   The fact that when the pores are small enough (such that the mean free path of the molecules becomes comparable to the dimension of the pore) the laws of molecular diffusion will no longer apply, but Knudsen diffusion will become important. In a physical sense, this means that in the Knudsen regime collisions between the gas molecules and the wall will become more frequent than collisions between gas molecules.

(4)   Under some circumstances, significant total pressure gradients may be established within the solid. Under these conditions, transfer due to a pressure gradient has to be taken into consideration, also.

The general qualitative conclusion that emerges from this discussion is that pore diffusion is more complex than molecular diffusion. Moreover, the rate of pore diffusion is smaller and may be much smaller than that of molecular diffusion, for comparable driving forces.

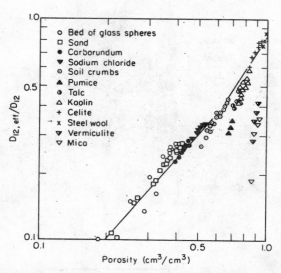

FIG. 2.4. A plot of the effective diffusivity/ordinary diffusivity ratio against porosity for various granular materials [94].

Pore diffusion is of considerable importance in a number of fields, including heterogenous catalysis, isotope separation, and drying, and for this reason, pore diffusion problems have been extensively studied.

A good discussion of the earlier work on pore diffusion is available in the texts by Satterfield [27] and Smith [28] and the review article by Mason and Marero [29]. The approach usually taken is to assume that the laws of molecular diffusion are obeyed in the porous medium and then to work in terms of an *effective diffusivity*. This effective diffusivity, which is smaller than the binary molecular diffusion coefficient, is then selected so as to incorporate the various factors that were mentioned earlier.

The nature of such an effective diffusivity is illustrated in Fig. 2.4 on a plot of the effective diffusivity/molecular diffusivity ratio against the porosity for various granular materials. Numerous methods have been proposed for both the estimation of the effective diffusivity and for the representation of pore diffusion through the use of this parameter. A detailed review of these would be beyond the scope of this monograph, so here we shall present a recent treatment proposed by Mason and coworkers [29, 30], who suggested that the diffusive flux of species A in a binary mixture in an isothermal porous solid may be represented by the expression

$$N_A = - D_{Aeff} \; \nabla C_A + X_A \, \delta_A N - X_A \gamma_A (C_T B_0 / \mu) \, \nabla p \qquad (2.3.1)$$

$$\underset{\substack{\text{diffusive} \\ \text{flux}}}{} \quad \underset{\substack{\text{effective} \\ \text{diffusivity}}}{} \quad \underset{\substack{\text{concentration} \\ \text{gradient}}}{} \quad \underset{\substack{\text{bulk flow} \\ \text{due to} \\ \text{diffusion}}}{} \quad \underset{\substack{\text{transport due to} \\ \text{gradient in} \\ \text{pressure}}}{}$$

and

$$N - \beta_A N_A = -(D_{BK}/RT)[1 + (B_0 p / D_K)] \, \nabla p \qquad (2.3.2)$$

where

$$N = N_A + N_B, \quad \text{i.e., the total flux} \qquad (2.3.3)$$

$$(D_{Aeff})^{-1} = (D_{AK})^{-1} + (D_{ABeff})^{-1} \qquad (2.3.4)$$

$$\delta_A = D_{Aeff}/D_{ABeff} \qquad (2.3.5)$$

$$\gamma_A = D_{Aeff}/D_{AK} \qquad (2.3.6)$$

$$\beta_A = 1 - (D_{BK}/D_{AK}) \qquad (2.3.7)$$

and

$$(D_K)^{-1} = (x_A/D_{AK}) + (x_B/D_{BK}) \qquad (2.3.8)$$

N_A is the flux of component A in moles per unit time per unit total (pore + solid) area normal to the flow, C_A the concentration of component A (moles per unit volume of pore), X_A the mole fraction of component A $(= C_A/C_T)$, C_T the total molar concentration $(= C_A + C_B)$, B_0 a parameter characteristic of the porous solid (but not of the diffusing–flowing gas), μ the viscosity of the gas mixture, p the total pressure, T the absolute temperature, D_{AK} the "Knudsen diffusivity" for component A, and D_{ABeff} the "effective molecular diffusivity" for the gas mixture AB.

An equation equivalent to (2.3.1) was derived independently by Gunn and King [31]. The governing equations (2.3.1)–(2.3.8) may seem quite complex, but Mason and coworkers have shown how this general expression may be simplified to more familiar forms for certain special conditions.

For uniform pressure

$$N_A/N_B = -(M_B/M_A)^{1/2} \qquad \text{(Graham's law of diffusion)} \qquad (2.3.9)$$

$$N_A = -\frac{D_{Aeff}\,\nabla C_A}{[1 - X_A\,\delta_A(1 - (M_A/M_B)^{1/2})]} \qquad (2.3.10)$$

For small pores, $D_{AK} \ll D_{ABeff}$

$$D_{Aeff} \approx D_{AK}, \qquad \delta_A \approx 0$$

and

$$N_A \approx -D_{AK}\,\nabla C_A \qquad (2.3.11)$$

which is the Knudsen diffusion equation.

For large pores, $D_{AK} \gg D_{ABeff}$

$$D_{ABeff} \approx D_{Aeff}, \qquad \delta_A \approx 1$$

$$N_A \approx \frac{-D_{ABeff}\,\nabla C_A}{[1 - X_A(1 - (M_A/M_B)^{1/2})]} \qquad (2.3.12)$$

which is simply Fick's first law written for the case

$$N_A/N_B = -(M_B/M_A)^{1/2} \qquad (2.3.13)$$

(this relationship has been found to hold true provided the pores are not excessively large).

For equimolar counterdiffusion $(N_A = -N_B)$

$$N_A = -\frac{C_T[1/x_A + 1/x_B]\,\nabla x_A}{(1/x_A\,D_{Aeff}) + (1/x_B\,D_{Beff})} \qquad (2.3.14)$$

$$dp/dx_A = p(D_{BK} - D_{AK})/(A_0 + A_1 p + A_2 p^2) \qquad (2.3.15)$$

where

$$A_0 = x_A\,D_{AK} + x_B\,D_{BK} \qquad (2.3.16)$$

$$A_1 = (D_{AK}\,D_{BK}/D_{ABeff}) + B_0/\mu \qquad (2.3.17)$$

$$A_2 = [(x_A\,D_{BK} + x_B\,D_{AK})/p D_{ABeff}]B_0/\mu \qquad (2.3.18)$$

It is seen that the r.h.s. of Eq. (2.3.15) tends to zero at low and high values of p.

For flow of pure gas A

$$N_A = [-D_{AK} - (pB_0/\mu)] \, \nabla p/RT \qquad \text{(Darcy's law)} \qquad (2.3.19)$$

Inspection of Eqs. (2.3.1)–(2.3.9) shows that diffusion and flow of a binary gas mixture AB through a porous medium is characterized by the parameters D_{ABeff}, D_{AK}, D_{BK}, μ, and B_0, the first three of which are dependent on both the gas mixture and the solid. A less rigorous, but certainly more convenient approach involves the postulates given by the following two equations:

$$D_{ABeff} = (\varepsilon/\tau)D_{AB} \qquad (2.3.20)$$

where ε is the porosity of the solid, τ a tortuosity factor characteristic of the solid, D_{AB} the (ordinary) molecular diffusivity for the pair AB, and

$$D_{AK} = \tfrac{4}{3}(8RT/\pi M_A)^{1/2}K_0 \qquad (2.3.21)$$

where the term in parentheses can be interpreted as the root mean square of the velocity of gas molecules A; K_0 is a parameter characteristic of the solid (with the dimensions of length).

Equations (2.3.20) and (2.3.21) merely serve as definitions of τ and K_0 and are based on the presumption (not necessarily correct) that as defined, these two quantities have the same value for different gases (but the same porous solid).

Flow and diffusion of gas in a porous medium is therefore now characterized by the parameters D_{AB}, M_A, M_B, μ, ε/τ, K_0, and B_0. M_A and M_B are readily determined, D_{AB} and μ may be found from published data or estimated by methods discussed previously. We note that further simplifications are possible under certain circumstances. *When the pores are large,* knowledge of ε/τ allows us to define D_{Aeff}. *For very small pores,* Knudsen diffusion will predominate; thus information on K_0 will allow us to define the problem.

Extensive measurements have been reported in the literature on tortuosity factors for commonly used catalyst pellets; some typical values of τ are shown in Fig. 2.5 and are within the range $1.5 \leq \tau \leq 10$.

Little information is available, however, on experimentally measured tortuosities and Knudsen diffusion coefficients for noncatalytic gas–solid reaction systems [32, 33]. The techniques available for the characterization of porous solids and the measurements of pore diffusion phenomena will be discussed in Chapter 6.

Methods are available for estimating the parameters ε/τ, K_0, and B_0 from first principles for certain idealized geometries; two of these will be briefly mentioned:

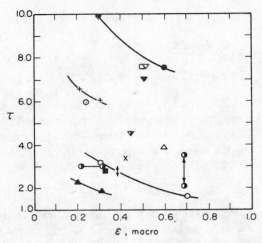

FIG. 2.5. A plot of tortuosity factor against macroporosity for several materials [94].

(a)   For a solid consisting of round pores, all of uniform pore diameter d, and all running parallel to the macroscopic concentration gradient, we have

$$\varepsilon/\tau = \varepsilon \tag{2.3.22}$$

$$K_0 = \varepsilon(d/4) \tag{2.3.23}$$

$$B_0 = \varepsilon(d^2/32) \tag{2.3.24}$$

(b)   Another idea of much broader potential practical application is the "dusty gas model" of Mason *et al.* [34], who proposed that for a solid matrix composed of uniform spherical grains we have

$$K_0^{-1} = (128/9)(n_d\tau/\varepsilon)r_g^2(1 + \pi/8) \tag{2.3.25}$$

where r_g is the radius of the spherical grains and $n_d = 3(1 - \varepsilon)/4\pi r_g^3$ is the number of solid grains per unit volume of porous solid. Let us illustrate the application of these two models by a simple example.

Example 2.3.1   (a)   A solid has uniform cylindrical pores all running parallel to each other. Calculate D_{AK} for $d = 2\ \mu m$, $\varepsilon = 0.5$, $T = 500°C$, and $M_A = 2$.

SOLUTION   From Eqs. (2.3.22)–(2.3.24), $\varepsilon/\tau = 0.5$; $B_0 = (0.5)(4 \times 10^{-8})/32 = 0.625 \times 10^{-9}$; $K_0 = (0.5)2 \times 10^{-4}/4 = 2.5 \times 10^{-5}$. From Eq. (2.3.21),

$$D_{AK} = \frac{4}{3}\left(\frac{8RT}{\pi M_A}\right)^{1/2} K_0 = \frac{4}{3}\left(\frac{8 \times 8.314 \times 10^7 \times 773}{3.14 \times 2}\right)^{1/2} 2.5 \times 10^{-5}$$

$$= 9.54\ cm^2/sec$$

(b)   Assume that a solid consists of spherical grains of diameter 0.2 μm. Calculate D_{AK} for the conditions of (a). It was found that the solid had $D_{ABeff} = 0.99$ for $H_2O - H_2$ and $D_{AB} = 5.15$.

SOLUTION   From Eq. (2.3.20),

$$\varepsilon/\tau = D_{ABeff}/D_{AB} = 0.99/5.15 = 0.192$$

From Eq. (2.3.25),

$$K_0^{-1} = \frac{128}{9} \frac{(1-0.5) \times 3}{4\pi(0.1 \times 10^{-4})(0.192)} \left(1 + \frac{\pi}{8}\right) = 12.3 \times 10^5$$

$$D_{AK} = \tfrac{4}{3}(8RT/\pi M_A)^{1/2} K_0 = 0.31 \text{ cm}^2/\text{sec}$$

It is thus seen that Knudsen diffusion would predominate for the small grains of (b) whereas molecular (bulk) diffusion (appropriately modified by tortuosity) would predominate for the large pores of (a).

Inspection of the form of Eqs. (2.3.4), (2.3.21), and (2.3.25) indicates the effect of the grain size on the pore diffusion coefficient, which is sketched in Fig. 2.6. It is seen that for large grains (i.e., large pores) D_{Aeff} is independent of the pore size; for small grains, i.e., small pores, D_{Aeff} is directly proportional to the grain size. Finally, there is an intermediate region between these two asymptotes.

For a given solid structure, r_g or the pore size is fixed; however, in many cases it is still possible to effect a change from bulk (molecular) diffusion to Knudsen diffusion or vice versa, by altering the total pressure of the system.

Figure 2.7 [44] shows a plot of the diffusive flux of ethylene in an ethylene–hydrogen mixture through a porous plug at various pressures. It is seen that at low pressures Knudsen diffusion predominates, whereas molecular (bulk) diffusion is the controlling factor at high pressures. This behavior is clearly consistent with physical reasoning, because the lower the pressure, the larger the mean free path of the gas molecules.

This dependence of the diffusive mechanism on the pressure of the system is of some practical importance because it provides a useful independent tool for discriminating between Knudsen and molecular diffusion by carrying out measurements at different pressures.

In order to illustrate the possible role of the viscous contribution [i.e., the last term on the r.h.s. of Eq. (2.3.1)] let us consider the following example.

Example 2.3.2   An anisotropic solid is made up of a large number of solid fibers lying parallel to each other and in contact. For practical purposes the interstices between fibers may be regarded as cylindrical pores, all of a uniform diameter of 0.15 μm. The porosity of the solid is 0.18. A pressure gradient of 0.1 atm/cm exists in the direction of the pores, at some region of

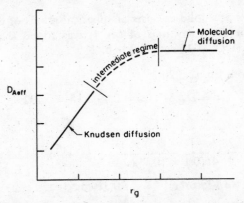

FIG. 2.6. A plot of effective diffusivity against grain size for diffusion in a granular porous medium.

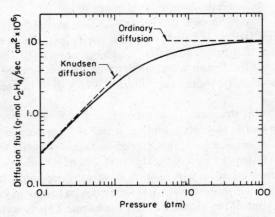

FIG. 2.7. The diffusive flux of ethylene in ethylene–hydrogen mixtures through a porous plug at various pressures [94].

the solid. Calculate the mass flux of component A in this region if there is no other gas component present. Temperature, $300°K$; pressure, 1 atm; gas viscosity, 2.0×10^{-4} g/cm sec; and M_A, 32.

From Eqs. (2.3.22)–(2.3.24),

$$K_0 = \varepsilon d/4 \qquad \text{and} \qquad B_0 = \varepsilon d^2/32$$

The two parameters characteristic of the solid are calculated as

$$K_0 = 0.18 \times 0.15 \times 10^{-4}/4 = 0.675 \times 10^{-6} \text{ cm}$$
$$B_0 = 0.18 \times 0.15^2 \times 10^{-8}/32 = 1.27 \times 10^{-12} \text{ cm}^2$$

From Eq. (2.3.21) we have

$$D_{AK} = \frac{4}{3}\left(\frac{8RT}{\pi M_A}\right)^{1/2} K_0 = \frac{4}{3}\left(\frac{8 \times 8.314 \times 10^7 \times 300°K}{3.14 \times 32}\right)^{1/2} \times 0.675 \times 10^{-6}$$

$$= \tfrac{4}{3}(19.9 \times 10^8)^{1/2} \times 0.675 \times 10^{-6} = 0.0402 \text{ cm}^2/\text{sec}$$

$$0.1 \text{ atm/cm} \equiv 1.01 \times 10^5 \text{ g/cm}^2 \text{ sec}^2$$

$$\nabla p/RT = 1.01 \times 10^{-2}/8.314 \times 300 \text{ mole/cm}^4$$

$$= 4.05 \times 10^{-6} \text{ mole/cm}^4$$

$$pB_0/\mu = 1.01 \times 10^{-6} \times 1.27 \times 10^{-12}/2 \times 10^{-4} = 0.0064 \text{ cm}^2/\text{sec}$$

Thus from Eq. (2.3.19),

$$N_A = (-0.0402 \text{ cm}^2/\text{sec} - 0.0064 \text{ cm}^2/\text{sec})4.05 \times 10^{-6} \text{ g-mole/cm}^4$$

$$= -0.0466 \times 4.05 \times 10^{-6} = -1.88 \times 10^{-7} \text{ g-mole/cm}^2 \text{ sec}$$

(the minus sign indicating that the flux is in the direction of decreasing pressure, as we would expect). We see that under the circumstances of this example the viscous flow contribution to the mass flux is small compared with the Knudsen flow contribution. It is left as an exercise for the reader to show that this is not the case if the mean pore size is 1.5 μm.

We stress that Eq. (2.3.1) is the rigorous starting point for describing diffusion in porous media for isothermal conditions. Simplified forms of this equation are used frequently by neglecting both the bulk flow and the viscous flow terms. The bulk flow term may be neglected only for equimolar counter-diffusion or when the mole fraction of the transferred substance is small.

The problem of viscous flow has been recently examined by Evans [35], and Evans and Song [36], who have shown that under certain circumstances errors may be introduced by neglecting the viscous flow term, which could be important for intermediate pore sizes and when there is a net generation of product gases.

MORE COMPLEX MODELS FOR PORE DIFFUSION

The previously described models for the structural parameters B_0, K, and τ, while straightforward, are oversimplified. More complex equations have been developed to describe diffusion in solids with broad pore size distribution. The model of Johnson and Stewart [37] (henceforth called the parallel-pore model) assumes that each pore makes its contribution to the total diffusive flux independently of the others. For isobaric conditions, the total flux is then given by

$$N_A = \int_0^\infty N_A'(r)\alpha(r)\,dr$$

$$= -\nabla C_A \int_0^\infty D'_{Aeff}(r)\alpha(r)\,dr + x_A \int_0^\infty \delta_A(r)N'(r)\alpha(r)\,dr \quad (2.3.26)$$

where $N_A'(r)$ is the flux of component A in pores of radius r, $\alpha(r)\,dr$ the volume fraction of pores with radius between r and $r + dr$, D'_{Aeff} the effective diffusivity for pores of radius r, $\delta_A'(r)$ the value of δ_A [see Eq. (2.3.5)] for pores of radius r, and $N'(r)$ the total flux in pores of radius r. In systems of practical interest, the integrals on the r.h.s. of Eq. (2.3.26) can usually be approximated by sums.

Satterfield and Cadle [38] determined the tortuosity factors for 17 commercially manufactured, pelleted catalysts and catalyst supports using the parallel-pore model. Except for two materials that had been calcined at very high temperatures, all tortuosity factors fell between 3 and 7. For about half the catalysts, the tortuosity factor was about 4, regardless of macroporosity or composition.

An alternative model for diffusion in porous media has been proposed by Wakao and Smith [39], who noted that many solids of interest have a bidisperse pore structure. That is, the solids consist of compacts of solid particles that are themselves porous. The solid therefore contains micropores (pores within the particles) and macropores (interstices between particles) and diffusion occurs through macropores, through micropores, and through micropores and macropores in series. For diffusion at constant pressure, we have

$$N_A = -\left[\varepsilon_a{}^2 D_a + (1 - \varepsilon_a)^2 D_i + 2\varepsilon_a(1 - \varepsilon_a)\left(\frac{2}{1/D_a + 1/D_i}\right)\right] C_T \nabla X_A \quad (2.3.27)$$

where ε_a is the void fraction of macropores and ε_i is the void fraction of micropores,

$$D_a = \frac{D_{AB}}{(1 - \alpha X_A) + (D_{AB}/D_{AKa})} \quad (2.3.28)$$

$$D_i = \frac{D_{AB}\,\varepsilon_i{}^2/(1 - \varepsilon_a)^2}{(1 - \alpha X_A) + (D_{AB}/D_{AKi})} \quad (2.3.29)$$

$$\alpha = 1 + (N_B/N_A) \quad (2.3.30)$$

D_{AKa} is the Knudsen diffusivity in the macropores,

$$D_{AKa} = \tfrac{2}{3}(8RT/\pi M_A)^{1/2} r_a \quad (2.3.31)$$

and D_{AKi} is the Knudsen diffusivity in the micropores. It is stressed that this random-pore model of Wakao and Smith is completely predictive, requiring no tortuosity factor or other adjustable parameters.

Cunningham and Geankoplis [40] extended the random-pore model to solids with tridisperse pore size distributions and provided experimental results showing a better fit for the extended model. Again, the model is predictive and involves no adjustable parameters. The parallel- and random-pore models (together with other models) have been the subject of a recent review by Youngquist [41], who concluded that

> No a priori method for predicting diffusion and flow fluxes for gases through porous solids is yet available. However, on the basis of the relatively meager experimental data available, it appears that simple capillary models coupled with elementary knowledge of the pore size distribution and a few flux measurements may be used with reasonable success to extrapolate data. For solids with rather broad pore size distributions, parallel pore models have given good correlations of data obtained under a wide range of pressure conditions. Random pore models in some cases have been used with considerable success in predicting fluxes through pellets made from pressed microporous powders.†

It should be noted that much of the work done on pore diffusion has been largely motivated by the relevance to heterogenous catalysis and adsorption phenomena (molecular sieves). In these systems, pore diffusion takes place in an essentially unchanging environment and in a reproducible solid structure.

Pore diffusion encountered in gas–solid reaction systems is generally rather more complex, because the solid structure may change in the course of the reaction. Moreover, the actual nature of the porous matrix may also be less well defined. In selecting pore diffusion models for the description of gas–solid reaction systems care should be taken that the sophistication of the model is consistent with the accuracy of the information available on the behavior of the system.

2.4 Adsorption and Chemical Reaction

Having described the transport of the gaseous reactants and products from the bulk of a moving gas stream to the solid surface and through the pores of the solid in the preceding sections, let us now consider the chemical

---

† Reprinted with permission from Youngquist, *Ind. Eng. Chem.* **62**, 52 (1970). Copyright by the American Chemical Society.

reaction itself together with the adsorption of the gaseous reactants and products, which necessarily accompanies this process.

We note that bulk (molecular) diffusion is well understood and its laws hold irrespective of the materials involved. Moreover, good general approximations may also be made regarding the pore diffusion of gases. In the consideration of adsorption and chemical reaction phenomena, matters will become highly specific to the nature of the substances involved.

In most of the work done up to the present on gas–solid reactions, relatively little attention has been paid to adsorption and to the actual chemical reaction phenomena, and very simple first-order rate expressions were generally postulated.

We hope that the brief discussion of adsorption and chemical kinetics that follows will stimulate further interest in the application of these concepts even to studies of gas–solid reaction systems of the engineering type.

ADSORPTION OF GASES ON SOLID SURFACES

When a gas (or a fluid in general) is brought into contact with a solid, some of the gas molecules will be adsorbed on the surface of the solid, because the surface atoms have no like atoms above the surface plane with which to form chemical bonds. Since a surface atom is capable of forming a similar number of bonds to those in the body of the solid, it may attract fluid molecules in order to satisfy the bonding capacity. There are two types of adsorption—physical adsorption and chemisorption. A detailed discussion of adsorption is beyond the scope of this book. For this reason, we shall examine only the salient features relevant to gas-solid reactions. For an extensive treatment of this topic, the reader is referred to various texts [42–46] and articles [47–51].

In *physical adsorption*, the adsorbed species are attracted to the surface by van der Waals or dispersion forces, which are much weaker than the forces involved in chemical bonding. The heat evolved in the process is rather small, generally in the range of 1–10 kcal/g-mole, which corresponds approximately to the heat of condensation of a vapor. In contrast, *chemisorption* takes place as a result of much stronger interaction; here the forces involved are of the same orders of magnitude as those for the formation of chemical bonds, and the heat of chemisorption is in the range of 10–150 kcal/g-mole. (Small values of the heat of chemisorption are known and even endothermic chemisorption has been shown to exist [52].) Since such valence forces diminish rapidly with distance, no more than a monolayer of adsorbed species can be formed by chemisorption. Physical adsorption, however, can form many adsorbed layers, because of the long-range nature of the forces involved. As the number of layers increases, the process approaches that of condensation, sometimes resulting in capillary condensation. Chemisorption is mainly responsible for gas–solid reactions and catalysis on solid surfaces, while physical adsorption does not usually play a role in chemical reactions because of the weak interaction involved. However, physical adsorption provides a valuable tool for determining the physical properties of porous solids, the most important of which is the measurement of the surface area and the pore size distribution.

Owing to its similarity to the liquefaction of a gas or the condensation of a vapor, physical adsorption may occur to a similar extent on different surfaces under a similar condition. Chemisorption, however, is specific to the particular gas and solid, just as a chemical reaction is. The adsorbed gas (adsorbate) and the solid surface (adsorbent) must meet exact requirements in terms of chemical forces and orientation, in order for chemisorption to take place.

Another significant difference between the two types of adsorption is the rate at which the process (adsorption or desorption) occurs. Physical adsorption, like condensation, requires no or very small activation energy and hence may be assumed to occur very fast. In contrast, chemisorption, may require a high activation energy, which in general implies a slow rate. However, chemisorption with a very small activation energy is known [41, 53], and physical adsorption on porous solids may take place slowly if diffusion limits the rate of adsorption.

In general, each type of adsorption is important in a different range of temperatures. Physical adsorption occurs to a significant extent only at temperatures below the boiling point of the adsorbate, and the equilibrium amount adsorbed becomes negligible above the critical temperature. In contrast, chemisorption generally occurs at a significant rate in a higher temperature range, although the amount chemisorbed at equilibrium decreases with temperature.

ADSORPTION ISOTHERMS AND RATES OF CHEMISORPTION AND DESORPTION

Langmuir [54] developed the first important treatment of adsorption by assuming that

(1) Adsorption takes place only onto certain "sites" within the solid surface and each site can accommodate only one adsorbed species.

(2) The coverage is a monolayer.

(3) The adsorbed molecules are localized, in that they are attached to definite sites.

(4) The surface is energetically uniform.

(5) There is no interaction between adsorbed species, that is, the adsorption on an empty site is not affected by the presence of other adsorbed species at neighboring sites.

The rate at which the gaseous species is adsorbed is proportional to the number of molecules colliding with the solid surface, which in turn is proportional to the partial pressure p_A, at a fixed temperature. The adsorption rate must also be proportional to the number of empty sites.† If the fraction of the sites covered is θ, then $\mathcal{R}_a$, the rate of adsorption per unit surface area, may be written as

$$\mathcal{R}_a = k_a p_A (1 - \theta) \tag{2.4.1}$$

$\mathcal{R}_d$, the rate of desorption of the adsorbed species, is proportional to the amount of adsorbate on the surface. Thus

$$\mathcal{R}_d = k_d \theta \tag{2.4.2}$$

In Eqs. (2.4.1) and (2.4.2), k_a and k_d are proportionality constants (or rate constants) for adsorption and desorption, respectively.

At equilibrium, the rates of adsorption and desorption are equal, and the extent of coverage at equilibrium, θ_e, may be obtained by equating $\mathcal{R}_a$ with $\mathcal{R}_d$. Thus we have

$$\theta_e = K p_A / (1 + K p_A), \qquad K = k_a / k_d \tag{2.4.3}$$

Equation (2.4.3) is the well-known Langmuir adsorption isotherm.

Although the Langmuir theory depicts a highly idealized situation that many real systems do not follow, it is nevertheless of great value in developing the kinetic interpretation of the heterogeneous reactions on solid surfaces. The concepts of the Langmuir theory have been extended to multilayer physical adsorption by Brunauer et al. [55]. The BET isotherm takes the form

$$p_A / v(p_0 - p_A) = 1/v_m a + (a - 1)p_A / a v_m p_0 \tag{2.4.4}$$

where v is the volume (expressed at standard conditions) of gas adsorbed on unit mass of porous solid, v_m the volume of gas adsorbed at monolayer coverage, p_A the partial pressure of the adsorbed species in the gas phase, p_0 the vapor pressure of the adsorbate, and a a constant, characteristic of the system.

The BET isotherm is often used for the experimental determination of the specific surface area (surface area per unit mass) of a porous solid [42, 45, 48, 55, 56]. This is done by carrying out a series of adsorption measurements at various partial pressures and then plotting $p_A/(p_0 - p_A)v$ against p_A/p_0. In accordance with Eq. (2.4.4) a straight line should be obtained. v_m may be calculated from the slope S and intercept I because

$$v_m = (I + S)^{-1} \tag{2.4.5}$$

---

† Within the framework of Langmuir's assumptions.

If the area occupied by one molecule of the adsorbed species A_m is known, the surface area per unit mass S_g may be calculated from

$$S_g = (v_m N/v_g)A_m \qquad (2.4.6)$$

where N is Avogadro's number and v_g the volume occupied by one mole of gas at standard conditions.

Example 2.4.1 Here we propose to determine the surface area of a given NiO particle by low-temperature nitrogen adsorption. Experiments were carried out and the following measurements were taken:

p_A (cm Hg):	21	19	16	13	10
v (cm^3):	1.50	1.41	1.27	1.13	0.98

v is the volume of gas absorbed at standard conditions per unit mass of solid and $p_0 = 76$ cm Hg. The area per nitrogen molecule $A_m = 14.2 \times 10^{-16}$ cm^2.

SOLUTION

$p_A/v(p_0 - p_A)$:	0.255	0.236	0.210	0.180	0.1555
p_A/p_0:	0.276	0.25	0.21	0.17	0.13

The plot of $p_A/v(p_0 - p_a)$ versus p_A/p_0 (Fig. 2.8) gives

intercept: $I = 0.0675$, slope: $S = 0.675$

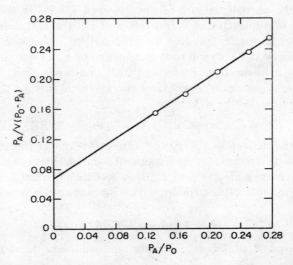

FIG. 2.8. Plot of Eq. (2.4.4). Slope, $S = 0.675$; intercept, $I = 0.0675$; ∘, experimental points.

From Eqs. (2.4.5)–(2.4.6) we have

$$v_m = (I + S)^{-1} = 1.347 \text{ cm}^3$$

$$S_g = \frac{v_m N A_m}{v_g} = \frac{(1.347)(6.023 \times 10^{23})(14.2 \times 10^{-16})}{22.4 \times 10^3}$$

$$= 5.32 \times 10^4 \text{ cm}^2/\text{g} = 5.32 \text{ m}^2/\text{g}$$

Among the assumptions made in deriving the Langmuir isotherm, the most questionable are those regarding the energy of adsorption and the interactions among the adsorbed species. Real surfaces are always nonuniform energetically, and even a noble gas shows considerable interaction between adsorbed molecules [54]. On a nonuniform surface, the energy of adsorption tends to decrease with coverage because the sites with higher energy of adsorption are occupied first. However, the interaction between the adsorbate molecules tends to cause the energy of adsorption to increase with coverage. In many cases, these two opposing effects cancel each other out approximately. The success of the Langmuir isotherm in describing the adsorption on many real surfaces may be attributed to these compensating factors.

It may be worthwhile to note closely the rate process of adsorption and desorption, in order to obtain a better understanding of the Langmuir theory. The rate at which gas molecules collide with a unit area of solid surface is given by

$$\angle = p_A (2\pi m \kappa T)^{-1/2} \tag{2.4.7}$$

where p_A is the partial pressure of the adsorbed species in the gas phase, m is the mass of a molecule of the species, and κ is the Boltzmann constant. Of the number of species colliding with the surface, only those striking the unoccupied portion $(1 - \theta)$ will have a chance to be adsorbed, since a monolayer coverage is assumed. Considering that the adsorption may be activated and only those molecules with sufficient energy would then be adsorbed, the rate of adsorption may be written as

$$\mathcal{R}_a = [\Omega/(2\pi m \kappa T)^{1/2}] p_A (1 - \vartheta) \exp(-E_a/RT) \tag{2.4.8}$$

where E_a is the activation energy for adsorption and Ω the condensation coefficient, which is defined as the fraction of the molecules colliding with the surface, with energies greater than E_a, that become adsorbed.

Similarly, the rate of desorption, which is always an *activated process*, is given by

$$\mathcal{R}_d = k_{d0} \theta \exp(-E_d/RT) \tag{2.4.9}$$

Again $\mathcal{R}_a$ and $\mathcal{R}_d$ are equal at equilibrium, and the isotherm may now be written

$$p_A = (k_{d0}/\Omega)(2\pi m \kappa T)^{1/2} [\theta_e/(1 - \theta_e)] \exp(-Q_a/RT) \tag{2.4.10}$$

where $Q_a \equiv -\Delta H_a = E_d - E_a$ is the heat of adsorption. Equation (2.4.10) reduces to the Langmuir equation, Eq. (2.4.3), if k_{d0}, Ω, and Q_a are independent of θ as assumed in the Langmuir derivation. On comparing Eqs. (2.4.3) and (2.4.10), the constant K in the Langmuir equation is seen to be given by

$$1/K = (k_{d0}/\Omega)(2\pi m\kappa T)^{1/2} \exp(-Q_a/RT) \quad (2.4.11)$$

If the gas molecules dissociate into n fragments on adsorption and each fragment occupies a site, then the rates may be written as

$$\mathscr{R}_a = k_a p_A (1 - \theta)^n \quad (2.4.12)$$

$$\mathscr{R}_d = k_d \theta^n \quad (2.4.13)$$

The adsorption isotherm now becomes

$$\theta_e = (Kp_A)^{1/n}/[1 + (Kp_A)^{1/n}] \quad (2.4.14)$$

If two or more different species are adsorbed simultaneously, the isotherm for nondissociative adsorption for each species may be given by

$$\theta_{ie} = K_i p_i/(1 + \sum_i K_i p_i) \quad (2.4.15)$$

where K_i and p_i are the Langmuir constant and the partial pressure respectively, of the ith species.

As pointed out earlier, the heat of adsorption is seldom independent of coverage and, in general, decreases with coverage. Two other widely used isotherms may be derived by considering this variation of the heat of adsorption with coverage.

The *Freundlich isotherm*, which originated as an empirical relation between coverage and pressure, may be derived by postulating the following relationship between the heat of adsorption and coverage [57, 58]†:

$$Q_a = Q_m \ln \theta \quad (2.4.16)$$

This assumption leads to the following relationship:

$$\theta_e = k_f p^{1/n} \quad (2.4.17)$$

where k_f is a proportionality constant and

$$n = Q_m/RT; \quad n > 1.0 \quad (2.4.18)$$

The *Temkin isotherm* is obtained by assuming that the heat of adsorption decreases linearly with coverage [60, 61]:

$$Q_a = Q_0(1 - \alpha\theta) \quad (2.4.19)$$

† Laidler [47] favors a slightly different derivation by Zeldowitch [59], which yields similar results.

where Q_0 is the heat of adsorption for $\theta = 0$ and α is a constant. This assumption gives the following:

$$\theta_e = k_T \ln(K_0 p) \tag{2.4.20}$$

where

$$k_T \equiv RT/\alpha Q_0 \tag{2.4.21}$$

and K_0 is the value of K in Eq. (2.4.11) with $Q_a = Q_0$.

Of the adsorption isotherms discussed above, the Langmuir equation is important because of its theoretical and conceptual contribution to the studies of adsorption and the kinetics of surface reactions. The Freundlich isotherm has found a wide use in correlating experimental data; the Temkin isotherm is obeyed by certain systems [62–64], but it is expected to apply only in the intermediate range of coverage due to the simplifying assumptions made in its derivation.

In deriving the expression for the *rate of adsorption on real surfaces*, one must again consider the changes of parameters with coverage. If E_a in Eq. (2.4.8) is assumed to change linearly with θ (say, $E_a = E_0 + \alpha\theta$), and the exponential term is considered to dominate the dependency on θ, we obtain the following expression, known as the Elovich equation [62, 65]:

$$d\theta/dt = a \exp(-\alpha\theta/RT) \tag{2.4.22}$$

where α is a constant. Usually the application of this expression is made in its integrated form:

$$\theta = (RT/\alpha) \ln[(t + t_0)/t_0] \tag{2.4.23}$$

where t_0 is a constant. A similar derivation may be made for the rate of desorption. Many systems have been found to obey the Elovich equation [51, 66] and a comprehensive review on the subject has recently appeared in the literature [51].

The theory of absolute reaction rates has also been applied to the prediction and interpretation of the rates of adsorption [42, 67]. The theory has proved useful in correlating and interpreting experimental data; however, due to the difficulty in describing accurately the chemical structure and energetics of the activated complex on the surface, the prediction of the rates has not been very successful so far.

We conclude our discussion of adsorption phenomena by stating that there exists a number of models for adsorption, which may be used for the interpretation of experimental measurements. The Langmuir equation and the further developments based on it provide a conceptually attractive picture of the dynamic equilibrium attained between the sorbate in the gaseous and the adsorbed state. From a practical viewpoint the BET equation is particularly helpful for the characterization of porous solids, through the determination of the surface area by physical adsorption studies.

The kinetics of adsorption processes is much less well understood and much further work needs to be done before the models currently available will be of immediate practical use.

THE KINETICS OF GAS–SOLID REACTIONS

The general description of the kinetics of gas–solid reactions is a very complex and difficult problem, especially for systems involving solid products, where not only the processes on the gas–solid interface but the advancement of the reaction interface between the reactant and product solids must also be taken into consideration.

The theory of the kinetics of heterogeneous catalytic reactions has been treated extensively elsewhere [68, pp. 119, 195; 69]. Here we shall only describe the basic principles underlying the derivation of the rate expression for gas–solid reactions using a simple system. Let us consider the reaction

$$A(g) + B(s) = C(g) + D(s) \tag{2.4.24}$$

The reaction process will involve the adsorption of the reactant gas A on the surface of solid B to form a surface complex X*, which may transform to another surface complex Y*, which then desorbs to give the gaseous product C and the solid product D. These steps, which are reversible in general, may be written

$$A + S \underset{k_{-1}}{\overset{k_1}{\rightleftharpoons}} X^* \tag{2.4.25}$$

$$X^* \underset{k_{-2}}{\overset{k_2}{\rightleftharpoons}} Y^* \tag{2.4.26}$$

$$Y^* \underset{k_3}{\overset{k_{-3}}{\rightleftharpoons}} C + S \tag{2.4.27}$$

where S designates the bare solid surface that is available to both A and C. When the reaction involves the complete gasification of solid, this idea poses no difficulty. When a solid product is formed, this implies that both the reactant and product gas compete for the same vacant site. The net rates of adsorption of A and desorption of C may be written with the aid of Eqs. (2.4.1) and (2.4.2):

$$\underset{\substack{\text{net rate of}\\\text{adsorption}\\\text{of A}}}{\gamma_1} = \underset{\substack{\text{rate of}\\\text{adsorption}\\\text{of A}}}{k_1 p_A \theta_S} - \underset{\substack{\text{rate of}\\\text{desorption}\\\text{of A}}}{k_{-1}\theta_{X^*}} \tag{2.4.28}$$

$$\underset{\substack{\text{net rate of}\\\text{desorption}\\\text{of C}}}{\gamma_3} = \underset{\substack{\text{rate of}\\\text{desorption}\\\text{of C}}}{k_{-3}\theta_{Y^*}} - \underset{\substack{\text{rate of}\\\text{adsorption}\\\text{of C}}}{k_3 p_C \theta_S} \tag{2.4.29}$$

where θ is the fraction of surface occupied by the various species indicated by the subscripts. The net rates at which the surface complexes are being produced may be written as

$$d\theta_{X\bullet}/dt = k_1 p_A \theta_S - k_{-1}\theta_{X\bullet} - k_2 \theta_{X\bullet} + k_{-2}\theta_{Y\bullet} \qquad (2.4.30)$$

$$d\theta_{Y\bullet}/dt = k_2 \theta_{X\bullet} - k_{-2}\theta_{Y\bullet} - k_{-3}\theta_{Y\bullet} + k_3 p_C \theta_S \qquad (2.4.31)$$

where the forward and reverse reaction (2.4.26) have been assumed to be first order with respect to the surface complex. Furthermore, we also have

$$\theta_S + \theta_{X\bullet} + \theta_{Y\bullet} = 1 \qquad \text{(overall balance on the sites)} \qquad (2.4.32)$$

The above rate expressions assume that the surface is energetically homogeneous, or average rate constants may be used.

The overall rate of reaction, expressed by the rate of production of C, may be obtained in terms of p_A by solving Eqs. (2.4.28)–(2.4.31). Strictly speaking, the θ's will change with time as p_A or p_C varies. Such a time-dependent complete solution could be obtained but the problem may be greatly simplified, without loss of much accuracy, if the steady state approximation is applied with regard to the number of surface sites. The physical implication of this assumption is that the rates of change of θ's are considered negligible compared with the overall rate of transformation of A into C, because the number of surface sites is much smaller than the number of molecules of A transformed through these sites (say $d\theta_{X\bullet}/dt, d\theta_{Y\bullet}/dt \ll k_1 p_A\theta_S$, $k_3 p_C\theta_S$). Therefore, there will be a dynamic steady state between the fluid phase concentration and the surface complexes.

In formulating the kinetic expression for a heterogeneous as well as a homogeneous reaction, this steady state approximation is widely used with regard to the rate of change of extremely reactive intermediates, which are the surface complexes. Mathematical justification of this assumption has been demonstrated for certain simple situations [70, 71]. Thus, on applying the steady state approximation, Eqs. (2.4.30) and (2.4.31) become

$$k_1 p_A \theta_S - k_{-1}\theta_{X\bullet} - k_2 \theta_{X\bullet} + k_{-2}\theta_{Y\bullet} = 0 \qquad (2.4.33)$$

$$k_2 \theta_{X\bullet} - k_{-2}\theta_{Y\bullet} - k_{-3}\theta_{Y\bullet} + k_3 p_C \theta_S = 0 \qquad (2.4.34)$$

From the steady state approximation it also follows that

$$\gamma_1 = \gamma_3 \qquad (2.4.35)$$

In principle we can obtain the general rate expression by considering the kinetics of all the steps shown in Eqs. (2.4.25)–(2.4.27); however, the resultant equation will contain many parameters. Such expressions may be used to obtain the overall rate if the values of these parameters are known or can be determined independently. Normally neither is the case, and the parameters must be evaluated by fitting the observed overall rate to the assumed reaction scheme. With many parameters to evaluate, the equation is too insensitive to test the validity of the reaction model. This difficulty can be alleviated if a particular step in the overall mechanism appears to control the overall rate. This will occur if the rate constants of one step are much smaller than those of the others, i.e., that step presents the major "resistance" to conversion of A into C. Since at steady state the net forward rates of all steps must be equal, the other reactions with larger rate constants will occur at near-equilibrium conditions. The step with the smallest rate constants is sometimes (somewhat erroneously) called "*the slow step*," although at steady state its net forward rate is the same as that of other steps. Let us now examine the effect of different rate-controlling mechanisms on the observable overall reaction rate.

SURFACE REACTION CONTROLLING

If the surface reaction is the slow step and hence controls the overall rate, the adsorption and desorption steps will be in near equilibrium. The equilibria may be calculated by setting $\gamma_1 = \gamma_3 = 0$ in Eqs. (2.4.28) and (2.4.29). We have

$$(\theta_{X\bullet})_{eq} = (k_1/k_{-1})p_A \theta_S = K_A p_A \theta_S \qquad (2.4.36)$$

and

$$(\theta_{Y\bullet})_{eq} = (k_3/k_{-3})p_C \theta_S = K_C p_C \theta_S \qquad (2.4.37)$$

where

$$K_A = k_1/k_{-1} \qquad (2.4.38)$$

$$K_C = k_3/k_{-3} \qquad (2.4.39)$$

The net forward rate of the surface reaction is then given by

$$\gamma_2 = k_2 \theta_{X\bullet} - k_{-2} \theta_{Y\bullet} \qquad (2.4.40)$$

Substituting Eqs. (2.4.36) and (2.4.37) into Eq. (2.4.40), we obtain

$$\gamma_2 = k_2 K_A p_A \theta_S - k_{-2} K_C p_C \theta_S \qquad (2.4.41)$$

The expression for θ_S may be obtained by combining Eqs. (2.4.32), (2.4.36), and (2.4.37) to give

$$\theta_S = 1/(1 + K_A p_A + K_C P_C) \qquad (2.4.42)$$

At steady state, γ_2 is also the same as the overall net forward rate, which is given by combining Eqs. (2.4.41) and (2.4.42):

$$\gamma = k_2 K_A \frac{p_A - (K_C/K_A K_S)p_C}{1 + K_A p_A + K_C p_C} \qquad (2.4.43)$$

where

$$K_S = k_2/k_{-2} \tag{2.4.44}$$

Since the reaction of surface complexes is not well understood, K_S is usually eliminated using its relation with the equilibrium constant for the overall reaction K, which may be written in terms of other equilibrium constants as follows:

$$K = \left(\frac{p_C}{p_A}\right)_{eq} = \left(\frac{\theta_{Y\ast}/K_C\,\theta_S}{\theta_{X\ast}/K_A\,\theta_S}\right)_{eq} = \frac{K_A}{K_C}\left(\frac{\theta_{Y\ast}}{\theta_{X\ast}}\right)_{eq} = \frac{K_A K_S}{K_C} \tag{2.4.45}$$

where we have made use of Eqs. (2.4.36) and (2.4.37) and the relationship $K_S = (\theta_{Y\ast}/\theta_{X\ast})_{eq}$. The expression for the overall rate of reaction now becomes

$$\gamma = k_2\,K_A\,\frac{p_A - p_C/K}{1 + K_A p_A + K_C p_C} \tag{2.4.46}$$

In deriving the rate expression we made a number of assumptions that are frequently used for reactions occurring at a gas–solid interface. For example, the Langmuir–Hinshelwood rate mechanism for heterogeneous catalysis assumes that the adsorption and desorption processes are in equilibrium and the rate of surface reaction is proportional to the concentration of surface complex formed by adsorption. On combining Eqs. (2.4.36) and (2.4.42), we have

$$\theta_{X\ast} = K_A p_A/(1 + K_A p_A + K_C p_C) \tag{2.4.47}$$

Then

$$\gamma = k_S \theta_{X\ast} = k_S[K_A p_A/(1 + K_A p_A + K_C p_C)] \tag{2.4.48}$$

When Eq. (2.4.46) is compared with Eq. (2.4.48), the analogy with the Langmuir–Hinshelwood mechanism is complete.

A rate expression similar to Eq. (2.4.46) has been applied to the reduction of iron oxide by hydrogen [72, 73], the reduction of nickel oxide by hydrogen [74], and the reaction of carbon with carbon dioxide and water vapor [75].

ADSORPTION OF REACTANT CONTROLLING

When the adsorption of A is the slow step, the surface reaction and the desorption of C may be assumed to occur at near equilibrium. Thus, $\theta_{Y\ast}$ is still given by Eq. (2.4.37), and by setting $\gamma_2 = 0$ in Eq. (2.4.40), we obtain

$$\theta_{X\ast} = \theta_{Y\ast}/K_S = (K_C K_S)p_C \theta_S \tag{2.4.49}$$

The net rate of adsorption of A, which again is equal to the overall rate of reaction, is then obtained from Eqs. (2.4.28) and (2.4.49). Thus we have:

$$\gamma = k_1 p_A \theta_S - k_{-1}(K_C/K_S)p_C \theta_S \tag{2.4.50}$$

On combining Eqs. (2.4.32), (2.4.37), (2.4.45), and (2.4.49), we obtain

$$\theta_S = \frac{1}{1 + (K_A/K + K_C)p_C} \qquad (2.4.51)$$

Substituting θ_S in Eq. (2.4.50) and rearranging the resultant expression, we have

$$\gamma = k_1 \frac{p_A - p_C/K}{1 + (K_A/K + K_C)p_C} \qquad (2.4.52)$$

DESORPTION OF PRODUCT CONTROLLING

In this case, the expression for γ should be obtained by using Eq. (2.4.29) and

$$\theta_{Y^\bullet} = K_S \theta_{X^\bullet} = K_S K_A p_A \theta_S \qquad (2.4.53)$$

$$\theta_S = 1/[1 + (K_A + K_C K)p_A] \qquad (2.4.54)$$

Thus we have

$$\gamma = k_3 K \frac{p_A - p_C/K}{1 + (K_A + K_C K)p_A} \qquad (2.4.55)$$

In all the rate expressions, Eqs. (2.4.46), (2.4.52), and (2.4.55), the rate becomes zero, and overall equilibrium is reached when $p_A = p_C/K$. The effect of solid species on equilibrium is through the equilibrium constant K, which is determined by the standard free energy change ΔF°, of the overall reaction:

$$K = \exp(-\Delta F^\circ/RT) \qquad (2.4.56)$$

Since the solid surface is usually nonuniform in many gas–solid systems, the formation of reaction nuclei at distinct surface locations and their propagation play important roles. The detailed description of the nucleation process is beyond the scope of this monograph. The reader is referred to other sources on the topic [76–78].

It may be worthwhile to summarize our discussion of the kinetic step in gas–solid reactions by stating that on the basis of the analogy with heterogeneous catalysis, the overall reaction process may be broken down into individual steps, namely, adsorption, the formation of a surface complex, and desorption. Each of these steps can be regarded as reversible and assigned both a forward and a reverse rate constant. The resultant overall rate expressions may then be simplified with the aid of the steady state approximation and a knowledge of the overall reaction equilibrium. This then results in characteristic rate expressions containing certain constants and the partial

pressures of the gaseous reactants and products in the bulk. These relationships may be very useful for the interpretation of experimental data with a
view of identifying the rate-controlling step. In closing, it should be remarked
that *the expressions for surface kinetics are generally nonlinear in the partial
pressure of the reactant gas*—notwithstanding that the assumption of linearity
has usually been made by most investigators of gas–solid reactions.

2.5 Heat Transfer

Many gas–solid reactions are sufficiently exothermic or endothermic that
the progress of the reaction is markedly affected by the temperature change
that accompanies the reaction. Moreover, many reactions will take place at
measurable rates only at elevated temperatures; it follows that both heat
transfer between a pellet and a moving gas stream and conductive heat
transfer within a pellet itself need to be discussed.

Since the subject of heat transfer is extensively covered in many engineering texts [1, 79, 80] only a very brief survey of topics that are of particular
relevance to gas–solid reaction systems will be presented here.

CONVECTIVE HEAT TRANSFER BETWEEN A SINGLE PARTICLE
AND A GAS STREAM

Let us consider a solid particle the surface temperature of which is
maintained at T_s in contact with a moving gas stream, which has a bulk
temperature T_0, as sketched in Fig. 2.9.

The heat flux from the solid surface may be related to T_0 and T_s in the
following manner:

$$q_y = h(T_s - T_0) \tag{2.5.1}$$

where q_y is the heat flux (rate of heat flow per unit area of external particle
surface), and h the heat transfer coefficient. The close analogy between Eqs.
(2.5.1) and (2.2.2) should be readily apparent. Just as in the case of the mass
transfer coefficients, in the majority of cases we have to rely on empirical
correlations for obtaining the heat transfer coefficient for various fluid–
particle systems. It may be shown [2–4] that these empirical correlations
should have the following form:

For forced convection

$$N_{Nu} = f(N_{Re}, N_{Pr}) \tag{2.5.2}$$

where $N_{Nu} = hL/k_g$ is the Nusselt number, $N_{Pr} = C_p \mu/k_g$ the Prandtl number,
and k_g the thermal conductivity of the gas.

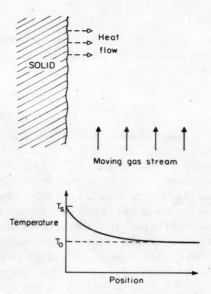

FIG. 2.9. Heat transfer from a solid into a moving gas stream.

For natural convection

$$N_{Nu} = f(N_{Gr}, N_{Pr})$$ (2.5.3)

where $N_{Gr} = gL^3\beta_H \Delta T/v^2$ is the Grashof number for heat transfer. Here $\beta_H = 1/T$ for ideal gases (with T measured on the absolute scale) and $\Delta T = |T_0 - T_s|$ is the characteristic temperature difference. Because of the close analogy between heat transfer and mass transfer involving a solid surface and a moving gas stream *the correlations for the heat transfer coefficient may be obtained from those previously given for mass transfer by replacing the Sherwood and Schmidt numbers with the Nusselt and Prandtl groups, respectively.*

Thus, the Ranz and Marshall correlations for heat transfer may be written as

$$N_{Nu} = 2.0 + 0.6 N_{Re}^{1/2} N_{Pr}^{1/3}$$ (2.5.4)

Moreover, the previously given relationship for j_D—the j factor for mass transfer [Eq. (2.2.12)]—may also be used for estimating the heat transfer coefficient by assuming that

$$j_D \simeq j_H$$ (2.5.5)

where

$$j_H = (hL/k_g)N_{Pr}^{-1/3}N_{Re}^{-1}$$

THERMAL RADIATION

At high temperatures thermal radiation may also become an important mechanism for heat transfer between a solid particle and its surroundings.

The rate at which a solid body emits radiation is given by the Stefan–Boltzmann law:

$$q_R = \tilde{\varepsilon}\sigma T_s^{\,4} \tag{2.5.6}$$

where q_R is the radiant flux emitted by the surface, $\tilde{\varepsilon}$ is the emissivity of the surface, σ is Boltzmann's constant, 1.37×10^{-12} cal/cm^2 sec $^\circ$K^4, and T_s is the surface temperature in absolute units. The calculation of radiant heat exchange between bodies or surfaces of irregular geometry is a fairly complex task because the procedure has to involve the establishment of heat balances in which account must be taken of the reflected and absorbed radiation.

For a detailed discussion of radiation, the reader is referred to the textbook literature [81–83]. We note here that a simple expression holds for describing the radiant flux received by a small pellet, located in a large cavity, the surface of which is isothermal, in the presence of nonabsorbing gas:

$$q_R' = \tilde{\varepsilon}\sigma(T_E^{\,4} - T_s^{\,4}) \tag{2.5.7}$$

where T_E is the temperature of the walls of the cavity and q_R' the net radiative flux at the solid surface. The situation described by Eq. (2.5.7) is often a good approximation for radiative heat transfer to single pellets held in a tubular furnace, an arrangement frequently used in kinetic studies.

If we wish to compare the relative importance of radiative and convective heat transfer, it is convenient to define a radiative heat transfer coefficient, h_r, in the following manner:

$$h_r = \tilde{\varepsilon}\sigma(T_E^{\,2} + T_s^{\,2})(T_E + T_s) \tag{2.5.8}$$

so that

$$q_R' = h_r(T_E - T_s) \tag{2.5.9}$$

Example 2.5.1   A burning coal particle 1 mm in diameter is exposed to air at 20°C. If the particle is at 1200°C and is moving at a velocity of 100 cm/sec and $\tilde{\varepsilon} = 0.53$, estimate:

(a)   the radiative heat transfer coefficient,
(b)   the convective heat transfer coefficient due to forced convection, and
(c)   the heat transfer coefficient due to natural convection.

ADDITIONAL DATA $C_p = 0.25$ cal/g $^\circ$K, $\mu = 0.0167$ cp, $k_g = 6.28 \times 10^{-5}$ cal/cm sec $^\circ$K, $\rho = 0.0012$ g/cm^3.

SOLUTION   (a) From Eq. (2.5.8),

$$h_r = (0.53)(1.37 \times 10^{-12})(293^2 + 1473^2)(293 + 1473)$$
$$= 0.00289 \text{ cal/cm}^2 \text{ sec } °K$$

(b) $N_{Pr} = C_p \mu/k_g = (0.25)(0.0167 \times 10^{-2})/6.28 \times 10^{-5} = 0.665$

$N_{Re} = L\rho/\mu = (0.1)(100)(0.0012)/0.0167 \times 10^{-2} = 71.86.$

From Eq. (2.5.4),

$$N_{Nu} = 2 + 0.6 N_{Re}^{1/2} N_{Pr}^{1/3} = 2 + 0.6(71.86)^{1/2}(0.665)^{1/3} = 6.64$$

$hL/k_g = 6.64,$ $h = (6.64)(6.28 \times 10^{-5})/(0.1) = 0.00417 \text{ cal/cm}^2 \text{ sec } °K$

(c) $N_{Gr} = gL^3\beta \, \Delta T/v^2 = (981) \times (0.1)^3 \times (1473 - 273)/\frac{1}{2}(293 + 1473)$

$\times (0.000167/0.0012)^2 = 67.69$

From Ranz and Marshall [6],

$$N_{Nu} = 2 + 0.60 N_{Gr}^{1/4} N_{Pr}^{1/3}$$
$$= 2 + 0.60 \times (67.69)^{1/4}(0.665)^{1/3} = 2.901$$
$$h = N_{Nu} k_g/L = 2.901 \times 6.28 \times 10^{-5}/0.1 = 0.00182 \text{ cal/cm}^2 \text{ sec } °K$$

CONDUCTIVE HEAT TRANSFER IN POROUS SOLIDS

The conduction of heat is described by Fourier's law, which states that the heat flux is proportional to the temperature gradient:

$$q_y = -k \, \partial T/\partial y \qquad \text{or} \qquad q = -k \, \nabla T \qquad (2.5.10)$$

Information is available on the thermal conductivities of gases, liquids, and dense solids [1, Chapter 8; 79; 84–86]. Moreover, the thermal conductivity of gases may be estimated by procedures analogous to those employed for the estimation of the diffusivity and the viscosity of gases.

In contrast, the thermal conductivity of porous solids is rather less well understood. The usual approach taken for the representation of porous solids is to consider them homogeneous from the macroscopic viewpoint, and then to relate the heat flux to the temperature gradient by means of an *effective conductivity*.

Ideally, one would wish to have experimentally measured values of the effective thermal conductivity. However, there are methods for estimating the conductivity of porous solids. In estimating the thermal conductivity, we can consider two extreme types of behavior, with real systems occupying an intermediate position.

(a) The material consists of a continuous solid phase containing numerous closed and isolated pores filled with a gas (or liquid) that has a thermal conductivity much lower than that of the solid phase. Commercially available refractories approximate this case and their thermal conductivity may be estimated by [87]

$$k_e = k_s(1 - \varepsilon) \qquad (2.5.11)$$

where k_e is the effective thermal conductivity of the porous solid, k_s is the thermal conductivity of the solid at zero porosity, and ε is the porosity of the solid. In such systems, the pores play a negligible part in transferring heat and, as might therefore be expected, the effective thermal conductivity is independent of the gas. A slightly different relationship was proposed by Maxwell [88] and Rayleigh [89] for solids of this type:

$$k_e = k_s[(1 - \varepsilon)/(1 + 0.5\varepsilon)] \qquad (2.5.12)$$

(b)   The material consists of a gaseous continuous phase (of relatively low thermal conductivity) containing individual solid particles that are in point contact with their nearest neighbors. Here the principal resistance to the overall rate at which heat is conducted through the matrix is in the vicinity of the points of contact of the particles, because heat is transferred from a particle to its neighbors by conduction through the gas adjacent to the point contacts. Under these conditions, the effective thermal conductivity of the porous solid depends on the thermal conductivity of the gas but not on the thermal conductivity of the solid. The effective thermal conductivity falls to zero as the gas within the pores is evacuated.

A practical example approaching such a case is a loosely compacted mass of metal powder. In such a composite, the particles have a finite area of contact with their nearest neighbors and the effective thermal conductivity is not zero under vacuum. The effective thermal conductivity of such real solids is dependent on the thermal conductivity of the gas, and to a lesser extent that of the solid, together with the extent of contact of adjacent particles, particle surface roughness, and other details of the solid geometry. Under such circumstances, prediction of the effective thermal conductivity is extremely difficult. The topic has been discussed by Woodside [90], Kunii and Smith [91], Luikov et al. [92], and many others [93]. Perhaps the safest procedure in this event is to measure the thermal conductivity.

It is of interest to note that sintering of a porous compact will change its character from case (b) toward case (a) and will thereby increase the effective thermal conductivity.

An extensive discussion of the thermal conductivity of porous catalyst particles is presented by Satterfield [94], who suggested that the effective thermal conductivity of most catalyst pellets falls in the range $8–4 \times 10^{-4}$ cal/cm sec °C.

COMBINED CONDUCTION AND CONVECTION

The reaction of solid particles is inherently an unsteady state process, so that the previously discussed gas to solid convection and intraparticle conduction phenomena have to be applied in a time-dependent manner.

In the general case, the unsteady state conduction within a solid particle where heat is generated or absorbed by a chemical reaction is given by Fourier's second law [95]:

$$\partial T/\partial t = \nabla \kappa_e \nabla T + [Q/(1-\varepsilon)\bar{\rho}_s C_p] \qquad (2.5.13)$$

where Q is the rate of heat generation (due to chemical reaction) per unit volume of pellet and κ is the effective thermal diffusivity. The convective heat transfer to or from the particles is then described by a boundary condition to Eq. (2.5.13), which would take the following form

$$q = -k_e \nabla T = h(T_s - T_0) \qquad \text{at the solid surface} \qquad (2.5.14)$$

where q is the heat flux out of the solid. The relative importance of convective heat transfer to the particle and conductive heat transfer within the particle may be assessed with the aid of the Biot number, defined as [79, 80]

$$N_{BiH} = hL/k_e \qquad (2.5.15)$$

It has been shown that when $N_{BiH} < 0.2$ the major resistance to transport of heat from the pellet interior to the surroundings is heat transfer between the pellet surface and the surrounding gas by convection (or radiation); under these conditions we may neglect temperature gradients within the solid particle. The rate of change of the temperature of the particle is given by

$$V_p \rho_s (1-\varepsilon) C_p \, dT_s/dt = V_p Q + hA_p(T_0 - T_s) \qquad (2.5.16)$$

where T_s is the (uniform) temperature of the particle. The use of Eqs. (2.5.15) and (2.5.16) will now be illustrated by a simple example.

Example 2.5.2 Prior to reacting an iron oxide pellet with hydrogen in a tubular furnace, we wish to preheat the specimen to the reaction temperature of 700°C by contacting it with a helium stream at 700°C. If the convective heat transfer coefficient is 10^{-4} cal/cm² °C sec, estimate the time required for the specimen to reach 650°C.

ADDITIONAL DATA $\bar{\rho}_s(1-\varepsilon)C_p = 0.21$ cal/cm³ (mean value over the temperature range); $k_e = 5 \times 10^{-4}$ cal/cm sec °C; initial pellet temperature, 20°C; diameter of spherical pellet, 1 cm.

SOLUTION The Biot number is

$$N_{BiH} = 10^{-4} \times 1.0/5 \times 10^{-4} \simeq 0.1$$

so that temperature gradients within the pellet may be neglected. Thus we may use Eq. (2.5.16) with $Q = 0$. Upon rearrangement and integration, we have

$$(T_0 - T_s)/(T_0 - T_{si}) = \exp[-hA_p t/V_p \rho_s C_p(1 - \varepsilon)]$$
$$50/680 = \exp[(-10^{-4} \times 3t)/0.12 \times 0.5)]$$

i.e., $t \simeq 920$ sec.

We note that for a practical case, at these temperatures, thermal radiation would have been quite significant, which would have reduced the heating time requirements substantially.

SIMULTANEOUS HEAT AND MASS TRANSFER

In some gas–solid reactions, the reactant gas and the solid surface may differ in temperature substantially. Thus, the heat and mass transfer processes described here will occur simultaneously.

When substantial temperature differences exist, care has to be taken in selecting the mean or "film" temperature at which the property values such as diffusivity and gas density are evaluated. A detailed discussion of this topic is available in Szekely and Themelis [96]. Here we shall note, however, that for most gaseous systems we may use the arithmetic mean of the bulk gas temperature and the solid surface temperature for evaluating the property values of the gas.

Another point worthy of note is the fact that when the temperature difference between the solid surface and the gas is very large, *thermal natural convection* may play a strong role in defining the flow conditions around the particle, so that this thermal natural convection would have to be taken into consideration in calculating *both* the heat transfer and the mass transfer coefficients.

2.6 Structural Changes in Gas–Solid Reactions

In Section 2.3 we discussed the diffusion of gaseous reactants and products through a porous matrix that could consist of the solid reactant, the solid product, or both. In the discussion, pore diffusion was assumed to be a steady state (or at least quasi-steady state) process occurring through a medium of unchanging structure. This assumption is usually correct for heterogeneous catalysis, investigation of which provided much of the motivation for work on pore diffusion. However, in the majority of gas–solid reaction systems, pore diffusion necessarily accompanies a chemical reaction, so that the solid matrix through which diffusion is taking place may undergo changes during the process.

Broadly speaking, one could envision two types of change, one leading to larger pores, or a more open structure, and the other causing densification, finer pores, and hence a lower effective diffusion coefficient.

The structural changes that may accompany gas–solid reactions can be quite complex but may be classified somewhat arbitrarily into the following categories:

structural modifications that are a direct result of the chemical change in the solid
  sintering
  swelling
  softening
  cracking

STRUCTURAL MODIFICATIONS DUE TO CHEMICAL CHANGE

For gas–solid reactions involving no solid product (e.g., the combustion of graphite or coal in air) the structural modification is evident and recent work by Schechter and Gidley [97] might be applied to these reactions. However, structural changes due to reactions occur even when a solid product exists. For example, the reduction of dense hematite or magnetite pellets at moderate temperatures yields a porous, rather than dense, iron pellet. If the gross dimensions of a reacting solid do not change during reaction, then a simple material balance yields the following relationship between the final and initial porosities of the solid:

$$\varepsilon_F = 1 - N(\rho_I/\rho_F)(1 - \varepsilon_I) \quad \text{for} \quad \varepsilon_F < 1 \quad (2.6.1)$$

where ε_F is the final porosity, ε_I is the initial porosity, ρ_I is the true molar density of solid reactant, and ρ_F is the true molar density of solid product, and N moles of solid product are formed by the reaction of one mole of solid reactant. Equation (2.6.1) can be manipulated into the form

$$\varepsilon_F - \varepsilon_I = (1 - \varepsilon_I)(1 - N\rho_I/\rho_F)$$

which implies [since $(1 - \varepsilon_I)$ is always positive] that the porosity will increase due to the reaction, provided $N\rho_I/\rho_F < 1$ and vice versa. The porosity will decrease if this same ratio is greater than one. A close analogy exists between this behavior and the behavior of solids undergoing dry corrosion, where the product solid either forms a pervious layer or an impervious layer, or spalls off according to the value of the Pilling–Bedworth ratio ($\bar{\rho}_I/\bar{\rho}_F$).

Structures developed for $N\rho_I/\rho_F < 1$ frequently have a large number of very small pores and as a consequence the solid has a large specific surface. This effect is exploited in the manufacture of high surface area heterogeneous catalysts. The development of such microporosity is a complex phenomenon dependent on (among other things) the method of preparation of the solid reactant [98]. The microporosity can be expected to disappear rapidly if sintering of the product solid is significant at the reaction temperature.

The variation of surface area accompanying the thermal decomposition of solids has been discussed by Nicholson [99], who took into consideration the possibility that the change from initial to final porosity might not occur instantaneously (on reacting a small particular region of the solid) but rather that a metastable lattice may form that changes into the solid product lattice with a certain characteristic time. Since the characteristic time cannot be determined a priori, in engineering calculations it is perhaps best to assume that the change occurs instantaneously.

SINTERING

Sintering in this context implies the phenomenon by which a porous compact increases its density as a consequence of being held at an elevated temperature, below its melting point.† Because gas–solid reactions are usually carried out at elevated temperatures, sintering may occur; this behavior may be of particular importance in the reduction of metal oxides where sintering of the porous metal product layer may markedly reduce the overall rate of reaction by limiting access by the reactant gas. Such findings have been reported for the reduction of nickel oxide with hydrogen above 750°C [100], and Fig. 2.10 shows a photomicrograph of a partially reduced nickel oxide pellet, where the sintered nickel layer is quite clearly seen.

Sintering is a complex phenomenon and, in spite of much careful investigation, it is still not possible to predict the rate at which the various structural characteristics (e.g., porosity, surface area) change under a given set of circumstances.

Activation energies for sintering are high; that is, the rate of sintering increases rapidly with increasing temperature. The increase is so rapid, in fact, that for a particular solid there is a temperature (known as the Tammann temperature) below which sintering is negligibly slow but above which it takes place quite rapidly. The Tammann temperature is usually 0.4–0.5 times the melting point on the absolute temperature scale.

The driving force for sintering is the surface energy of the compact. The surface area of the unsintered compact being large, a considerable reduction in free energy takes place on forming a solid with a smaller surface area. Several mechanisms have been proposed for the sintering process:

† The term *sintering* is also used to describe the agglomeration of ores as a means of feed preparation for the iron (or lead) blast furnace. This operation, which actually increases the porosity of the feed material, is not the one discussed here.

FIG. 2.10. A photograph of a partially reduced and sectioned nickel oxide pellet, showing surface layers of dense sintered nickel. Diagonal striations were caused by the tool used to section the pellet.

(a)   *Viscous (plastic) flow*   This is generally discredited as a sintering mechanism although Kingery and Berg [101] have shown that glass sinters by this route.

(b)   *Evaporation–condensation*   The vapor pressure above a convex solid surface exceeds that above a concave surface and consequently there is a tendency for evaporation from most of the particle surfaces and condensation in the necks between particles. Kingery and Berg suggested that sodium chloride sinters in this manner.

(c)   *Volume diffusion*   This mechanism proceeds by a diffusion of atoms through the particles to the necks. This is synonymous with stating that vacancies diffuse away from the necks. The sinks for the vacancies may be either the particle surfaces remote from the neck or grain boundaries separating the two particles. It is difficult to see how shrinkage can occur in the former case.

(d)   *Surface diffusion*   This mechanism involves the migration of atoms across the particle surface to the necks.

(e)   *Grain boundary diffusion*   Vacancies diffuse away from the necks via the grain boundaries and the grain boundaries act as sinks for the vacancies. Most solids are thought to sinter by grain boundary or volume diffusion although surface diffusion may be significant at low temperatures or small particle sizes [102].

Mechanisms (b) and (d) cannot lead to gross shrinkage of the compact.

The majority of the interpretations developed for sintering phenomena postulate the existence of three stages for the sintering process, as shown in Fig. 2.11.

The initial stage involves the growth of the small areas of contact between the grains. Since these "necks" represent the weakest part of the solid, this initial stage is usually accompanied by a great increase in mechanical strength. This stage together with the intermediate stages, usually involves the overall shrinkage of the solid. As seen in Fig. 2.11 the *intermediate stage* involves the intersection of the "necks" until ultimately closed pores are formed. *The final stage* of sintering involves the elimination of these closed pores.

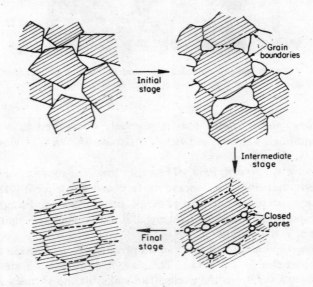

FIG. 2.11. Schematic representation of the three stages of sintering.

The literature of sintering has grown appreciably in recent years. While a great deal of useful work is being done, the development of the field has not yet reached the stage where these models can be applied directly to the modeling of gas–solid reactions in which structural changes are important.

The reviews [103–105] and journal articles [106–110] constitute potentially useful reading for gaining a better understanding of sintering phenomena.

SWELLING

Pellets occasionally present problems in iron blast furnaces because of swelling during reduction. Most iron oxide pellets and ores swell to some extent during reduction but some exhibit what has become known as a "catastrophic" swelling [111]. Balon and coworkers [112] studied the effect of swelling in an experimental blast furnace and concluded that no difficulties were encountered with pellets swelling less than 20% (by volume), but pellets swelling between 20 and 40% could be used only up to a maximum of 65% of the charge. With pellets swelling more than 40% the blast furnace was found to work in an unstable manner, even if the burden contained less than 65% pellets. Lecomte *et al.* [113] have devised a laboratory test procedure for measuring the swelling of ores or pellets.

The volume changes accompanying catastrophic swelling are large, sometimes approaching 300%. The swelling results in the production of a mass of metallic wires or whiskers, rather than the fairly dense sponge of the normal reduction, and causes severe problems in blast furnace operation. Chang *et al.* [114] studied the swelling of seventeen production pellets from various U.S. steel plants in 1967 and concluded that none of these exhibited catastrophic swelling. The maximum swelling they observed varied from 11–32% and occurred at a degree of reduction varying from 31–40%. No obvious differences were found between the behavior of hematites and magnetites.

Vom Ende and coworkers [115] have demonstrated that small additions of alkalis (Na_2CO_3 or K_2CO_3) can result in the catastrophic swelling of otherwise normal pellets. The adverse effect could be prevented by the addition of a fine-grained acid gangue to form stable alkali silicates.

Bleifuss [116] has suggested calcium oxide as a cause of catastrophic swelling. Wenzel and Gudenau [117] were able to reduce the swelling by indurating the pellets at a higher temperature. These investigators were able to bring about similar benefits by carrying out the reduction at above 1100°C after heating to this temperature in an inert or oxidizing atmosphere [118].

According to Bleifuss the catastrophic swelling involved in the reduction of calciferous ores is due to the formation of a surface layer of lime saturated with iron. A wire filament nucleates at the surface and additional iron migrates to the nucleation point by solid state diffusion and surface diffusion.

Normal swelling is probably caused by the crystallographic transformations that occur during reduction [119].

SOFTENING

In the iron blast furnace, the descending burden loses most of its mechanical strength at a temperature somewhat below its melting point. Burden descent is improved and pressure drop across the bosh region is reduced if the charge has a high softening temperature and a melting temperature only a little above the softening temperature. This softening has been the subject of recent papers by Lecomte and coworkers [113, 120].

As far as the effects on gas–solid reactions are concerned, softening may be considered as an extreme form of sintering. At softening temperatures, pores must close extremely rapidly, which in turn prevents diffusion of gaseous reactant into the solid and thus severely hinders the reaction. If reaction is carried out in a packed or fluidized bed, deformation of the solid under the mechanical load of the superimposed solid (in the packed bed) or agglomeration will make the modeling of the reaction very difficult. Moreover, softening of the material may cause undesirable agglomeration in fluidized beds.

CRACKING

There have been many publications on the cracking of iron ores during reduction, recent examples of which are [121–124]. Cracking of the solid is usually undesirable for gas–solid reactions systems carried out on an industrial scale, because the resultant breakage of the solids may cause operational difficulties, e.g., due to the increased resistance to the flow of gases through the reactor.

However, cracking of a pellet or ore lump may increase the overall rate of reaction (for that particular pellet) because it provides an easy access by the gaseous reactant to the solid by reducing the diffusional resistance. It is important that investigators be aware of cracking phenomena in kinetic studies because these might lead to the falsification of the kinetic data and could also signal possible difficulties for the operation of industrial or even pilot scale units.

2.7 Concluding Remarks

In this chapter we have introduced the reader to the elements that are involved in the reaction between a gas and a solid particle. We have seen that in many instances even these individual steps themselves are quite complex and inadequately understood, so that the task of integrating this information into workable models for describing the overall rate of gas–solid reactions seems a formidable undertaking. Indeed, this state of affairs precludes the development of entirely general "all purpose" models for gas–solid reactions that incorporate all the phenomena described here.

The reaction systems that are well understood tend to be dominated by one of the simpler steps (e.g., mass transfer or pore diffusion) so that the interpretation is then quite feasible. Nevertheless, it is thought very important to generate awareness of the principal mechanisms that could control the overall reaction rate because in future work on more complex systems their effects will have to be taken into consideration.

Notation

A_p	pore surface area of pellet
a, b	stoichiometric coefficients
B_0	geometric parameter characteristic of porous solid
C_{A0}, C_{As}	bulk, surface, concentration of gaseous reactant A
C_p	specific heat
d	pore diameter
D	diffusivity
D_{AB}	ordinary diffusivity of gas pair A–B
D_{ABeff}	effective ordinary diffusivity of gas pair A–B
D_a	defined by Eq. (2.3.28)
D_{Aeff}	effective diffusivity of gas A
$D_{Aeff(r)}$	effective diffusivity of gas A in pores of radius r
D_{AK}	Knudsen diffusivity of gas A
D_i	defined by Eq. (2.3.29)
D_K	defined by Eq. (2.3.8)
E_a, E_d	activation energy for adsorbtion (or reaction), desorption
ΔF°	standard free energy of reaction
ΔH_a	heat of adsorption
h	heat transfer coefficient
h_D	mass transfer coefficient
h_r	radiative heat transfer coefficient
j_D, j_H	Chilton–Colburn j factors for mass transfer, heat transfer
K	Langmuir constant
K_E	equilibrium constant for reaction
K_0	geometric parameter characteristic of porous solid
k	heterogeneous reaction rate constant
k_a, k_d	rate constant for adsorption, desorption
ℓ_g	thermal conductivity
L	characteristic length
M	molecular weight
N	Avogadro's number
N_A, N_C	flux of species A, C
N_{Pr}, N_{Re}, etc.	Prandtl number, Reynolds number, etc.
n_d	number of grains per unit volume of porous solid
p_0	saturation vapor pressure of the adsorbate
p, p_A	total pressure, partial pressure of A
Q	rate of heat generation per unit volume
q	heat flux
q_r	radiative heat flux
R	gas constant
$\mathscr{R}, \mathscr{R}_a, \mathscr{R}_d$	rate of reaction, adsorption, desorption

r_g	radius of grain
S_g	pore surface area per unit mass of solid
T, T_s, T_E	temperature, surface temperature, environment temperature
t	time
U	fluid velocity
V_p	volume of pellet
$\bar{v}$	molar volume
v, v_m	volume of gas adsorbed per unit mass of porous solid, at monolayer coverage
v_g	volume of a mole of gas at standard conditions
X_A	mole fraction of A
α	defined by Eq. (2.3.30)
$\alpha(r)\,dr$	volume fraction of pores with radius between r and $r + dr$
β_A	defined by Eq. (2.3.7)
β_D	defined following Eq. (2.2.5)
β_H	coefficient of thermal expansion (volumetric)
γ_A	defined by Eq. (2.3.6)
γ_1, γ_{-1}	net forward rate, backward rate, of reaction 1
δ_A	defined by Eq. (2.3.5)
$\varepsilon, \varepsilon_I, \varepsilon_F$	porosity of solid, initial porosity, final porosity
ε_A	Lennard-Jones 6-12 parameter
$\varepsilon_a, \varepsilon_i$	macropore porosity, micropore porosity
$\tilde{\varepsilon}$	emissivity
θ, θ_e	fractional occupation of surface adsorption sites at equilibrium
θ_x	fraction of surface adsorption sites occupied by species x
κ_e	effective thermal diffusivity
κ	Boltzmann constant
μ	viscosity
ν	kinematic viscosity
ρ	density
σ	Stefan–Boltzmann constant
σ_A	Lennard-Jones 6-12 parameter
τ	tortuosity factor characteristic of porous solid
Ω	condensation coefficient
Ω_{AB}	collision integral

Subscripts

A, B, etc.	pertaining to chemical species A, B, etc.
0	pertaining to the bulk
s	pertaining to the surface

References

1. R. B. Bird, W. E. Stewart, and E. N. Lightfoot, "Transport Phenomena." Wiley, New York, 1960.
2. J. Szekely and N. J. Themelis, "Rate Phenomena in Process Metallurgy." Wiley, New York, 1971.
3. H. L. Langhaar, "Dimensional Analysis and the Theory of Models." Wiley, New York, 1951.
4. R. E. Johnstone and M. W. Thring, "Pilot Plants, Models and Scale-up Methods in Chemical Engineering." McGraw-Hill, New York, 1957.
5. W. E. Ranz and W. R. Marshall, Jr., *Chem. Eng. Prog.* **48**, 141, 173 (1952).
6. M. Frössling, *Beitr. Geophys.* **52**, 170 (1938).
7. P. N. Rowe and K. T. Claxton, *Trans. Inst. Chem. Eng.* **43**, T 231 (1965).
8. R. L. Steinberger and R. E. Treybal, *AIChE J.* **6**, 227 (1960).
9. S. Evnochides and G. Thodos, *AIChE J.* **7**, 78 (1961).
10. S. K. Friedlander, *AIChE J.* **7**, 347 (1961).
11. R. B. Keey and J B. Glen, *Can. J. Chem. Eng.* **42**, 227 (1964).
12. G. A. Hughmark, *AIChE J.* **13**, 1219 (1967).
13. K. Lee and H. Barrow, *Int. J. Heat Mass Trans.* **11**, 1013 (1968).
14. W. J. Lavender and D. C. T. Pei, *Int. J. Heat Mass Trans.* **10**, 529 (1967).
15. J. O. Hirschfelder, C. F. Curtiss, and R. B. Bird, "Molecular Theory of Gases and Liquids." Wiley, New York, 1956.
16. T. R. Marrero, Ph.D. Thesis, Brown Univ., Providence, Rhode Island, 1970.
17. J. C. Slattery and R. B. Bird, *AIChE J.* **4**, 137–142 (1958).
18. R. A. Svehla, Tech. Rep. R-132, Lewis Res. Center, NASA, Cleveland, Ohio (1962). (N63-22862).
19. R. S. Brokaw, *IEC Proc. Design Develop.* **8**, 240 (1969).
20. L Monchick and E. A. Mason, *J. Chem. Phys.* **35**, 1676 (1961).
21. R. D. Nelson, D. R. Lide, and A. A. Margott, "Selected Values of Electric Dipole Moments for Molecules in the Gas Phase, NSRDS." Nat. Bur. Std. 10, (Sept. 1, 1967).
22. E. N. Fuller, P. D. Schettler, and J. C. Giddings, *Ind. Eng. Chem.* **58** (5), 19 (1966).
23. J. H. Bae and T. M. Reed, III, *Ind. Eng. Chem. Fundam.* **10**, 36 (1971).
24. J. H. Bae and T. M. Reed, III, *Ind. Eng. Chem. Fundam.*, **10**, 269 (1971).
25. N. J. Themelis and J. C. Yannopoulos, *Trans. AIME*, **236**, 414 (1966).
26. R. J. Fruehan and L. J. Martonik, *Met. Trans.* **4**, 2789–2792, 2793–2974 (1973).
27. C. N. Satterfield, "Mass Transfer in Heterogeneous Catalysis," Chapter 1. M.I.T. Press, Cambridge, Massachusetts, 1970.
28. J. M. Smith, "Chemical Engineering Kinetics," 2nd ed., Chapter 11. McGraw-Hill, New York, 1970.
29. E. A. Mason and T. R. Marrero, *Advan. At. Mol. Phys.* **6**, 155-232 (1970).
30. E. A. Mason, A. P. Malinauskas, and R. B. Evans, *J. Chem. Phys.* **46**, 3199-3216 (1967).
31. R. D. Gunn and C. J. King *AIChE J.* **15**, 507 (1969).
32. R. G. Olsson and W. M. McKewan, *Trans. AIME*, **236** 1518 (1966).
33. J. Szekely and H. Y. Sohn, *Trans. Inst. Min. Met.* **82**, C 92 (1973).
34. E. A. Mason, A. P. Malinauskas, and R. B. Evans, *J. Chem. Phys.* **46**, 3199 (1967).
35. J. W. Evans, *Can. J. Chem. Eng.* **50**, 811 (1972).
36. J. W. Evans and S. Song, *Can. J. Chem. Eng.* **51**, 616 (1973).
37. M. F. L. Johnson and W. E. Stewart, *J. Catal.* **4**, 248 (1965).
38. C. N. Satterfield and J. P. Cadle, *IEC Process Design Develop.* **7**, 257 (1968).

39. N. Wakao and J. M. Smith *Chem. Eng. Sci.* **17**, 825 (1962).
40. R. D. Cunningham and C. J. Geankoplis, *Ind. Eng. Chem. Fundam.* **7**, 535 (1968).
41. G. R. Youngquist, *Ind. Eng. Chem.* **62**, 52 (1970).
42. J. M. Thomas and W. J. Thomas, "Introduction to the Principles of Heterogeneous Catalysis." Academic Press, New York, 1967.
43. J. H. de Boer, "The Dynamical Character of Adsorption." Oxford Univ. Press, London and New York, 1953.
44. S. Ross and J. P. Olivier, "On Physical Adsorption." Wiley (Interscience), New York, 1964.
45. D. M. Young and A. D. Crowell, "Physical Adsorption of Gases." Butterworths, London and Washington, D.C., 1962.
46. D. O. Hayward and B. M. W. Trapnell, "Chemisorption," 2nd ed. Butterworths, London and Washington, D.C., 1962.
47. K. J. Laidler, *in* "Catalysis" (P. H. Emmett, ed.), Vol. 1, p. 75. Van Nostrand-Reinhold, Princeton, New Jersey, 1954.
48. S. Brunauer, L. E. Copeland, and D. L. Kantro, *in* "The Solid-Gas Interface" (E. A. Flood, ed.), Vol. 1. Dekker, New York, 1967.
49. T. L. Hill, *Advan. Catal.* **4**, 211 (1952).
50. J. H. de Boer, *Advan. Catal.* **8**, 17 (1956).
51. C. Aharoni and F. C. Tompkins, *Advan. Catal.* **21**, 1 (1970).
52. H. J. de Boer, *Advan. Catal.* **9**, 472 (1972).
53. G. Padberg and J. M. Smith, *J. Catal.* **12**, 111 (1968).
54. I. Langmuir, *J. Amer. Chem. Soc.* **40**, 1361 (1918).
55. S. Braunuer, P. H. Emmett, and E. Teller, *J. Amer. Chem. Soc.* **60**, 309 (1938).
56. P. H. Emmett, *in* "Catalysis" (P. H. Emmett, ed.,) Vol. 1, p. e1. Van Nostrand-Reinhold, Princeton, New Jersey, 1954.
57. H. Galsey and H. S. Taylor, *J. Chem. Phys.* **15**, 624 (1947).
58. G. Halsey, *Advan. Catal.* **4**, 259 (1952).
59. J. Zeldowitch, *Acta Physicohcim. URSS*, **1**, 961 (1935).
60. A. Slygin and A. Frumkin, *Acta Physicochim. URSS* **3**, 791 (1936).
61. S. Brunauer, K. S. Love, and R. G. Keenan, *J. Amer. Chem. Soc.* **64**, 751 (1952).
62. P. H Emmett and S. Brunauer, *J. Amer. Chem. Soc.* **56**, 35 (1934).
63. M. I. Temkin and V. Pyzhev, *Acta Physicochim. URSS*, **12**, 327 (1940).
64. A. S. Porter and F. C. Tompkins, *Proc. Roy. Soc.* **A217**, 544 (1953).
65. S. Yu Elovich and G. M. Zhabrova, *Zh. Fiz. Khim.* **13**, 1761, 1775 (1939).
66. M. J. D. Low, *Chem. Rev.* **60**, 267 (1960).
67. G. C. Bond, "Catalysis by Metals." Academic Press, New York, 1962.
68. K. J. Laidler, *in* "Catalysis" (P. H. Emmett, ed.), Vol. 1. Van Nostrand-Reinhold, Princeton, New Jersey, 1954.
69. O. A. Hougen and K. M. Watson, "Chemical Process Principles," Part 3, Kinetics and Catalysis. Wiley, New York, 1947.
70. E. E. Petersen, "Chemical Reaction Analysis," Chapter 3. Prentice-Hall, Englewood Cliffs, New Jersey, 1958.
71. M. Boudart, "Kinetics of Chemical Processes," Chapter 3. Prentice-Hall, Englewood Cliffs, New Jersey, 1965.
72. J. M. Quets, M. E. Wadwsorth, and J. R. Lewis, *Trans. Met. Soc. AIME* **221**, 1186 (1961).
73. W. M. McKewan, *Trans. Met. Soc. AIME*, **224**, 387 (1962).
74. T. Kurosawa, R. Hasegawa, and T. Yagihashi, *Nippon Kinzoku Gakkaishi.* **34**, 481 (1970).
75. P. L. Walker, Jr., F. Rusinko, Jr., and L. G. Austin, *Advan. Catal.* **XI**, 133 (1959).

76. B. Delmon, "Introduction a la Cinétique Hétérogene." Éditions Technip, Paris, 1969.
77. G. Pannetier and P. Souchay, "Chemical Kinetics" (translated by H. D. Gesser and H. H. Emond). Elsevier, Amsterdam, 1967.
78. D. A. Young, "The International Encyclopedia of Physical Chemistry and Chemical Physics," (F. C. Tompkins, ed.), Topic 21, Solid and Surface Kinetics, Vol. 1, Decomposition of Solids. Pergamon, Oxford, 1966.
79. W. H. McAdams, "Heat Transmission." McGraw-Hill, New York, 1954.
80. B. Gebhart, "Heat Transfer." McGraw-Hill, New York, 1971.
81. E. M. Sparrow and R. D. Cess, "Radiation Heat Transfer." Brooks/Cole Publ., Belmont, California 1966.
82. H. Hottel and A. F. Sarofim, "Radiation Heat Transfer." McGraw-Hill, New York, 1967.
83. J. Szekely and N. J. Themelis, "Rate Phenomena in Process Metallurgy." Chapter 9. Wiley, New York, 1971.
84. Y. S. Toulokian (ed.), "Thermophysical Properties of High Temperature Solid Materials." McMillan, New York, 1967.
85. R. C. Weast (ed.), "Handbook of Chemistry and Physics." Chemical Rubber Co., Cleveland, Ohio, 1971.
86. International Critical Tables.
87. J. Francl and W. D. Kingergy, J. Amer. Ceram. Soc. 37, 99 (1954).
88. C. Maxwell, "Treatise on Electricity and Magnetism," Vol. I, p. 365. Oxford Univ. Press, London and New York, 1873.
89. W. R. Rayleigh, Phil. Mag. 5, 418 (1892).
90. W. Woodside, Can. J. Phys. 36, 815 (1958).
91. D. Kunii and J. M. Smith, AIChE J. 6, 71 (1960).
92. A. V. Luikov, A. G. Shashkov, L. L. Vasillies and Y. E. Fraimon, Int. J. Heat. Mass Trans. 11, 117 (1968).
93. J-H. Huang and J. M. Smith, J. Chem. Eng. Data 8, 437 (1963).
94. C. Satterfield, "Mass Transfer in Heterogeneous Catalysis." pp. 169 et seqq. M.I.T. Press, Cambridge, Massachusetts, 1970.
95. H. S. Carslaw and J. C. Jaeger, "Conduction of Heat in Solids," pp. 11–12. Oxford Univ. Press, London and New York, 1957.
96. J. Szekely and N. J. Themelis, "Rate Phenomena in Process Metallurgy," Chapter 13. Wiley, New York, 1971.
97. R. S. Schechter and J. L. Gidley, AIChE J. 15, 339 (1969).
98. P. G. Fox, J. Ehretsmann, and C. E. Brown, J. Catal. 20, 67–73 (1971).
99. D. Nicholson, Trans. Faraday Soc. 61, 990–998 (1965).
100. J. Szekely and J. W. Evans, Chem. Eng. Sci. 26, 1901 (1971).
101. W. D. Kingery and M. Berg, J. Appl. Phys. 26, 1205 (1955).
102. H. E. N. Stone, Mettalurgia 74, 151 (1966).
103. G. X. Kuczynski, N. A. Hooton and C. F. Bibbon (eds.), "Sintering and Related Phenomena." Gordon and Breach, New York, 1967.
104. M. M. Ristic, Phys. Sinter. 1, A1–A14 (1969).
105. M. M. Ristic, Phys. Sinter. 2, 49–61 (1969).
106. D. L. Johnson, J. Appl. Phys. 40, 192 (1969).
107. H. H. Bache, J. Amer. Ceram. Soc. 53, 1205 (1955).
108 R. L. J. Coble, Appl. Phys. 32, 79 (1967).
109. R. L. J. Coble, Appl. Phys. 36, 2327 (1965).
110. R. L. J. Coble and M. C. Flemings, Met. Trans. 2, 409 (1971).
111. S. Watanabe and M. Yoshinaga, Sumitomo J. 77, 323–330 (1965).
112. R. Balcon, L. Baty, L. Hovent, and A. Poos, J. Metals 93 (June 1967).

113. P. Lecomte, R. Vidal, A. Poos, and A. Decker, C.N.R.M., No. 16, (Sept. 1968).
114. M. C. Chang, J. Vlanty, and D. W. Kestner, Paper, *Ironmaking Conf.*, *26th AIME*, *Chicago* (1967).
115. H. Vom Ende, J. Grege, S. Thomalla, E. E. Hofmann, *Stahl Eisen* **90** (13) 667–676 (1970). (Ger.).
116. R. L. Bleifuss, *Trans. AIME* **247**, 225–231 (1970).
117. W. Wenzel, H. W. Gudenau, *Stahl Eisen* **90** (13) 689–697 (1970) (Ger.).
118. W. Wenzel and H. W. Gudenau, Germ. Offen. 1, 922, 440 (1970).
119 B. B. L. Seth and H. V. Ross, *Trans. AIME* **233** 180–185 (1965).
120. P. Lecomte, R. Vidal, A. Poos, and A. Decker, C.N.R.M., No. 21, (December 1969).
121. H. Brill-Edwards, H. E. N. Stone, B. L. Daniell, *J. Iron Steel Inst.*, *London* **207** (Pt 12) 1565–1577 (1969).
122. H. E. N. Stone and B. L. Daniell, *J. Iron Steel Inst. London* **208** (Pt 4), 406 (1970).
123. N. Surtees, H. E N. Stone, and B. L. Daniell, *J. Iron Steel Inst. London* **208** (Pt 7), 669–674 (1970).
124. C. Offroy, *Rev. Met.* (*Paris*) **66** (7-8) 491–517 (1969). (Fr.).
125. F. A. Schwertz and J. E. Brow, *J. Chem. Phys.* **19**, 640 (1951).

Chapter 3 | **Reactions of Nonporous Solids**

3.1 Introduction

In many gas–solid reactions, the solid is initially nonporous. The reaction may or may not involve gaseous reactants, and the product may be a gas, a solid, or a mixture of both. The porosity of the solid product, which depends on the particular reaction system, greatly influences the characteristics of the reaction.

A long list of examples of reactions involving nonporous solids can be found elsewhere [1].

The majority of the models for gas–solid reactions have been based on the assumption of a nonporous reactant—even when the solid reactant has considerable porosity. This has often led to difficulties in the interpretation of experimental results. This will be discussed in detail in the next chapter, which deals with the reaction of porous solid reactants.

In this chapter we shall concern ourselves with systems wherein the solid reactant is nonporous and the reaction occurs at a sharp interface between two phases—either gas–solid or solid–solid.

It is convenient to classify gas–solid reactions according to the geometry that the system exhibits in the course of reaction. The reaction of non-porous solids can be divided into three types of geometrical groups:

1. A shrinking particle, which results from products being gaseous or solid products flaking off the surface of the reactant solid as soon as they are formed.

2. A particle whose overall size is unchanged, with a product or "ash" layer remaining around the unreacted core (hence the term "shrinking un-reacted-core" model).

3.   A particle whose overall size is changed due to the difference in the volume of the solid before and after the reaction.

An important common characteristic of the reaction of nonporous solids is that the steps of chemical reaction and mass transport are coupled in series. Since the chemical reaction occurs at the plane surface of a solid, the surface always appears as one of the boundary conditions in the mass transport equations. This makes the analysis of nonporous solid systems much easier than that of porous reactant solid systems—a reason why most previous models assumed a nonporous solid reactant.

In the following sections we will consider these systems for constant-bulk-fluid conditions throughout the duration of the reaction. Extensions can easily be made to accomodate the changing bulk conditions in systems such as batch, fixed-bed, or moving-bed reactors.

A remark to be made at this point is that for some gas–solid reactions, nucleation is an important process. When the growth of reaction nuclei controls the overall rate, the relationship between the conversion and time is complex, involving the period of accelerating rate.

In this monograph we will deal mostly with interfacial chemical reactions on a sharp boundary which retains its initial shape (i.e., parallel plates, coaxial cylinders, or concentric spheres, depending on the original shape) for successive times. This also applies to systems requiring nucleation, if the temperature of reaction is raised, in which case nucleation usually occurs rapidly over the entire solid surface [2,3].

3.2   Shrinking Nonporous Particles

The simplest system in fluid–solid reactions is that of a shrinking nonporous particle forming no solid product layer:

$$A(g) + bB(s) \longrightarrow cC(g)$$

Examples of this type often occur in chemical and metallurgical processes: the combustion and gasification of carbon and coal, the decomposition of solids into gases, the dissolution of solids in liquids, formation of metal carbonyls, fluorination and chlorination of metals, and some electrochemical processes.

Figure 3.1 shows the situation schematically. The particles may have different shapes [4], such as flat disks, long cylinders, or spheres. Of these an approximately spherical shape is the most commonly encountered, and most real particles are represented by equivalent spheres.

The important characteristic dimensions, so far as reaction with gases is concerned, are the thickness for a flat disk, and the radius for a long cylinder and for a sphere.

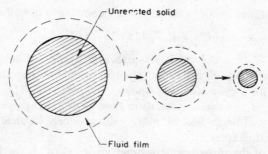

FIG. 3.1. A nonporous solid participating in a gas–solid reaction (gasification) in which no solid product is formed.

We shall first consider the limiting cases—chemical reaction control and mass transport control—and then study the combined effect of these two mechanisms on the overall reaction rate of the shrinking particles with the surrounding fluid.

3.2.1   Chemical Reaction Control

This is a simple case and the overall rate can be expressed as the rate of disappearance of A by the surface chemical reaction as

$$\mathscr{R}_s = k[C_{As}^n - (C_{Cs}^m/K_E)] \qquad (3.2.1)$$

where $\mathscr{R}_s$ is the rate of reaction (moles of A per unit time per unit surface area of B), k the heterogeneous rate constant, K_E the equilibrium constant, and C_{As} and C_{Cs} denote the gaseous reactant and product concentrations, respectively, at the reaction surface. In the present case we shall assume that the overall rate is controlled by chemical kinetics, so that the resistance due to mass transfer is negligible. Under these conditions $C_{As} = C_{A0}$ and $C_{Cs} = C_{C0}$, i.e., the concentrations of these components at the reaction surface are the same as in the bulk of the gas stream.

The rate of reaction of A may be equated to the rate of disappearance of B:

$$b\mathscr{R}_s = \rho_s \, dr_c/dt \qquad (3.2.2)$$

where ρ_s is the molar density of the solid B, and r_c is the distance coordinate perpendicular to the solid surface. We note at this point that Eqs. (3.2.1) and (3.2.2) apply to any particle geometry.

Combining the two equations and rearranging the resulting equation in dimensionless form, we obtain

$$d\xi/dt^* = -1 \qquad (3.2.3)$$

where

$$\xi \equiv (A_p/F_p V_p)r_c \qquad (3.2.4)$$

(r_c now is the distance from the center of the geometry to the solid surface) and

$$t^* \equiv (bk/\rho_s)(A_p/F_p V_p)[C_{A0}^n - (C_{C0}^m/K_E)]t \tag{3.2.5}$$

where A_p and V_p are the original surface area and volume, respectively, of the solid particle, and F_p is the shape factor, which has the values 1, 2, and 3 for flat plates, long cylinders, and spheres, respectively. The characteristic dimension $F_p V_p/A_p$ is a useful quantity in generalizing the analysis for different geometries. Its usefulness will be more evident when we deal with porous pellets in the next chapter.

In the cases of the three simple geometries, $F_p V_p/A_p$ is the half-thickness for a flat plate, and the radius for a long cylinder or a sphere.

The initial condition for Eq. (3.2.3) is

$$\xi = 1 \qquad \text{at} \quad t^* = 0 \tag{3.2.6}$$

Integrating Eq. (3.2.3) with the initial condition, we obtain the changing size of the solid with time:

$$
\begin{aligned}
\xi &= 1 - t^* &&\text{for} \quad t^* \leq 1 \\
\xi &= 0 &&\text{for} \quad t^* \geq 1
\end{aligned}
\tag{3.2.7}
$$

Strictly speaking, this integration is valid only for those particles that maintain their original shapes in the course of reaction. However, most particles without thin horn- or thornlike structures are expected to maintain their original shapes throughout most of the reaction period.

The overall extent of reaction is given by

$$X = 1 - \xi^{F_p} \tag{3.2.8}$$

Substituting Eq. (3.2.7) in Eq. (3.2.8), we obtain the relationship between the conversion and time [disregarding the trivial solution given by the second equality in Eq. (3.2.7)]:

$$t^* = (bk/\rho_s)[C_{A0}^n - (C_{C0}^m/K_E)](A_p/F_p V_p)t = 1 - (1 - X)^{1/F_p} \tag{3.2.9}$$

from which the time for complete conversion is found to be

$$t_{X=1} = \frac{\rho_s(F_p V_p/A_p)}{bk[C_{A0}^n - (C_{C0}^m/K_E)]} \tag{3.2.10}$$

We see that in the case of chemical reaction control, $t_{X=1}$ is proportional to the initial size of the particle. For subsequent use we define a conversion function in the absence of diffusional resistance:

$$g_{F_p}(X) \equiv 1 - (1 - X)^{1/F_p} \tag{3.2.11}$$

The shape factor F_p of large particles can be estimated from their visual appearance. If the particles are too small to be easily observed visually, F_p

can be obtained as the value that gives a straight line between the experimental values of $g_{F_p}(X)$ and the reaction time t under the absence of external mass transfer resistance. Furthermore, from the slopes of such straight lines obtained with different bulk concentrations and temperatures, one can determine the reaction order n, the activation energy, and the preexponential factor.

Examples of reactions obeying this type of kinetics can be found elsewhere in the literature [1].

3.2.2 Mass Transport Control

In many gas–solid systems with fast chemical reactions, the overall rate is found to be controlled by mass transport between the reaction surface and the bulk fluid. Moreover, the rate of mass transport is an important quantity in heterogeneous systems in that it represents the maximum overall rate attainable, and hence provides a useful guideline for reactor design and performance.

When mass transport controls the overall process, the rate can be expressed as

$$-\frac{\rho_s}{b}\frac{dr_c}{dt} = h_D\left(C_{A0} - \frac{C_{C0}}{K_E}\right)\left(\frac{K_E}{1 + K_E}\right) \tag{3.2.12}$$

where h_D is the mass transfer coefficient, which depends on particle size, the gas velocity, and the properties of the gas.

For single spheres or approximately spherical particles, or for reactions occurring in fixed or fluidized beds, the correlations discussed in Chapter 2 or 7 may be used to calculate h_D. These correlations take the form

$$N_{Sh} = f(N_{Sc}, N_{Re}) \tag{3.2.13}$$

For the reaction of individual particles in a moving gas stream where the particle size decreases as reaction proceeds, Eq. (3.2.13) can be put in the form†

$$N_{Sh} = h_D r_c/D = 1 + a' N_{Sc}^{1/3}(N_{Re}^0)^{1/2}(r_c/r_p)^{1/2} \tag{3.2.14}$$

where r_p is the initial radius of the particle, $N_{Re}^0 = 2r_p u\rho/\mu$, and a' is a numerical coefficient. Substituting Eq. (3.2.14) into Eq. (3.2.12) and rearranging into a dimensionless form, we obtain

$$d\xi/dt^+ = -\tfrac{1}{2}(1 + a\xi^{1/2})/\xi \tag{3.2.15}$$

† We note that in Eq. (3.2.14) and throughout Chapters 3 and 4, we have chosen to define the Sherwood number by using the particle radius as the characteristic length, rather than the particle diameter, in order to conform to earlier published work.

where

$$\xi = r_c/r_p \tag{3.2.16}$$

$$a = a' N_{Sc}^{1/3} (N_{Re}^0)^{1/2} \tag{3.2.17}$$

and

$$t^+ = \frac{2bD}{\rho_s r_p^2} \left(C_{A0} - \frac{C_{C0}}{K_E} \right) \left(\frac{K_E}{1 + K_E} \right) t \tag{3.2.18}$$

The initial condition for Eq. (3.2.15) is

$$\xi = 1 \quad \text{at} \quad t^+ = 0 \tag{3.2.19}$$

Upon integrating Eq. (3.2.15) with this initial condition, we obtain

$$t^+ = \frac{4}{3a}(1 - \xi^{3/2}) - \frac{2}{a^2}(1 - \xi) + \frac{4}{a^3}(1 - \xi^{1/2}) - \frac{4}{a^4} \ln \left(\frac{1 + a}{1 + a\xi^{1/2}} \right) \tag{3.2.20}$$

When $a = 0$, i.e., for a single sphere in a stagnant atmosphere, Eq. (3.2.20) reduces to†

$$t^+ = 1 - \xi^2 \tag{3.2.21}$$

From Eqs. (3.2.20) and (3.2.21) we see that for a large initial Reynolds number N_{Re}^0, the time for complete conversion is proportional to $r_p^{1.5}$ whereas it is proportional to r_p^2 for a small values of N_{Re}^0.

3.2.3   Overall Rate Determined by Both Chemical Reaction and Mass Transport

When chemical reaction and mass transport present comparable resistances to the progress of reaction, the contributions of these processes must be considered simultaneously. Assuming pseudosteady state, the overall rate is identical to the rate of interfacial chemical reaction and also to that of mass transport:

$$-(\rho_s/b)(dr_c/dt) = k(C_{As}^n - C_{Cs}^m/K_E) \tag{3.2.22}$$

$$= h_D(C_{A0} - C_{As}) = h_D(C_{Cs} - C_{C0}) \tag{3.2.23}$$

---

† Equation (3.2.21) may be obtained directly by integrating Eq. (3.2.15) with $a = 0$, or from Eq. (3.2.20) using the following procedure: Rewrite Eq. (3.2.20) as

$$t^+ = [\tfrac{4}{3}a^3(1 - \xi^{3/2}) - 2a^2(1 - \xi) + 4a(1 - \xi^{1/2}) - 4\ln(1 + a) + 4\ln(1 + a\xi^{1/2})]/a^4$$

As $a \to 0$, t^+ becomes indeterminate. Therefore, using L'Hôpital's rule,

$$t^+ = [4a^2(1 - \xi^{3/2}) - 4a(1 - \xi) + 4(1 - \xi^{1/2}) - 4(1 + a)^{-1} + 4\xi^{1/2}(1 + a\xi^{1/2})]/4a^3$$

t^+ still becomes indeterminate at $a \to 0$. Using L'Hôpital's rule three more times and substituting $a = 0$, we obtain

$$t^+ = 1 - \xi^2$$

These equations† may be used to solve the for unknowns C_{Cs} and C_{As}, and by substituting these into either equation, we obtain the overall rate as a function of the changing size. We shall do this for the simple case of a first-order interfacial reaction.

On combining Eqs. (3.2.22) and (3.2.23) for $n = 1$, $m = 1$, we have

$$\left(C_{As} - \frac{C_{Cs}}{K_E}\right) = \frac{h_D}{k(1 + 1/K_E) + h_D}\left(C_{A0} - \frac{C_{C0}}{K_E}\right) \tag{3.2.24}$$

The overall rate is now given in terms of C_{A0} and C_{C0}:

$$-\frac{\rho_s}{b}\frac{dr_c}{dt} = \frac{(C_{A0} - C_{C0}/K_E)}{[(1 + 1/K_E)/h_D] + 1/k} \tag{3.2.25}$$

In dimensionless form, Eq. (3.2.25) can be written

$$\frac{d\xi}{dt^*} = \frac{-1}{1 + k(1 + K_E)/(h_D K_E)} \tag{3.2.26}$$

where ξ and t^* are dimensionless variables defined in Eqs. (3.2.4) and (3.2.5), respectively.

For spherical particles, h_D may be found with the aid of Eq. (3.2.14) to give

$$dt^*/d\xi = -[1 + 2\sigma_0{}^2\xi/(1 + a\xi^{1/2})] \tag{3.2.27}$$

The initial condition is again

$$t^* = 0, \quad \text{at} \quad \xi = 1 \tag{3.2.28}$$

Integrating Eq. (3.2.27), we obtain

$$t^* = (1 - \xi) + \sigma_0{}^2 q(\xi) \tag{3.2.29}$$

where

$$\sigma_0{}^2 \equiv kr_p(1 + 1/K_E)/h_D$$

and $q(\xi)$ represents the r.h.s. of Eq. (3.2.20).

In Eq. (3.2.29) we see that *the time required to reach a certain ξ (and hence a certain conversion) is the sum of two terms: the time required to reach the same ξ in the absence of mass transport resistance* [Eq. (3.2.7)] *and that for pure mass transport control* [Eq. (3.2.20)]. This is an important result that applies to any gas–solid reaction consisting of first-order rate processes in series. Subsequently, we will show that this finding will also apply to other nonporous solid systems.

---

† It is implicitly assumed in the statement of Eq. (3.2.23) that the mass transfer coefficient is the same for all the diffusing species. This assumption will be made routinely throughout and is valid for any binary diffusion system involving equimolar counterdiffusion.

We note that when $\sigma_0{}^2$ is small (chemical "resistance" much greater than diffusional resistance), the system reduces to the chemical-reaction-controlled case and when $\sigma_0{}^2$ is large, to the diffusion-controlled case.

The relative importance of chemical kinetics and external mass transfer is illustrated in the following example.

Example 3.2.1 Spherical graphite particles of $r_p = 1$ mm are burned in a stagnant atmosphere of 10% O_2 at 900°C and 1 atm total pressure. Assume that the reaction is

$$C + O_2 \longrightarrow CO_2$$

irreversible and of first order with respect to oxygen.

Calculate the time necessary for complete conversion. What can be said about the controlling step?

The following information is given: density of graphite = 2.26 g/cm^3, $k = 20$ cm/sec, $D = 2$ cm^2/sec.

Repeat the calculation for $r_p = 0.1$ mm.

SOLUTION

$$\sigma_0{}^2 = (20)(0.1)/(2)(2) = \tfrac{1}{2}$$

From this value of $\sigma_0{}^2$ we know that mass transport and chemical reactions are of comparable importance.

Since $a = 0$, $q(\xi) = 1 - \xi^{1/2}$,

$$t^* = (bk/\rho_s r_p)(p_{A0}/R_g T)t = \frac{(1)(20)(0.1)t}{(2.26/12)(0.1)(82.06)(900 + 273)} = 1.1 \times 10^{-3}t$$

Substituting in Eq. (3.2.29), we have

$$t_{X=1} = (1 + \tfrac{1}{2})/1.1 \times 10^{-3} = 1364 \text{ sec} = 22.7 \text{ min}$$

For $r_p = 0.1$ mm, $\sigma_0{}^2 = 0.05$ and the resistance to mass transport is much smaller; the overall rate is controlled by the rate of chemical reaction. Again from Eq. (3.2.29),

$$t_{X=1} = (1 + 0.05)/1.1 \times 10^{-2} = 95 \text{ sec} = 1.6 \text{ min}$$

3.2.4 Nonisothermal Behavior

So far we have examined the isothermal behavior of shrinking nonporous particles. When there exists a significant difference in the surface and the bulk temperatures, one must consider the transport of heat as well as that of mass.

The complete governing equations in such cases must allow for transient heat conduction inside the solid particle, and hence will be a rather involved problem to solve. In simple cases where the heat capacity of the solid is small and its thermal conductivity large, the solid at any instant is at a uniform temperature. The problem is then considerably simplified.

The changing particle size, however, will alter the relative importance of the rates of chemical reaction and the transport of heat and mass. It is conceivable then that there may occur a switch from transport-controlled regime to kinetic-controlled regime during the reaction of nonporous particles even when conditions external to the particle remain constant throughout the reaction period.

This effect is usually more likely when the rates of mass and heat transport depend strongly on the extent of reaction. Therefore, we will discuss this problem in more detail for the case in which the product forms an "ash layer" around the reactant solid.

Gas–solid reactions of the type disucssed in this section are of obvious importance in the combustion of coal and other solid fuels. Considerable experimental work has been done on this subject and there is an extensive literature. No attempt will be made to review this work here; instead the reader is referred to a recent review article by Mulcahy and Smith [45].

3.3   Systems Displaying a Shrinking Unreacted Core

In many gas–solid reaction systems, the product solid forms an "ash layer" around the nonporous reactant solid:

$$A(g) + bB(s) \longrightarrow cC(g) + dD(s)$$

The examples of such a system are frequently encountered in chemical and metallurgical processes, such as the reduction of metal oxides, the oxidation of metals, the roasting of ores, and the decomposition of metal compounds. The production of lime and the combustion of ashy coal are further examples.

The progress of reaction of a nonporous solid to ash is shown schematically in Fig. 3.2. As the reaction proceeds, the unreacted core diminishes in size. The overall size of the solid may or may not change, depending on the relative

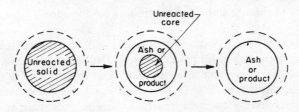

FIG. 3.2. A gas–solid reaction proceeding with the formation of a shrinking unreacted core.

densities of solid reactant and product. The reaction occurs at a sharp interface or within a narrow region between the unreacted core and the product layer. The product may or may not be porous. In either case, the gaseous reactant must be able to penetrate through the product layer in order for the reaction to proceed to completion. This is not the case when the product forms a "protective film" as in the case of the oxidation of aluminum, where the growth of oxide film virtually stops after the film attains a certain thickness.

As we can see in Fig. 3.2, the overall process consists of chemical reaction at the interface, and the diffusion of gaseous reactants and products through the solid-product layer and through the boundary layer at the external surface of the solid. The overall rate may be controlled by the rate of chemical reaction or by the rate of diffusion. In other cases these two steps may present comparable resistances to the progress of reaction and may both influence the overall process. Most models proposed previously assume either of the first two extremes. This is not justified for many systems, nor is it necessary, because they are merely two asymptotes of the general case of both chemical reaction and diffusion determining the overall rate. The solution for the general case encompasses the extreme cases, as we shall see later.

External mass transport through the gas boundary layer usually provides a negligible resistance to the progress of reaction and hence has only a small effect on the overall rate.

We shall again discuss the extreme cases first and then the general case of mixed control including the external mass transfer will be considered. Finally, the more complex case of a nonisothermal system will be discussed.

3.3.1 Control by Chemical Reaction

The overall rate is controlled by the rate of chemical reaction at the interface, when it is the principal resistance to the progress of reaction. The solution becomes identical to that for shrinking particles reacting under conditions of chemical control. The conversion is again given by the following equation:

$$t^* = t/t_{X=1} = g_{F_p}(X) \tag{3.3.1}$$

which applies to infinite slabs, long cylinders, and spheres ($F_p = 1, 2,$ and 3, respectively).

Equation (3.3.1) for spherical particles ($F_p = 3$) has been extensively used for describing many gas–solid reactions [1, 5]. However, a certain amount of caution is necessary in using this equation. The existence of a straight-line relationship according to Eq. (3.3.1) does not necessarily mean that the process is controlled by chemical reaction, because this straight line is very difficult to distinguish from the curve for pure diffusion control, when plotted on the same coordinates (Fig. 3.3). This has frequently led to an incorrect

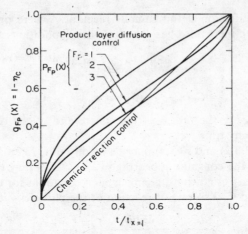

FIG. 3.3.   Thickness of product layer as a function of normalized time for different controlling mechanisms and geometries.

assumption of chemical control, when in fact reaction was controlled by ash layer diffusion or was under mixed control, and subsequently led to calculation of an incorrect value of the activation energy for the intrinsic chemical reaction [6].

3.3.2   Control by Diffusion through Product Layer

When the chemical reaction at the interface presents a negligible resistance to the progress of reaction compared with diffusion through the ash layer, the overall rate is controlled by the latter. In order to obtain the rate of diffusion (hence the overall rate), we make the pseudosteady state approximation [7, 8]; as far as diffusion is concerned the reaction interface may be considered stationary at any time due to the high density of a solid compared to that of a gas.†

We shall obtain the solution for spherical particles. (The solutions for infinite slabs and long cylinders can be obtained similarly.) The rate of diffusion through the product layer of a spherical particle is given by

$$- 4\pi r^2 D_e \, dC_A/dr = \text{const} \tag{3.3.2}$$
$$= 4\pi r^2 D_e \, dC_C/dr \tag{3.3.3}$$

In writing these equations it is assumed that the diffusion in the ash layer is equimolar counterdiffusion. Equation (3.3.2) would also apply to the case of

----

† Alternatively we may state that the rate of accumulation of the gaseous species in the ash layer is negligibly small compared to the diffusive fluxes.

nonequimolar diffusion if the gaseous reactant A were present as a dilute component in an inert third gas. The boundary conditions are

$$C_A = C_{A0} \quad \text{and} \quad C_C = C_{C0} \qquad \text{at} \quad r = r_0 \tag{3.3.4}$$

and

$$C_A = C_C/K_E \qquad \text{at} \quad r = r_c \tag{3.3.5}$$

where r_0 and r_c are the radius of the sphere and of the reaction interface, respectively, at any time.

Equations (3.3.2) and (3.3.3) may be readily integrated. The constants of integration and the constant appearing in Eq. (3.3.2) may then be evaluated from the boundary conditions, giving explicit equations for the concentration profiles of A and C. The rate of reaction of the particle is then given by

$$b4\pi r^2 D_e \frac{dC_A}{dr} = \frac{4\pi b D_e}{1/r_c - 1/r_0} \left(C_{A0} - \frac{C_{C0}}{K_E} \right) \frac{K_E}{1 + K_E} \tag{3.3.6}$$

PARTICLES OF UNCHANGING SIZE

First we consider the case where the total size of the particle is unchanged with reaction; r_0 is constant and equal to $F_p V_p/A_p$. The case of changing size will be discussed later.

It is noted that r_c changes with time as the reaction proceeds through the solid. The rate of reaction of particle may also be expressed as

$$\text{rate of reaction of particle} = -4\pi r_c^2 (\rho_s) \, dr_c/dt \tag{3.3.7}$$

Equating Eqs. (3.3.6) and (3.3.7) and integrating for the initial condition

$$r_c = F_p V_p/A_p \qquad \text{at} \quad t = 0 \tag{3.3.8}$$

we obtain

$$\left(\frac{K_E}{1 + K_E} \right) \left(C_{A0} - \frac{C_{C0}}{K_E} \right) \frac{6bD_e}{\rho_s} \left(\frac{A_p}{F_p V_p} \right)^2 t = 1 - 3\eta_c^2 + 2\eta_c^3 \tag{3.3.9}$$

where

$$\eta_c = (A_p/F_p V_p) r_c \tag{3.3.10}$$

Thus, η_c is the dimensionless position of the reaction front. In terms of conversion X, Eq. (3.3.9) can be written as

$$\left(\frac{K_E}{1 + K_E} \right) \left(C_{A0} - \frac{C_{C0}}{K_E} \right) \frac{6bD_e}{\rho_s} \left(\frac{A_p}{F_p V_p} \right)^2 t = 1 - 3(1 - X)^{2/3} + 2(1 - X) \tag{3.3.11}$$

where

$$X = 1 - \eta_c^{F_p} \tag{3.3.12}$$

For flat plates and long cylinders the relationships, whose derivations will be left to the reader as an exercise, are as follows:

For infinite slabs

$$\frac{K_E}{1 + K_E}\left(C_{A0} - \frac{C_{C0}}{K_E}\right)\frac{2bD_e}{\rho_s}\left(\frac{A_p}{F_p V_p}\right)^2 t = (1 - \eta_c)^2 \qquad (3.3.13)$$

$$= X^2 \qquad (3.3.14)$$

For long cylinders

$$\frac{K_E}{1 + K_E}\left(C_{A0} - \frac{C_{C0}}{K_E}\right)\frac{4bD_e}{\rho_s}\left(\frac{A_p}{F_p V_p}\right)^2 t = 1 - \eta_c^2 + \eta_c^2 \ln \eta_c^2 \qquad (3.3.15)$$

$$= X + (1 - X)\ln(1 - X) \qquad (3.3.16)$$

The relationship for different geometries can be written by a single equation as

$$t^+ \equiv t^*/\sigma_s^2 = p_{F_p}(X) \qquad (3.3.17)$$

where $p_{F_p}(X)$ represents conversion functions given in the r.h.s. of Eqs. (3.3.14), (3.3.16), and (3.3.11) for $F_p = 1, 2,$ and 3, respectively; t^* is defined by Eq. (3.2.9); and

$$\sigma_s^2 = (k/2D_e)(V_p/A_p)(1 + 1/K_E) \qquad (3.3.18)$$

which represents the ratio of the capacities of the shrinking unreacted-core system for chemical reaction and for diffusion. This ratio is a special form of Damköhler group II, defined for shrinking unreacted-core systems. We shall henceforth call σ_s the *shrinking-core reaction modulus*. The usefulness of this quantity will be seen below, where we shall discuss the combined effect of chemical reaction and diffusion.

Examples of the application of Eq. (3.3.17) to spherical particles include the work of Carter [9] for the oxidation of nickel, Kawasaki et al. [10] for the reduction of iron oxides, and Weisz and Goodwin [11] for the combustion of coke deposits on catalysts. In the latter two cases, the solid was initially a porous pellet; when diffusion controls the overall rate, however, the relationships above may be used. We will discuss this in more detail when the system of a porous reactant solid is presented. Many studies on the oxidation of metals have been made in one-dimensional geometries [12,13]—hence the term parabolic law for the rate of progress of such oxidation reactions under conditions of diffusion control [see Eq. (3.3.14)]. Hutchins [14] has verified Eq. (3.3.16) experimentally using a cylindrical system.

It was noted in the previous section on chemical control that it is difficult to determine the controlling mechanism by simply plotting the conversion data according to an assumed mechanism and finding a reasonable straight line. Figure 3.3 shows the comparison between the conversion functions $g_{Fp}(X)$ for chemical reaction control and $p_{Fp}(X)$ for diffusion control. In principle, the controlling mechanism could be determined from plots of this type. In reality, however, the scatter of experimental points makes it very difficult. Furthermore, when both chemical reaction and diffusion contribute significantly in determining the overall rate, an assumption of either extreme mechanism would give an equally good straight line.

In the case of an infinite slab, the distinction is easier than in the cases of long cylinders and spheres. Therefore, one-dimensional geometry is recommended if one must estimate the controlling mechanism from conversion versus time data only, without additional supporting evidence [15]. Most experimental data have been obtained with spherical particles for which the distinction is most difficult. This makes dubious some of the conclusions made by previous investigators with regard to the controlling mechanism based only on the type of plot shown in Fig. 3.3. The rate-controlling mechanism can be determined accurately only by a careful and comprehensive investigation of the effect of important variables affecting the overall rate, such as particle size, temperature, and concentration of the gaseous reactant.

PARTICLES OF CHANGING SIZE

When the external dimensions of the solid particle change upon reaction, the overall size [such as r_0 in Eq. (3.3.6) for a sphere] is no longer constant but given by

$$(F_p V_p/A_p)^{F_p}_{\text{anytime}} = Z(F_p V_p/A_p)^{F_p}_{\text{orig}} + (1 - Z)r_c^{F_p} \qquad (3.3.19)$$

where Z is the volume of the product formed from unit volume of reactant.

Using this relationship instead of a constant overall size, we obtain the following counterparts of Eq. (3.3.17):

$$t^+ = ZX^2 \qquad \text{for} \quad F_p = 1 \qquad (3.3.20)$$

$$t^+ = \frac{[Z + (1 - Z)(1 - X)]\ln[Z + (1 - Z)(1 - X)]}{Z - 1} + (1 - X)\ln(1 - X)$$

$$\text{for} \quad F_p = 2 \qquad (3.3.21)$$

$$t^+ = 3\left\{\frac{Z - [Z + (1 - Z)(1 - X)]^{2/3}}{Z - 1} - (1 - X)^{2/3}\right\}$$

$$\text{for} \quad F_p = 3 \qquad (3.3.22)$$

The time for complete conversion for each geometry is given by

$$t^+_{X=1} = Z \qquad\qquad \text{for} \quad F_p = 1 \qquad (3.3.23)$$

$$t^+_{X=1} = Z \ln Z/(Z - 1) \qquad \text{for} \quad F_p = 2 \qquad (3.3.24)$$

$$t^+_{X=1} = 3(Z - Z^{2/3})/(Z - 1) \qquad \text{for} \quad F_p = 3 \qquad (3.3.25)$$

Figure 3.4 shows the effect of Z on conversion compared with the case of unchanging particle size ($Z = 1$). For infinite slabs the relationship between conversion and the normalized time ($t/t_{X=1}$) remains identical. For long cylinders and spheres, the shape of the curve changes somewhat with Z. However, in the majority of cases the accuracy of experimental data is unlikely to be sufficient to show this difference. Therefore, for all practical purposes, the curve for $Z = 1$ can be used for systems of changing size with $0.5 \leq Z \leq 2$. The approximate equation is then given by

$$t^+/t^+_{X=1} = p_{Fp}(X) \tag{3.3.26}$$

The net effect of this approximation is as follows: for the prediction of the conversion with known D_e and Z, Eq. (3.3.17) can be used simply by replacing D_e by $D_e/t^+_{X=1}$. Conversely, when the effective diffusivity is to be determined from experimental data, a plot of $p_{Fp}(X)$ against time will yield $D_e/t^+_{X=1}$, instead of D_e as in the case of unchanging particles.

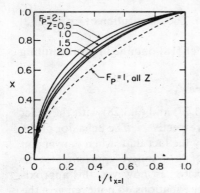 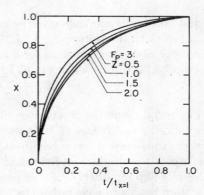

FIG. 3.4. The effect of volume change due to reaction on the relationship between conversion and normalized time.

Comments on some of the approximate theories

It is of some interest at this point to make a few comments on the so-called Jander equation, which has been used often for representing diffusion-controlled gas–solid reactions [1,9,16]. This equation is satisfactory only for the reaction of slablike particles ($F_p = 1$) or for the initial stages of the reaction of particles of other geometries. The equation was put forward by Jander [17] in 1927 for a spherical solid reacting under diffusion control:

$$3[1 - (1 - X)^{1/3}]^2 = t/t_{X=1} \tag{3.3.27}$$

This equation is obtained by assuming that the product layer around the spherical particle is flat, and using the relationship for the thickness of product layer

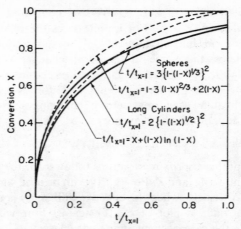

FIG. 3.5. Comparison of Jander equations with exact solutions.

and time for infinite slabs given by Eq. (3.3.13). The substitution of Eq. (3.3.12) in Eq. (3.3.13) results in Eq. (3.3.27).

The equivalent equation for a long cylinder [18] obtained by substituting Eq. (3.3.12) for $F_p = 2$ in Eq. (3.3.13) is

$$2[1 - (1 - X)^{1/2}]^2 = t/t_{X=1} \qquad (3.3.28)$$

Figure 3.5 shows the comparison of Eqs. (3.3.27) and (3.3.28) with the exact solutions given by Eqs. (3.3.11) and (3.3.16), respectively. The behavior of the approximate equations is readily explained by the fact that, when conversion is small, the product layer around a cylinder or a sphere approximates an infinite slab reasonably well; as conversion increases, however, this approximation becomes inaccurate. In fact, the Jander equations do not even give the correct time for complete conversion.

The approximate agreement at low conversion is the apparent reason why the Jander equation is still used by some investigators. The experimental data to which it is applied rarely extend beyond 60% conversion [16]. Using the data on the oxidation of nickel particles, Carter [9] has clearly shown the inapplicability of the Jander equation at conversions higher than 60%. Therefore, instead of the Jander equations, the exact equations given by Eqs. (3.3.11) and (3.3.16) should be used.

Various investigators [19,20] have occasionally attempted to interpret their data in terms of the unsteady state diffusion of matter into a sphere, which is described by the following equation:

$$1 - X = (6/\pi^2) \sum_{n=1}^{\infty} (1/n^2) \exp(-n^2 k^+ t) \qquad (3.3.29)$$

where $k^+ = \pi^2 D_e/r_p{}^2$ and r_p is the radius of the sphere.

This equation is applicable to the saturation of a sphere with another species by diffusion in the absence of chemical reaction. When diffusion occurs toward a sharp boundary where chemical reaction takes place, the situation becomes entirely different. For such reacting systems, the agreement is usually limited to less than 60–70 % conversion, when this equation is forced upon the experimental data. Moreover, the equation is very cumbersome to use.

3.3.3 Overall Rate Determined by Both Chemical Reaction and Diffusion

If neither chemical reaction nor diffusion controls the overall process completely, both steps should be taken into consideration simultaneously in obtaining the expression for the overall rate. To make the analysis complete, we will also consider the effect of external mass transport.

To obtain the general rate expression, we can start from the governing differential equation for diffusion-through the product layer with interfacial chemical reaction and external mass transport providing the boundary conditions. Since each step occurs in series and is independent of the others, we can make use of the results already obtained in the previous two sections.

First, we will consider the case where external mass transport is fast. The overall rate of reaction can be expressed in terms of the rate of chemical reaction or the rate of diffusion through the product layer.

From Eq. (3.3.1), the following equation is obtained by replacing C_{A0} with C_{Ac}, the concentration of A at the reaction interface, and also replacing C_{C0} with C_{Cc} in the definition of t^*:

$$\frac{dX}{dt} = \frac{bk(C_{Ac}^n - C_{Cc}^m/K_E)}{\rho_s g'_{F_p}(X)} \left(\frac{A_p}{F_p V_p}\right) \tag{3.3.30}$$

By differentiating Eq. (3.3.17) with $(C_{A0} - C_{C0}/K_E)$ in the definition of σ_s replaced by the new driving force $[(C_{A0} - C_{C0}/K_E) - (C_{Ac} - C_{Cc}/K_E)]$, we have

$$\frac{dX}{dt} = \frac{2bD_e[K_E/(1 + K_E)][(C_{A0} - C_{C0}/K_E) - (C_{Ac} - C_{Cc}/K_E)]}{F_p \rho_s p'_{F_p}(X)} \left(\frac{A_p}{V_p}\right)^2 \tag{3.3.31}$$

where $g'_{F_p}(X)$ and $p'_{F_p}(X)$ are the derivatives of $g_{F_p}(X)$ and $p_{F_p}(X)$, respectively, with respect to X.

Upon assuming pseudosteady state conditions, the above two expressions can be equated to solve for $(C_{Ac} - C_{Cc}/K_E)$ and hence the overall rate expression can be obtained. We shall consider the simple case of a first-order interfacial reaction, when $(C_{Ac} - C_{Cc}/K_E)$ is given by

$$\frac{(C_{Ac} - C_{Cc}/K_E)}{(C_{A0} - C_{C0}/K_E)} = \frac{g'_{F_p}(X)}{(k/2D_e)(1 + 1/K_E)(V_p/A_p)p'_{F_p}(X) + g'_{F_p}(X)} \tag{3.3.32}$$

Substituting $(C_{Ac} - C_{Cc}/K_E)$ back into Eq. (3.3.30) and integrating, we obtain

$$t^* = g_{Fp}(X) + \sigma_s^2 p_{Fp}(X) \tag{3.3.33}$$

This general expression, which applies to infinite slabs, long cylinders, and spheres, is analogous to Eq. (3.2.29) obtained for shrinking particles. It is again seen that the time required to reach a certain conversion is the sum of the time to reach the same conversion in the absence of diffusional resistance [Eq. (3.3.1)] and that for pure diffusion control [Eq. (3.3.17)].

For particles of changing size, $p_{F_p}(x)$ in Eq. (3.3.33) can be replaced by the terms given in the r.h.s. of Eqs. (3.3.20)–(3.3.22).

Lu [21] obtained equations similar to Eq. (3.3.33) for infinite slabs and spheres, and Seth and Ross [22] used the equation for spherical particles.

The use of Eq. (3.3.33) for the interpretation of experimental data is illustrated by the following two examples.

Example 3.3.1   The reduction of flat particles of high-density nickel oxide in hydrogen for two different sizes of thin and thick samples gave the following results under the identical conditions but in the absence of resistance due to external mass transfer:

thin samples

0.05 cm thick	t (min):	0.5	1	1.5	2	2.5
	X:	0.1	0.2	0.3	0.4	0.5
0.1 cm thick	t (min):	1	2	3	4	5
	X:	0.1	0.2	0.3	0.4	0.5

thick samples

1 cm thick	t (min):	15	60	135	240
	X:	0.1	0.2	0.3	0.4
2 cm thick	t (min):	60	240	375
	X:	0.1	0.2	0.25

Using these data, predict the conversion versus time relationship for the reduction of high-density spherical particles of nickel oxide of 0.3-cm radius. Neglect external mass transfer and consider the conditions to remain unchanged.

SOLUTION   For the results with thin samples we test Eq. (3.3.1), which for $F_p = 1$ may be rewritten

$$X = (kC_{A0}/\rho_s r_p)t$$

From the plot of X versus t (Fig. 3.6), according to this equation, we obtain straight lines with

$$\text{slope} = kC_{A0}/\rho_s r_p = 0.2 \quad \text{for} \quad r_p = 0.025 \text{ cm}$$

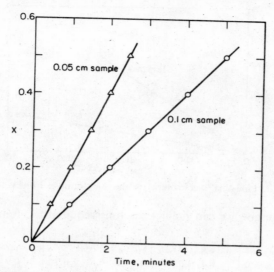

FIG. 3.6. Plot of extent of reaction, X, versus time for the thin samples of nickel oxide.

and

$$\text{slope} = kC_{A0}/\rho_s r_p = 0.1 \qquad \text{for} \quad r_p = 0.05 \text{ cm}$$

Therefore, $kC_{A0}/\rho_s = 0.01$ for both samples. The fact that we obtain straight lines that give identical k values indicates chemical reaction control.

For the thick sample we apply Eq. (3.3.14), which for $F_p = 1$ may be re-written

$$(2C_{A0}/\rho_s)D_e(t/r_p^2) = X^2$$

The plots of X^2 versus t (Fig. 3.7) give straight lines with

$$\text{slope} = 2C_{A0} D_e/\rho_s r_p^2 = 6.67 \times 10^{-4} \qquad \text{for} \quad r_p = 0.5 \text{ cm}$$

and

$$\text{slope} = 2C_{A0} D_e/\rho_s r_p^2 = 1.667 \times 10^{-4} \qquad \text{for} \quad r_p = 1 \text{ cm}$$

From both plots we get

$$2C_{A0} D_e/\rho_s = 1.667 \times 10^{-4}$$

indicating diffusion control. For the spherical pellet using the general relationship given by Eq. (3.3.33), we get

$$t = 30[1 - (1 - X)^{1/3}] + 180[1 + 2(1 - X) - 3(1 - X)^{2/3}]$$

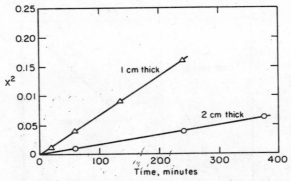

FIG. 3.7. Plot of X^2 versus time for the thick samples of nickel oxide.

From this equation we can predict the time required to attain certain conversions:

t (min):	1.668	4.85	9.64	17	26	39.54	56	80	116	210
X:	0.1	0.2	0.3	0.4	0.5	0.6	0.7	0.8	0.9	1.0

Example 3.3.2 The estimation of k and D_e (and thus σ_s^2) can best be made by varying the experimental condition such that the diffusion through product layer either has no effect on or completely controls the overall rate. When such data cannot be obtained one must make use of Eq. (3.3.33) in order to predict the X versus t relationship under conditions different from those of the experiment.

Seth and Ross [22] proposed to determine the values of σ_s^2 and τ (where $t^* = t/\tau$) by using two experimental points of X–t data with Eq. (3.3.33) and solving simultaneously. This would require rather precise data. And sometimes the two-point method may give inconsistent results. An improvement on this method may be the least-squares method, using as many points as convenient. Find τ and σ_s from the data given below [22] for the reaction of a spherical particle:

t (hr):	0.08	0.1	0.12	0.16	0.22	0.32	0.45
X:	0.2	0.3	0.4	0.6	0.7	0.8	0.9

SOLUTION   Equation (3.3.33) may be rewritten

$$t_i = \tau g_{F_p}(X_i) + \tau \sigma_s^2 p_{F_p}(X_i)$$

Let $\tau = a_1$ and $\tau \sigma_s^2 = a_2$. The least-squares method requires the minimization of the expression

$$\sum_{i=1}^{n} (\text{error})_i^2 = \sum_{i=1}^{n} [a_1 g_{F_p}(X_i) + a_2 p_{F_p}(X_i) - t_i]^2$$

with respect to a_1 and a_2, where n is the number of data points to be used.

The computer solution to the above problem gives

$$\tau = 0.533, \qquad \sigma_s^2 = 0.999$$

from which one can determine k and D_e.

Example 3.3.3 Using the values of σ_s^2 and τ obtained in the preceding example, make a plot of $dX/dt^*|_{act}$ and $dX/dt^*|$ against X in the absence of diffusional resistance. Observe how the effect of diffusion varies with conversion.

SOLUTION The actual rate and the rate in the absence of diffusional resistance are given as

$$dX/dt^*|_{act} = (3)(1 - X)^{2/3}/\{1 + 6\sigma_s^2(1 - X)^{1/3}[1 - (1 - X)^{1/3}]\}$$
$$dX/dt^*|_{no\ diff} = (3)(1 - X)^{2/3}$$

In Figs. 3.8 and 3.9 plots of X versus rate are given. As can be seen, in the case of spherical pellets (Fig. 3.8) the diffusional effect first increases and than decreases to zero. But for flat plates (Fig. 3.9) it increases until $X = 1$. Later in Figs. 3.11 and 3.12 similar comparisons are made in terms of $-d\eta_c/dt^*$, the rate of advancement of the reaction interface.

Now that we have the equation for combined interfacial reaction and diffusion through the product layer, let us generalize it to include the effect of external mass transport. The overall rate can be expressed in terms of the rate of conversion within the particle or the rate of external mass transport:

$$\text{rate of reaction of particle} = \rho_s V_p \, dX/dt \qquad (3.3.34)$$
$$= bA_p h_D(C_{A0} - C_{As}) \qquad (3.3.35)$$
$$= b A_p h_D(C_{Cs} - C_{C0})$$

where C_{A0} is the concentration of A in the bulk.

Assuming pseudosteady state, the term dX/dt can be obtained from Eq. (3.3.33) by replacing C_{A0} and C_{C0} with the surface concentrations C_{As} and C_{Cs};

$$\frac{dX}{dt} = \frac{bk(C_{As} - C_{Cs}/K_E)}{\rho_s}\left(\frac{A_p}{F_p V_p}\right)[g'_{F_p}(X) + \sigma_s^2 p'_{F_p}(X)]^{-1} \quad (3.3.36)$$

Substituting Eq. (3.3.36) into Eq. (3.3.34) and setting equal to Eq. (3.3.35), we obtain

$$\frac{C_{As} - C_{Cs}/K_E}{C_{A0} - C_{C0}/K_E} = \frac{[g'_{F_p}(X) + \sigma_s^2 p'_{F_p}(X)]}{2\sigma_s^2/N_{Sh}^* + [g'_{F_p}(X) + \sigma_s^2 p'_{F_p}(X)]} \qquad (3.3.37)$$

where

$$N_{Sh}^* \equiv \frac{h_D}{D_e}(F_p V_p/A_p) = (D/D_e)N_{Sh} \qquad (3.3.38)$$

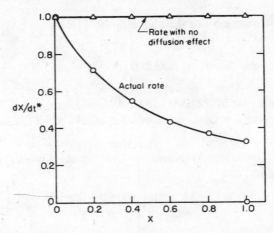

FIG. 3.8. Reaction rate versus conversion for spherical pellet.

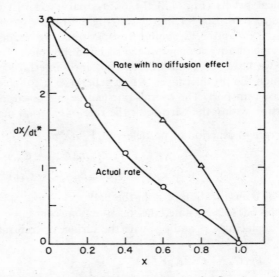

FIG. 3.9. Reaction rate versus conversion for flat plate pellet.

is the modified Sherwood number. Finally, on combining Eqs. (3.3.36) and (3.3.37) and integrating, we obtain the general expression between conversion and time:

$$t^* = g_{F_p}(X) + \sigma_s^2[p_{F_p}(X) + 2X/N_{Sh}^*] \qquad (3.3.39)$$

It can be seen in this equation that, when σ_s^2 is small, i.e., when chemical reaction controls, the effect of N_{Sh}^* is negligible, and that even when σ_s^2 is large, external mass transport is of only secondary importance compared with the diffusion through the product layer.

A similar expression has been obtained for the particular case of a spherical particle by Shen and Smith [23] and by St. Clair [24].

Now the general expression for the time for complete reaction in isothermal, first-order gas–solid reactions, showing the contributions of various steps is given by

$$t_{X=1}^* = 1 + \sigma_s^2(1 + 2/N_{Sh}^*) \tag{3.3.40}$$

From the above discussion we see that the shrinking-core reaction modulus σ_s provides a simple but general criterion for the relative importance of chemical reaction and diffusion. The effect of σ_s^2 is shown in Fig. 3.10 for

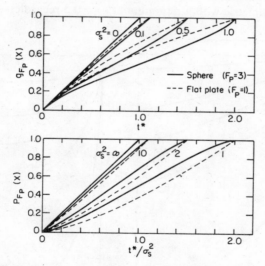

FIG. 3.10. Effect of the reaction modulus σ_s on the conversion versus time relationships.

$N_{Sh}^* = \infty$. When σ_s^2 is smaller than 0.1, the system can be considered to be controlled entirely by chemical reaction. On the other hand, when σ_s^2 is larger than 10, the system is controlled by the diffusion through the product layer and by external mass transfer. The maximum deviation from both extremes occurs when $\sigma_s^2 = 1$. These criteria are general, regardless of geometry.

It is seen in Fig. 3.10 that the relationship between $g_{F_p}(X)$ and time is approximately linear even for σ_s^2 as large as 0.5, i.e., when the resistances due to chemical reaction and diffusion are of comparable magnitude. This indicates that the existence of an approximate linear relationship between $g_{F_p}(X)$ and time from experimental results does not necessarily mean that the system is controlled by chemical reaction. One may obtain an approximately linear plot even when there is a substantial resistance due to diffusion in the product layer.

Figures 3.11 and 3.12 show the rates of regression of the reaction interface as a function of its position for an infinite slab and a sphere, respectively In the case of an infinite slab the effect of diffusion becomes progressively larger as the diffusion path is increased. In the case of a spherical particle, however, an interesting phenomenon occurs. In the early stages of reaction, the effect of diffusion increases with conversion due to the same reason. As the reaction interface regresses further, the converging geometry makes the rate of diffusion increase with the conversion. In fact, the rate of regression of the reaction interface is symmetric with a minimum at $\eta_c = 0.5$ regardless of σ_s.

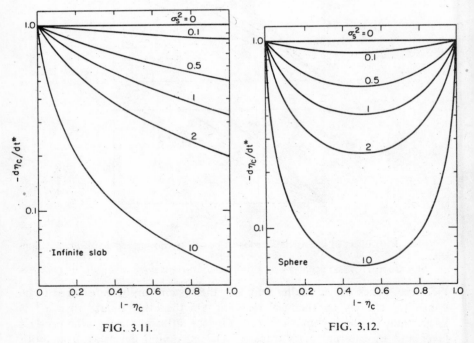

FIG. 3.11.   FIG. 3.12.

FIG. 3.11. Rate of regression of the reaction interface as a function of its position for the case of an infinite slab and first-order reaction.

FIG. 3.12. Rate of regression of reaction interface as a function of its position for the case of a sphere and first-order reaction.

3.4 Effect of Reaction Order on Gas–Solid Reactions

In the previous analyses of the combined effects of chemical reaction and diffusion, we have used first-order kinetics for the interfacial reaction. In this section we will examine the effect of reaction order with respect to the concentration of gaseous reactant (n, henceforth to be called simply, the "reaction order"). We shall do this for the shrinking unreacted-core system without external-mass-transport resistance, and for irreversible reactions ($K_E \to \infty$).

Equation (3.3.30) may be rewritten, with the aid of Eqs. (3.2.5), (3.2.11), and (3.3.12), as [44]

$$d\eta_c/dt^* = -\psi^n \qquad (3.4.1)$$

where

$$\psi \equiv C_{Ac}/C_{A0} \qquad (3.4.2)$$

Rewriting Eq. (3.3.31),

$$d\eta_c/dt^* = (1 - \psi)/\sigma_s^2 p'_{F_p}(\eta_c) \qquad (3.4.3)$$

where $p_{F_p}(\eta_c)$ is given by the r.h.s. of Eqs. (3.3.13), (3.3.15), and (3.3.9) for $F_p = 1$, 2, and 3, respectively.

Equations (3.4.1) and (3.4.3) are solved simultaneously by eliminating ψ to obtain the rate or the position of the reaction interface as a function of time [44].

The results are shown in Figs. 3.13–3.15 for infinite slabs and spheres, and for $n = 0$, $\frac{1}{2}$, 1, and 2. Long cylinders are expected to exhibit behavior that is intermediate between those of infinite slabs and spheres.

The results indicate that the use of Eq. (3.3.33), which is exact for first-order kinetics, for other reaction orders may introduce a serious error. This is in contrast to the result for the heterogeneous catalysis in a porous pellet or on a solid surface, in which the difference between the effectiveness factors for different reaction orders is much smaller, when plotted against the Thiele modulus generalized for different reaction order [25, 26].

The reason for the greater error in a noncatalytic system is that the gaseous reactant must first diffuse through the product layer to reach the reaction interface. The concentration at the interface is smaller than in the bulk, making the rate of chemical reaction more sensitive to the reaction order. (Generalization is made in terms of the bulk concentration.)

The error becomes largest when the relative importance of chemical reaction and diffusion are of the same magnitude ($\sigma_s^2 = 1$). As the asymptotic regime of either chemical reaction control ($\sigma_s^2 < 0.1$) or diffusion control ($\sigma_s^2 > 10$) is approached, the error becomes small.

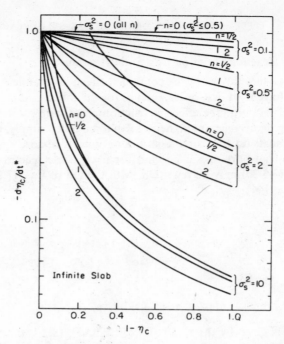

FIG. 3.13. The effect of reaction order n on the rate of regression of the reaction interface for various values of the reaction modulus σ_s [44].

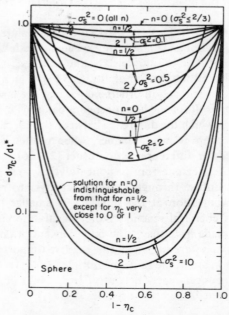

FIG. 3.14. The effect of reaction order n on the rate of regression of the reaction interface for various values of the reaction modulus σ_s [44].

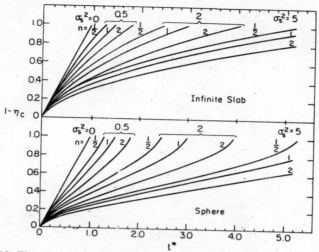

FIG. 3.15. The effect of reaction order n on the progress of reaction for various values of the reaction modulus σ_s [44].

3.5   Gas–Solid Reactions Combined with Solid State Diffusion

When the solid product is nonporous and forms a dense layer around the reactant solid, the transport of matter across the product layer occurs by solid state diffusion. This is usually a slow process with a high activation energy.

Solid state diffusion has been found to control, under certain conditions, the overall rate of the reduction of metal oxides [27, 28] and the oxidation of metals [29].

Landler and Komarek [27] studied the partial reduction of wüstite with hydrogen:

$$\text{Fe}_{1-y_1}\text{O} \longrightarrow \text{Fe}_{1-y_2}\text{O}$$

(wüstite is a nonstoichiometric compound) and interpreted their results in terms of a model in which the overall rate was determined by chemical reaction at the wüstite surface and the solid state diffusion of iron within the wüstite. The interaction of these steps is shown in Fig. 3.16. It had been shown previously [29, 30] that oxygen was not the diffusing species in the partial reduction or oxidation of wüstite.

Assuming the solid to be an infinite slab with a finite thickness L, the unidirectional diffusion of iron within the solid is described by

$$\partial c/\partial t = D_s\, \partial^2 c/\partial y^2 \tag{3.5.1}$$

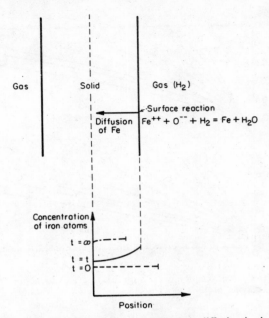

FIG. 3.16. Combined surface reaction and solid state diffusion in the partial reduction of wüstite.

with the boundary conditions

$$c = c_0 \qquad \text{at} \quad t = 0 \qquad (3.5.2)$$

$$D_s \, \partial c / \partial y = k(c - c_{eq}) \qquad y = L/2 \qquad (3.5.3)$$

$$\partial c / \partial y = 0 \qquad \text{at} \quad y = 0 \qquad (3.5.4)$$

The boundary condition at the gas–solid interface, Eq. (3.5.3), assumes that the surface reaction is reversible and of first order with respect to iron concentration and that the gas phase composition remains constant.

The solution for iron concentration as a function of y and t is given by [31]

$$\frac{c - c_{eq}}{c_0 - c_{eq}} = \sum_{n=1}^{n=\infty} 2A_n \cos \frac{2\beta_n y}{L} \exp\left(-\frac{4D_s t}{L^2} \beta_n^2\right) \qquad (3.5.5)$$

where

$$A_n = [(kL/2D_s + (2D_s/kL)\beta_n^2 + 1) \cos \beta_n]^{-1} \qquad (3.5.6)$$

and β_n is defined by

$$\cos \beta_n = \beta_n (2D_s/kL) \qquad (3.5.7)$$

The overall conversion can be calculated using Eq. (3.5.5) to give

$$X = 1 - \frac{1}{L} \int_{-L/2}^{L/2} \left(\frac{c - c_{eq}}{c_0 - c_{eq}} \right) dy$$

$$= 1 - \sum_{n=1}^{n=\infty} 2 \exp\left(-\frac{4D_s t}{L^2} \beta_n^2 \right) \left[\beta_n^2 + \left(\frac{2D_s \beta_n^2}{kL} \right)^2 + \frac{2D_s}{kL} \beta_n^2 \right]^{-1} \quad (3.5.8)$$

It is seen that conversion is given as a function of reduced time, $4D_s t/L^2$ with kL/D_s as a parameter. This is analogous to Eq. (3.3.33) except that the relationship is more complicated. The values of the sum in Eq. (3.5.8) are tabulated as a function of $(kL/2D_s)$ and $(4D_s t/L^2)$ in the literature [31].

Although Landler and Komarek determined the diffusivity of iron using Eq. (3.5.8) and experimental conversion data, a more valid use of this equation would be in predicting the conversion from separately obtained information on surface reaction and diffusion. The solid state diffusion can be studied independently, as shown by Edstroem and Bitsianes [30] and Himmel and coworkers [32].

3.6 Nonisothermal Shrinking Unreacted-Core Systems

The preceding analyses of the shrinking unreacted-core systems assume the solid to be isothermal during the reaction. When the chemical reaction involves heat of reaction, a temperature gradient may exist within the pellet. If the reaction is exothermic, the reaction rate is enhanced by the higher temperature in the interior of the solid. This temperature difference may, however, cause some difficulties such as sintering in the reduction of metal oxides and deactivation in the case of the regeneration of catalyst pellets. Furthermore, exothermic gas–solid reactions may present the interesting and often important problems of multiple steady states and geometrical and thermal instability. If the reaction is endothermic, the system is always stable with a single steady state.

3.6.1 Temperature Rises in a Diffusion-Controlled Shrinking Unreacted-Core System

For a system that undergoes an exothermic reaction and is controlled by the diffusion through the product layer, Luss and Amundson [33] have calculated the maximum temperature rise in the solid as affected by the various parameters.

The solution for a spherical particle is given in the following. The reader is referred to the original article of Luss and Amundson for detailed mathematics. The temperature at any position in the pellet when the reaction front is at

a given position is given by

$$\theta = \tfrac{2}{3}A \sum_{n=1}^{\infty} f(\alpha_n)[\sin(\alpha_n\eta)/\eta] \int_{\eta_c}^{1} \eta^* \sin(\alpha_n\eta^*)$$

$$\times \exp\left\{-\frac{A\alpha_n{}^2}{3Nu^*}\left[\frac{1}{3}\left(1-\frac{1}{N_{Sh}^*}\right)(\eta_c{}^3 - \eta^{*3}) + \frac{1}{2}(\eta^{*2} - \eta_c{}^2)\right]\right\} d\eta^* \quad (3.6.1)$$

where

$$\theta \equiv [h(T_s - T_0)/b(-\Delta H)_B D_e C_{A0}](F_p V_p/A_p) \tag{3.6.2}$$

$$A \equiv (h/\rho c_s)(\rho_s F_p/b D_e C_{A0})(F_p V_p/A_p) \tag{3.6.3}$$

$$f(\alpha_n) \equiv [(N_{Nu}^* - 1)^2 + \alpha_n{}^2]/[N_{Nu}^*(N_{Nu}^* - 1) + \alpha_n{}^2] \tag{3.6.4}$$

$$N_{Nu}^* \equiv (h/\lambda_e)(F_p V_p/A_p) = (\lambda/\lambda_e)N_{Nu} \tag{3.6.5}$$

$$\eta \equiv (A_p/F_p V_p)r \tag{3.6.6}$$

and α_n are the positive roots of

$$\alpha_n \cot \alpha_n = 1 - N_{Nu}^* \tag{3.6.7}$$

A typical temperature profile within the particle as a function of the location of the interface between the reactant and the product layers is shown in Fig. 3.17. The maximum temperature at a given value of η_c increases initially as the reaction front moves into the solid. Eventually, due to the reduction of reaction rate by the increasing diffusional resistance, the maximum temperature starts to decrease.

Figure 3.18 shows the effect of N_{Nu}^* on the variation of the maximum temperature. The value of the highest θ_{max} increases and occurs for larger values of η_c, as N_{Nu}^* increases.

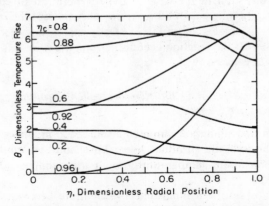

FIG. 3.17. Temperature profiles within a reacting particle as a function of the location of the interface between reactant and product layers. $A = 100$, $N_{Sh}^* = 100$, $N_{Nu}^* = 1$ [33].

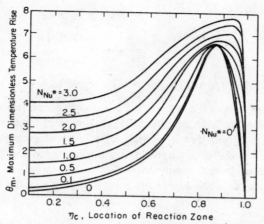

FIG. 3.18. Effect of the Nusselt number on the maximum temperature as a function of the position of the reaction zone. $A = 100$, $N_{Sh}^* = 100$ [33].

The condition of $N_{Nu}^* = 0$, i.e., very large thermal conductivity of the solid and hence uniform temperature within the solid, is shown to be well approached when $N_{Nu}^* < 0.1$. It is also seen that the value of the highest θ_{max} does not vary greatly for $0 < N_{Nu}^* < 3$. Therefore, when the maximum temperature rise in the solid at any time during the reaction is of interest, the much simpler solution for $N_{Nu}^* = 0$ can be used to obtain a reasonable estimate. In such a case, the temperature difference $T = T_s - T_0$ is determined by the relationship

$$\rho c_s V_p \, dT/dt + h A_p T = Q \qquad (3.6.8)$$

The position of the reaction front can be determined by Eq. (3.3.39) with $\sigma_s \to \infty$ and Eq. (3.3.12). The rate of heat generation Q can be expressed in terms of $d\eta_c/dt$. For a spherical particle, these are given by

$$d\eta_c/dt = (b D_e C_{A0}/\rho_s)(A_p/F_p V_p)^2 \{\eta_c^2(1 - 1/N_{Sh}^*) - \eta_c\}^{-1} \qquad (3.6.9)$$

and

$$Q = -\rho_s(-\Delta H)_B A_p \eta_c^{F_p - 1}(F_p V_p/A_p) \, d\eta_c/dt \qquad (3.6.10)$$

Combining Eqs. (3.6.8)–(3.6.10) and rearranging in dimensionless form one obtains

$$d\theta/d\eta_c + A[\eta_c^2(1 - 1/N_{Sh}^*) - \eta_c]\theta = -A\eta_c^2 \qquad (3.6.11)$$

with boundary condition

$$\theta = 0 \qquad \text{at} \quad \eta_c = 1 \qquad (3.6.12)$$

The solution to Eq. (3.6.11) is

$$\theta = A \exp[f(\eta_c)] \int_{\eta_c}^{1} \eta_c^2 \exp[-f(\eta_c)] \, d\eta_c \tag{3.6.13}$$

where

$$f(\eta_c) \equiv A\eta_c^2[\tfrac{1}{2} - \tfrac{1}{3}(1 - 1/N_{Sh}^*)\eta_c] \tag{3.6.14}$$

The solution given by Eq. (3.6.13) still requires numerical evaluation. An analytical solution has been obtained [34] that is exact for a infinite slab and is approximately correct for a particle of a general geometry. From this analytical solution, useful asymptotic and approximate solutions have been obtained for the maximum temperature rise.

For an infinite slab, the equations corresponding to Eqs. (3.6.8)–(3.6.11) are given as

$$\rho c_s V_p \, dT/dt + hA_p T = Q \tag{3.6.15}$$

$$d\eta_c/dt = (bD_e C_{Ab}/\rho_s)(A_p/F_p V_p)^2[\eta_c - (1 + 1/N_{Sh}^*)]^{-1} \tag{3.6.16}$$

$$Q = -\rho_s(-\Delta H)_B A_p(F_p V_p/A_p) \, d\eta_c/dt \tag{3.6.17}$$

and

$$d\theta/d\delta + A(\delta + 1/N_{Sh}^*)\theta = A \tag{3.6.18}$$

where

$$\delta \equiv 1 - \eta_c \tag{3.6.19}$$

The boundary condition is

$$\theta = 0 \quad \text{at} \quad \delta = 0 \tag{3.6.20}$$

Equation (3.6.18) is identical to that obtained for a spherical particle by linearizing Eq. (3.6.11) for small values of δ. In view of the fact that for a spherical particle the maximum temperature rise occurs for a small value of δ [33], one can anticipate that the solution to Eq. (3.6.18) will be approximately correct for a spherical particle.

Integration of Eq. (3.6.18) yields

$$\theta = \sqrt{2A} \, e^{-x^2}[D(x) - D(\alpha)] \tag{3.6.21}$$

where

$$x \equiv \sqrt{A/2}(\delta + 1/N_{Sh}^*) \tag{3.6.22}$$

$$\alpha \equiv (1/N_{Sh}^*)\sqrt{A/2} \tag{3.6.23}$$

and $D(x)$ is defined by the so-called Dawson's integral

$$e^{-w^2}D(w) \equiv e^{-w^2} \int_0^w e^{t^2} \, dt \tag{3.6.24}$$

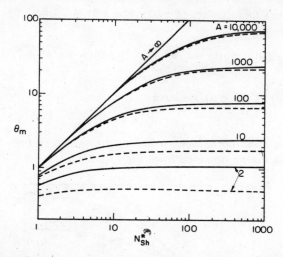

FIG. 3.19. Comparison between exact and approximate solutions for the maximum temperature rise. ——— Exact solution for slablike pellets or approximate solution for pellets of other geometries. - - - Exact solution for spherical pellets [34].

The value of this function is tabulated and its series expansion is also available in the literature [34–36].

Figure 3.19 shows the comparison for the maximum temperature rise θ_m obtained from Eqs. (3.6.21) and (3.6.13). It is seen that there is a good agreement for $A > 10$. For $A < 10$ the solution differs for particles of different shapes. In this region, however, the temperature rise is small and is of no great practical significance. Therefore, θ_m obtained from Eq. (3.6.21), which is exact for an infinite slab, can be used for a spherical particle (or for particles of other geometries).

From the solution given by Eq. (3.6.21), some asymptotic and approximate solutions can be obtained for θ_m as a function of A and N_{Sh}^* [34].

As N_{Sh}^* becomes large, i.e., $\alpha \to 0$, Eq. (3.6.21) reduces to

$$\theta = \sqrt{2A}\, e^{-x^2} D(x) \qquad (3.6.25)$$

from which we obtain [35]

$$\theta_{m,\,N_{Sh}^* \to \infty} = 0.765\sqrt{A} \qquad (3.6.26)$$

which occurs at

$$\delta_{N_{Sh}^* \to \infty} = 1.31/\sqrt{A} \qquad (3.6.27)$$

This asymptotic solution is recommended for $\alpha < 0.1$.

For small but finite values of α $(0.1 < \alpha < 1)$ the following approximate solution applies:

$$\theta_{m,\alpha} = \sqrt{2A}\{\exp-[(\alpha + \sqrt{\alpha^2 + 2})/2]^2\}\{D[(\alpha + \sqrt{\alpha^2 + 2})/2] - D(\alpha)]$$

(3.6.28)

For large values of $\alpha(1 < \alpha)$, we have

$$\theta_{m,\alpha \to \infty} = N_{Sh}^*(1 + 1/\alpha^2)^{-1}$$

(3.6.29)

The asymptotic and approximate solutions are shown in Fig. 3.20 together with the exact solution obtained from Eq. (3.6.21).

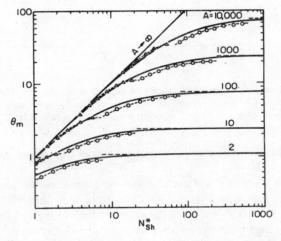

FIG. 3.20. Asymptotic and approximate analytical solutions for the maximum temperature rise. — Exact solution, - - - Asymptotic solution $(\alpha \le 0.1)$, - - $\bigcirc$ - - Approximate solution $(0.1 < \alpha < 1)$, - - $\triangle$ - - Approximate solution $(1 \le \alpha)$ [34].

It has been shown above that very large temperature rises may occur during diffusion-controlled gas–solid reactions. When these elevated temperatures cause structural changes such as sintering the situation becomes even more complex and is beyond the reach of present-day models. The treatment developed here may be used to assess whether substantial local temperature rises are likely to occur for the conditions considered. This is illustrated by the following example.

Example 3.6.1   Luss and Amundson [33] give the following data for the combustion of coke in a typical cracking catalyst. Predict the maximum temperature rise in the pellet during the combustion. Consider the catalyst spherical.

$$D_e = 5 \times 10^{-3} \text{ cm}^2/\text{sec}, \qquad D = 2 \times 10^{-1} \text{ cm}^2/\text{sec}$$

$$C_{A0} = 3 \times 10^{-6} \text{ g-mole/cm}^3, \qquad \rho_s = 4.5 \times 10^{-3} \text{ g-mole/cm}^3$$

$$\lambda = 8 \times 10^{-5} \text{ cal/cm sec } ^\circ\text{C}, \qquad \lambda_e = 8 \times 10^{-4} \text{ cal/cm sec } ^\circ\text{C}$$

$$\rho = 1.0 \text{ g/cm}^3, \qquad c_s = 0.3 \text{ cal/g } ^\circ\text{C}$$

$$b = 1, \qquad (-\Delta H)_B \doteq 83{,}000 \text{ cal/g mole}, \qquad N_{Nu} = N_{Sh} = 2.5$$

From the above data,

$$h(F_p V_p/A_p) = \lambda N_{Nu} = 2 \times 10^{-4} \text{ cal/cm } ^\circ\text{C sec}$$

$$N^*_{Nu} = (\lambda/\lambda_e)N_{Nu} = 0.25, \qquad N^*_{Sh} = (D/D_e)/N_{Sh} = 100$$

Substituting these values in Eq. (3.6.3) we obtain A = 600, and from Fig. 3.19, $\theta_m = 14$. Using Eq. (3.6.2) we get $\Delta T_{max} = (T_s - T_0)_{max} = 87^\circ\text{C}$.

3.6.2 Multiple States and Thermal Instability

In the general case of nonisothermal shrinking-core systems controlled both by chemical reaction and diffusion, the thermal effect of the reaction may bring about multiple steady states and instability due to sudden transition of rate-controlling steps during the reaction. The problem of thermal instability in noncatalytic gas–solid reactions was first pointed out by Cannon and Denbigh [37] and has been discussed by Shen and Smith [23] and Wen and coworkers [38, 39].

In order to analyze the nonisothermal systems, an energy balance must be added to the mass balance within the solid particle. Although this is not essential in the development that follows, we shall assume for the sake of simplicity that both D_e and the concentration profiles are unaffected by temperature. These assumptions should be reasonable when the temperature difference across the product layer is small.

Then the expression for $C_{Ac} - C_{Cc}/K_E$, the chemical driving force, will be the same as that obtained in the isothermal case. Thus combining Eqs. (3.3.32) and (3.3.37), we obtain for the first-order irreversible chemical reaction

$$\frac{C_{Ac}}{C_{A0}} = [\sigma_s(T_0)]^2 \left[\frac{4/N^*_{Sh} + p'_{F_p}(X)}{g'_{F_p}(X)}\right] \exp\left[\gamma\left(1 - \frac{T_0}{T_c}\right)\right]^{-1} \qquad (3.6.30)$$

where σ_s is based on T_0, the bulk temperature, and

$$\gamma \equiv E/RT_0 \qquad (3.6.31)$$

When the heat capacity of the solid is small,† the energy balance within the

---

† Wen and Wang [39] discussed the unsteady state heat transfer problem when the heat capacity is not small.

product layer yields the following relationship:

$$\frac{T_c}{T_0} - \left\{ \beta[\sigma_s(T_0)]^2 \frac{C_{Ac}}{C_{A0}} \left[\frac{4/N_{Nu}^* + p'_{F_p}(X_c)}{g'_{F_p}(X)} \right] \exp\left[\hat{\gamma}\left(1 - \frac{T_0}{T_c} \right) \right] \right\}^{-1} = 0 \quad (3.6.32)$$

where

$$\beta \equiv [(-\Delta H)D_e/\lambda_e T_0]C_{A0} \quad (3.6.33)$$

In the above derivation, the physical and transport parameters were assumed to be independent of temperature.

Equations (3.6.30) and (3.6.32) give the relationship between the conversion and C_{Ac}. Substitution of this relationship back into Eq. (3.3.30) yields the relationship between conversion and time.

For an exothermic reaction, Eqs. (3.6.30) and (3.6.32) may have three solutions at a given conversion value. One of the solutions represents an unstable condition. The analysis of thermal instability is best demonstrated by comparing the rates of heat generation and heat loss as functions of temperature.

These rates per unit surface area of reaction interface are given by

$$q_{gen} = (-\Delta H)k(T_c)C_{Ac} \quad (3.6.34)$$

$$q_{loss} = \frac{\lambda_e(A_p/V_p)(T_c - T_0)}{[1/N_{Nu}^* + p'_{F_p}(X)/2]/g'_{F_p}(X)} \quad (3.6.35)$$

However, since the conversion changes during the reaction in the shrinking unreacted-core systems, the location of the heat generation curve and the slope of the heat loss curve will change even for constant bulk conditions. If these changes occur such that the operating rate-controlling step is kinetic or diffusional, the system will undergo a gradual change and will not experience sudden transitions. Thermal instability occurs when these changes cause a sudden transition in the operating regime from a high-temperature stable state to a low-temperature stable state or vice versa.

These phenomena are shown in Fig. 3.21. The reaction starts at point A, where the slow rate of chemical reaction controls the overall rate. As the conversion increases, point B is reached and a sudden jump to B' occurs, exhibiting an "ignition" process. As the reaction progresses further, or for a reaction that started at point B', the operating conditions may go through point C and reach point D, where a sudden shift to point D' occurs representing an "extinction" process.

It must be stressed to the reader that the preceding discussion was based on the (simultaneous) steady state solution of the thermal and material transport equations. In a real physical situation the onset of ignition or extinction phenomena is inherently an unsteady state process; it follows that the transition to ignition or extinction will take place at a finite rate.

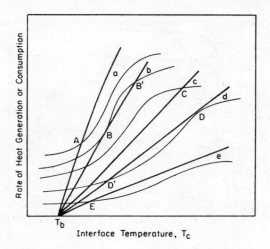

FIG. 3.21. Thermal balance at the reaction interface. Locations of q_{gen} and q_{loss} as reaction progresses: $a \to b \to c \to d \to e$; operating conditions: $A \to B/B' \to C \to D/D' \to E$.

Various criteria have been developed and discussed to predict the possibility of a thermal instability in gas–solid reactions [23, 40, 41], assuming pseudosteady state regarding the heat transfer within the solid and negligible heat capacity of the solid. Wen and Wang [39] discussed the effect of various factors on the thermal instability and have extended their analysis to study the unsteady state heat transfer within the particle when the heat capacity of the solid is not negligible compared to the heat of reaction.

3.7   The Effect of Structural Changes on Reaction

In an earlier section we discussed the effects of a change in the gross volume of the particle during reaction on the progress of reaction. At elevated temperatures the porous ash layer through which the gaseous reactant and product are diffusing may undergo sintering, which will affect the effective diffusivities and hence the rate of reaction. Figure 3.22 is a plot of experimental data [42] on the reduction of nickel oxide spheres by hydrogen at elevated temperatures.† At temperatures above approximately 750°C sintering of the nickel layer markedly reduces the rate of reaction, the effect being greater at higher temperatures, where sintering is more rapid.

Modification of the shrinking-core model to accommodate structural

---

† The nickel oxide pellets were actually porous, but reduction was carried out under conditions where the shrinking-core model was applicable (see Chapter 4).

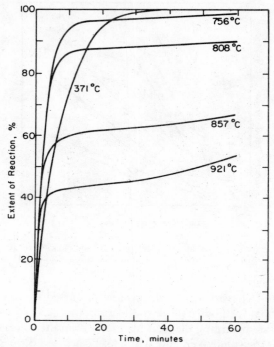

FIG. 3.22. Experimental results on the reduction of nickel oxide by hydrogen at temperatures where sintering of the nickel occurs [42].

changes of this nature is possible [43]. For an isothermal reaction of a non-porous spherical particle we have

$$\frac{dX}{dt} = -\frac{3r_c^2}{r_0^3}\frac{dr_c}{dt} = \frac{3h_D(C_{A0} - C_{C0}/K_E)}{r_0\rho_s[1 + (1/K_E) + h_D r_0^2 I(1 + 1/K_E)]} \qquad (3.7.1)$$

where

$$I = \int_{r_c}^{r_0} dr/D_e(r, t)r^2 \qquad (3.7.2)$$

Integration of Eq. (3.7.1) presents some difficulty since our present knowledge of sintering kinetics is insufficient to allow the prediction of the effective diffusivity as a function of time and position in the ash layer.

Figure 3.23 presents the results of numerical calculations [42] made using equation (3.7.1) and an assumed exponential dependence of D_e on time of the form

$$D_e(r, t) = D_{e0} \exp[-(t - t_c)/\tau_{SI}] \qquad (3.7.3)$$

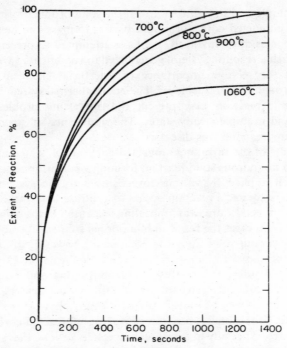

FIG. 3.23. Prediction of the progress of a gas–solid reaction proceeding at temperatures where sintering of the solid product occurs [42].

where D_{e0} and τ_{SI} are constants and t_c is the time at which the reaction front reaches the radius r. τ_{SI}, the characteristic time for sintering, is assumed to be related to temperature by

$$\tau_{SI} = \tau_{SO} \exp[-E_{AS}/RT] \qquad (3.7.4)$$

where τ_{SO} is a constant, E_{AS} an activation energy for sintering, and T the absolute temperature.

This rather simplistic model leads to a predicted behavior which is at least in qualitative agreement with the experimental results depicted in Fig. 3.22. The refinement of models of this type must await the development of a better understanding of sintering kinetics.

3.8   Concluding Remarks

In this chapter we have discussed gas–solid reactions involving nonporous solid reactants. When such a reaction produces gaseous products or solid products that are immediately removed from the reaction interface, the analysis is simple and straightforward. If the solid product forms an ash layer around the unreacted core, forming a barrier to mass and heat transfer, the analysis becomes more involved. We have attempted to present a systematic picture of such a reaction system, generalized to encompass various geometric factors and the relative importance of interfacial chemical reaction and diffusion through the product layer. The nonisothermal system that undergoes an exothermic reaction may present the interesting problem of thermal instability and multiple steady states. The signficance of structural changes accompanying reaction was discussed.

The results of the shrinking unreacted-core model, which are valid to the reaction of a nonporous solid reactant forming a porous product layer around it, have been applied by various investigators to the system of a porous reactant where chemical reaction occurs in a diffuse zone rather than a sharp interface. This procedure was appealing because of the simplicity of the model; in certain cases the reaction of a porous solid can be approximated to that of a nonporous solid. In many other cases, however, this indiscriminate application of the shrinking-core model to the reaction of porous solids introduced difficulties and confusion in interpreting and extending experimental data, in calculating reaction and diffusion parameters, and in determining the dependency of overall rate on various physical parameters.

Therefore, one should use the shrinking unreacted-core model with the utmost care, especially in extrapolating results beyond the range of parameters employed in obtaining experimental data.

In the next chapter we will discuss the problems involved in the reaction of porous solids and will show the salient comparisons discussed above.

Notation

a, a'	constants defined in Eqs. (3.2.17) and (3.2.14), respectively
A	dimensionless quantity defined by Eq. (3.6.3)
A_p	external surface area of the particle
b, c, d	stoichiometric coefficients
c_s	specific heat of the particle including solid product and inert solid, if any
C_i	molar concentration of species i
D	molecular diffusivity
D_e	effective diffusivity
D_s	solid state diffusivity
$D(x)$	function defined by Eq. (3.6.24)
F_p	particle shape factor ($=1, 2,$ and 3 for infinite slab, long cylinder, and sphere, respectively)
$g_{F_p}(X)$	conversion function defined by Eq. (3.2.11)
h	heat transfer coefficient

h_D	mass transfer coefficient
$(-\Delta H)_B$	molar heat of reaction of solid B
k	reaction rate constant
K_E	equilibrium constant
m, n	reaction orders with respect to gaseous reactants
$N_{Nu}, N_{Re}, N_{Sc}, N_{Sh}$	Nusselt, Reynolds, Schmidt, and Sherwood numbers, respectively
N_{Nu}^*, N_{Sh}^*	modified Nusselt and Sherwood numbers defined by Eqs. (3.6.5) and (3.3.38), respectively
$p_{Fp}(X)$	conversion function defined by Eqs. (3.3.17), (3.3.14), (3.3.16), and (3.3.11)
$q(\xi)$	function defined in Eq. (3.2.29)
r	distance coordinate
r_c	position of reaction interface
r_0	position of external surface at any time
r_p	original position of external surface
R_s	rate of reaction at gas–solid interface
S	surface area
t	time
t^*	dimensionless time defined by Eq. (3.2.5)
t^+	dimensionless time defined by Eq. (3.3.17)
T	temperature ($=T_s - T_0$)
T_c	temperature at reaction interface
T_0	temperature of the bulk fluid
T_s	temperature of the particle
V_p	volume of the particle
X	fractional conversion of the solid
Z	volume of solid product formed from unit volume of initial solid
α	dimensionless quantity defined by Eq. (3.6.23)
β	dimensionless quantity defined by Eq. (3.6.33)
γ	dimensionless group defined by Eq. (3.6.31)
δ	dimensionless variable defined by Eq. (3.6.19)
η	dimensionless distance defined by Eq. (3.6.6)
η_c	dimensionless position of reaction interface defined by Eq. (3.3.10)
θ	dimensionless temperature defined by Eq. (3.6.2)
θ_m	maximum value of θ
λ	thermal conductivity of gas
λ_e	effective thermal conductivity of the particle
ξ	dimensionless position of external surface of a shrinking non-porous particle, defined by Eq. (3.2.4)

ρ	density of the particle including solid product and inert solid, if any
ρ_s	molar concentration of solid reactant
σ_0^2	dimensionless parameter defined in Eq. (3.2.29)
σ_s^2	shrinking-core reaction modulus defined by Eq. (3.3.18)
ψ	dimensionless concentration defined by Eq. (3.4.2)

Subscripts

A	gas A
B	solid B
c	value at reaction interface
C	gas C
D	solid D
0	bulk property
p	value for the original particle
s	property of solid reactant, or value at the external surface

References

1. F. Habashi, "Extractive Metallurgy," Vol. 1, General Principles, Chapter 8. Gordon and Breach, New York, 1969.
2. D. A. Young, "Decomposition of Solids." Pergamon, Oxford, 1966.
3. G. Pannetier and P. Souchay, "Chemical Kinetics" (H. D. Gesser and H. H. Edmon, trans.), Chapter 9. Elsevier, Amsterdam, 1967.
4. J. S. Hirshhorn, "Introduction to Powder Metallurgy," p. 46. Amer. Powder Metallurgy Inst., 1969.
5. N. J. Themelis and W. H. Gauvin, *Trans. Met. Soc. AIME* **227**, 290 (1963).
6. J. Szekely and H. Y. Sohn, *Trans. Inst. Mining Met.* **82**, C92 (1973).
7. K. B. Bischoff, *Chem. Eng. Sci.* **18**, 711 (1963); **20**, 783 (1965).
8. D. Luss, *Can. J. Chem. Eng.* **46**, 154 (1968).
9. R. E. Carter, *in* "Ultrafine Particles" (W. E. Kuhn, ed.), p. 419. Wiley, New York, 1963; *J. Chem. Phys* **35**, 1137 (1961).
10. E. Kawasaki, J. Sanscrainte, and T. J. Walsh, *AIChE J.* **8**, 48 (1963).
11. P. B. Weisz and R. D. Goodwin, *J. Catal.* **2**, 397 (1963).
12. O. Kubaschewski and B. E. Hopkins, "Oxidation of Metals and Alloys." Butterworths, London and Washington, D.C., 1962.
13. P. Kofstad, "High Temperature Oxidation of Metals," Chapter V. Wiley, New York, 1966.
14. J. R. Hutchins, III, *Trans. Met. Soc. AIME* **239**, 1990 (1967).
15. W. K. Lu and G. Bitsianes, *Can. Met. Quart.* **7**, 3 (1968).
16. P. P. Budnikov and A. M. Ginstling, "Principles of Solid State Chemistry, Reaction in Solids" (K. Shaw, ed. and translator), Chapter 5. MacLaren and Sons, London, 1968.
17. W. Jander, *Z. Anorg. Allgem. Chem.* **163**, 1 (1927).
18. F. E. Massoth and W. E. Hensel, Jr., *J. Phys. Chem.* **65**, 636 (1961).
19. D. E. Y. Walker, *J. Appl. Chem.* **15**, 128 (1965).

20. J. Hlavac, *Proc. Int. Symp. Reactivity Solids 4th* (J. H. DeBoer, ed.), 129. Elsevier, Amsterdam, 1961.
21. W. K. Lu, *Trans. Met. Soc. AIME* **227**, 203 (1963).
22. B. B. L. Seth and H. U. Ross, *Trans. Met. Soc. AIME* **233**, 180 (1965); *Can. Met. Quart.* **5**, 315 (1966).
23. J. Shen and J. M. Smith, *Ind. Eng. Chem. Fundamentals* **4**, 293 (1965).
24. H. W. St. Clair, *Trans. Met. Soc. AIME* **233**, 1145 (1965).
25. E. E. Petersen, "Chemical Reaction Analysis," Chapter 4. Prentice Hall, Englewood Cliffs, New Jersey, 1965.
26. H. Y. Sohn, R. P. Merrill, and E. E. Petersen, *Chem. Eng. Sci.* **25**, 399 (1970).
27. P. F. J. Landler and K. L. Komarek, *Trans. Met. Soc. AIME* **236**, 138 (1966).
28. H. O. Lien, A. E. El-Mehairy, and H. U. Ross, *J. Iron and Steel Inst.*, **209**, 541 (1971).
29. M. H Davies, M. T. Simnad, and C. E. Birchenall, *Trans Met. Soc. AIME* **191**, 889 (1951).
30. J. O. Edstroem and G. Bitsianes, *Trans. Met. Soc. AIME* **203**, 760 (1955).
31. A. B. Newman, *Trans. AIChE* **27**, 203 (1931).
32. L. Himmel, R. R. Mehl, and C. E Birchenhall, *Trans. Met. Soc. AIME* **197**, 827 (1953).
33. D. Luss and N. R. Amundson, *AIChE J.* **15**, 194 (1969).
34. H. Y. Sohn, *AIChE J.* **19**, 191 (1973); **20**, 416 (1974).
35. M. Abramowitz and I. A. Stegun, "Handbook of Mathematical Functions." Dover, New York, 1965.
36. J. B. Rosser, Theory and Application of $\int_0^z e^{-x^2} \, dx$ and $\int_0^z e^{-p^2y^2} \, dy \int_0^y e^{-x^2} \, dx$, Mapleton House, Brooklyn, New York, 1948.
37. K. J. Cannon and K. G. Denbigh, *Chem. Eng. Sci.* **6**, 145, 155 (1957).
38. M. Ishida and C. Y. Wen, *Chem. Eng. Sci.* **23**, 125 (1968).
39. C. Y. Wen and S. C. Wang, *Ind. Eng. Chem.* **62** (8), 30 (1970).
40. R. Aris, *Ind. Eng. Chem. Fundamentals* **6**, 315 (1967).
41. G. S. G. Beveridge and P. J. Goldie, *Chem. Eng. Sci.* **23**, 913 (1968).
42. J. Szekely and J. W. Evans, *Chem. Eng. Sci.* **26**, 1901 (1971).
43. J. W. Evans, J. Szekely, W. H. Ray, and Y. K. Chuang, *Chem. Eng. Sci.* **28**, 683 (1973).
44. H. Y. Sohn and J. Szekely, *Can. J. Chem. Eng.* **50**, 674 (1972).
45. M. F. R. Mulcahy and I. W. Smith, *Rev. Pure Appl. Chem.* **19**, 81 (1969).

Chapter 4 | **Reactions of Porous Solids**

4.1 Introduction

In Chapter 3 we discussed gas–solid reactions involving nonporous solids and showed how different reaction steps that are coupled in series interact with each other. In many gas–solid reactions the solid is porous, allowing diffusion and chemical reaction to occur simultaneously throughout the solid; therefore, the reaction occurs in a diffuse zone rather than at a sharp boundary. The reaction of a porous solid with a gas has not been investigated as extensively as that of a nonporous solid due to the difficulties in analyzing the experimental data. Furthermore, even the analysis of the results of experiments on a porous solid has often been based on the shrinking unreacted-core model.

Since many solid reactants have some initial porosity and the simple shrinking unreacted-core model is often inapplicable to such systems, there have been recent efforts to find valid models for these reaction systems. A review of these will be presented.

Heterogeneous reactions involving a porous solid and a gas generally include the following steps:

1. external mass transport of reactant from the bulk fluid to the outside surface of the porous solid;
2. diffusion of the reactant gases within the pores of the solid;
3. chemical reaction of the solid surface including the adsorption of reactants and the desorption of products;
4. diffusion of the gaseous product through the porous solid;
5. mass transfer of the gaseous product into the bulk gas stream.

As in the case of nonporous reactant solids, it is very important to understand the relative significance of these steps in order to understand the overall kinetics of the reaction. The external mass transport and the intrapellet diffusion may affect significantly the overall rate, the apparent activation energy, and the apparent order of reaction.

4.2   Complete Gasification of Porous Solids

Many gas–solid reactions are of the type

$$A(g) + bB(s) \longrightarrow \text{gaseous products}$$

Specific examples are the combustion of carbon or graphite

$$O_2 + 2C \longrightarrow 2CO$$

the water–gas reaction

$$C + H_2O \longrightarrow CO + H_2$$

and the formation of carbonyls

$$4CO + Ni \longrightarrow Ni(CO)_4$$

Here we assume that solid B is sufficiently pure that complete gasification takes place and that negligible quantities of solid residue remain. Cases where appreciable quantities of solid residue are left after reaction (e.g., the combustion of coals of high ash content) are treated in a subsequent section.

It was indicated in Chapter 3 that the reaction of a shrinking nonporous solid with a gas may be divided into two regimes, i.e., those controlled by chemical reaction and by mass transport, with an intermediate regime where both steps affect the overall rate.

In the gasification of a porous solid, diffusion of gaseous reactants within the porous solid adds another step; thus the reaction may be divided into three main regimes as shown in Fig. 4.1. Wicke [1] and Walker et al. [2] have applied such classification to the combustion and gasification of porous carbon.

At low temperatures, where the intrinsic reactivity of the solid is low, a molecule of gaseous reactant entering the porous solid has a high probability of diffusing deeply into the pellet before finally reacting with the pore surface. In this region, which we shall call regime I, the concentration of gaseous reactant is essentially uniform throughout the porous solid and equal to that in the bulk gas stream. The overall rate is controlled by the intrinsic chemical reaction that takes place with the gaseous reactant concentration at the level in the bulk gas stream. The activation energy of the overall reaction is the same as the intrinsic value. All other kinetic parameters are the intrinsic values for the reaction and the rate does not depend on the size of the sample.

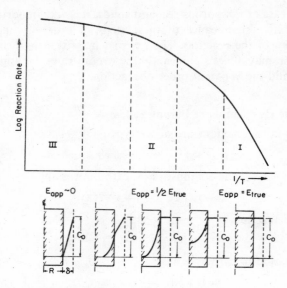

FIG. 4.1. Kinetic regimes for the gasification of a porous solid [2].

Since the concentration is uniform everywhere in the system, the reaction takes place uniformly throughout the solid. The pores will be enlarged inside the solid but the overall size will remain unchanged until the solid is almost completely consumed.

At somewhat higher temperatures, where the intrinsic solid reactivity is greater (regime II), the probability of a gaseous reactant molecule *penetrating deeply into the pellet before reacting is small*. Most of the reaction occurs in a zone near the external surface of the pellet. Diffusion within the pores limits the overall rate of reaction but does not " control " it because both chemical reaction and pore diffusion exert an influence on the progress of reaction. Thus, even in this regime, increasing the intrinsic solid reactivity (for example, by raising the temperature) still increases the overall rate, as does increasing the rate of diffusion. The temperature dependence of the overall rate is also affected by that of diffusion (which is very small), as we shall see subsequently. Measured values such as activation energy and the order of reaction do not generally correspond to the intrinsic values. In this regime reaction will proceed with reduction of the external dimensions of the pellet, while the center of the pellet will remain relatively unchanged until the final stages of reaction.

At still higher temperatures the intrinsic solid reactivity will be so high that the gaseous reactant molecules will react with the solid as soon as they have crossed the boundary layer enveloping the pellet. In this regime (regime

III) the concentration of the gaseous reactant *at the external surface of the pellet is near zero and the progress of reaction is controlled by external mass transfer*. Since mass transport increases only slightly with temperature, the apparent activation energy is small. In this regime all the reaction takes place at the external surface and the interior of the pellet is unaffected until it is exposed to gaseous reactant by the gasification of solid above it.

There is a gradual transition from one regime to the next on increasing the solid temperature. The nature of these transitions and the ranges of temperature over which the three regimes exist will depend on such parameters as the pore diffusivity, pellet size, and activation energy for reaction.

In the following we shall derive mathematical equations describing the gasification of a solid in each of the three regimes.

PARTICLES WITH ENLARGING PORES (REGIME I)

One of the first models for the gasification of solids in which account was taken of the changes in pore structure by the consumption of solid is due to Petersen [4], who assumed that the solid contains uniform cylindrical pores with random intersections.

The rate of reaction per unit volume of porous solid for such a system is given by

$$(\rho_s S_v/b) \, dr/dt = (\rho_s/b) \, d\varepsilon/dt \tag{4.2.1}$$

where S_v is the surface area per unit volume of the pellet, r the radius of the pores, ε porosity, ρ_s the molar density of pore-free solid and b is a stoichiometric coefficient.

From Eq. (4.2.1) it is seen that

$$S_v = d\varepsilon/dr \tag{4.2.2}$$

We also have

$$dr/dt = (bk/\rho_s)C_A{}^n \tag{4.2.3}$$

where k is the reaction rate constant and C_A the concentration of gaseous reactant.

In order to solve Eq. (4.2.1) a relationship must be obtained between S_v and r. Let us consider that the porous solid may be represented by the idealized network of cylindrical pores shown in Fig. 4.2. For a constant concentration of the gaseous reactant, the pore radius increases uniformly from the initial value r_0 to r. The total length L of the pore system per unit volume is defined as the sum of the center-line distances of the pores

$$L = \overline{ab} + \overline{bc} + \overline{cd} + \overline{de} + \cdots \tag{4.2.4}$$

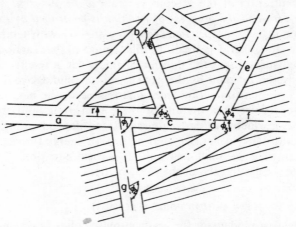

FIG. 4.2. Model of a porous solid with random-pore intersections [4].

If the pores enlarge uniformly, the surface area of the smooth pores of radius r is given by

$$S_v = 2\pi r\left[L - r\sum_{i=1}^{n}(1/\sin \varphi_i)\right] - r^2\sum_{i=1}^{n}\beta(\varphi_i) \qquad (4.2.5)$$

where n is the number of pore intersections per unit volume. The second term on the r.h.s. of Eq. (4.2.5) accounts for the area of the openings in the pore walls and $\beta(\varphi_i)$ is a shape factor of the ith intersection based on the angle φ_i. The term $\sum_{i=1}^{n}(1/\sin \varphi_i)$ allows for the shortening of the actual length of the pore as the intersection enlarges. If no new pore intersections are produced, the terms $\sum_{i=1}^{n}(1/\sin \varphi_i)$ and $\sum_{i=1}^{n}\beta(\varphi_i)$ are constant for a given solid. Equation (4.2.5) may then be rewritten

$$S_v = 2\pi rL - Kr^2 \qquad (4.2.6)$$

The value of the characteristic constant K depends on the number of pore intersections per unit volume and the angles at which the poles intersect each other.

Upon combining Eqs. (4.2.2) and (4.2.6) we obtain

$$\varepsilon = \int_{0}^{r}(2\pi rL - Kr^2)\,dr = \pi Lr^2 - (K/3)r^3 \qquad (4.2.7)$$

We also have that $\varepsilon = \varepsilon_0$ when $r = r_0$. Therefore,

$$\frac{\varepsilon}{\varepsilon_0} = \frac{\pi Lr^2 - (K/3)r^3}{\pi Lr_0{}^2 - (K/3)r_0{}^3} = \zeta^2\left(\frac{G - \zeta}{G - 1}\right) \qquad (4.2.8)$$

where

$$\zeta = r/r_0 \qquad (4.2.9)$$

$$G = 3\pi L/Kr_0 \qquad (4.2.10)$$

To evaluate G, Eq. (4.2.8) is assumed valid from $\varepsilon = 0$ to $\varepsilon = 1$. We know that $S_v \to 0$ as $\varepsilon \to 1$. Combining Eqs. (4.2.2) and (4.2.8) we obtain

$$S_v = d\varepsilon/r_0 \, d\zeta = (\varepsilon_0/r_0)(2G - 3\zeta)\zeta/(G - 1) \qquad (4.2.11)$$

from which we have

$$\zeta\Big|_{\varepsilon=1} = \tfrac{2}{3}G \qquad (4.2.12)$$

Substituting Eq. (4.2.12) into Eq. (4.2.8) written for $\varepsilon = 1$ we get

$$(4/27)\varepsilon_0 \, G^3 - G + 1 = 0 \qquad (4.2.13)$$

We see that G is a function only of ε_0 and may be obtained by solution of the cubic Eq. (4.2.13). With the value of G thus obtained Eqs. (4.2.1), (4.2.3), and (4.2.11) give the rate of reaction per unit volume of solid:

$$kS_v C_A{}^n = k(\varepsilon_0/r_0)[(2G - 3\zeta)\zeta/(G - 1)]C_A{}^n \qquad (4.2.14)$$

If we are only interested in the initial rate (in many instances this initial rate may be used for calculating the reaction rate constant) this may be calculated using Eq. (4.2.14) combined with the diffusion equation for the gaseous reactant. A particular case of negligible diffusional resistance is illustrated by the following example.

Example 4.2.1   Consider the gasification of graphite with carbon dioxide at 1100°C. The reaction is first order with respect to carbon dioxide and may be considered isothermal (see Example 4.2.4). The graphite sample has an initial porosity and pore radius of 0.3 and 2 μm, respectively.

ADDITIONAL DATA

$$k = 10^{-3} \text{ cm/sec}, \qquad C_{Ab} = 10^{-5} \text{ g-mole/cm}^3$$

SOLUTION   From Eq. (4.2.13) with $\varepsilon_0 = 0.3$, we obtain $G = 4.1$. The initial value of S_v, surface area for unit volume of solid, may be obtained from Eq. (4.2.11):

$$S_v = 0.25 \times 10^4 \text{ cm}^2/\text{cm}^3$$

The initial rate in the absence of diffusional resistance is therefore

$$-dN_A/dt = kS_v C_{Ab} = 0.25 \times 10^{-4} \text{ g-mole/sec cm}^3 = 1.08 \text{ g/hr cm}^3$$

For the system controlled by chemical reaction, the concentration of gaseous reactant C_A in Eq. (4.2.14) remains constant throughout the porous solid for all times at the bulk value. Consequently, Eq. (4.2.3) can be integrated to give

$$\zeta = r/r_0 = 1 + t/\tau_c \qquad (4.2.15)$$

where ·

$$\tau_c = r_0\, \rho_s / bk C_A{}^n \qquad (4.2.16)$$

The extent of reaction of the solid is

$$X = (\varepsilon - \varepsilon_0)/(1 - \varepsilon_0) \qquad (4.2.17)$$

which may be reexpressed, using (4.2.8) and (4.2.15), as

$$X = \frac{\varepsilon_0}{(1 - \varepsilon_0)}\left[\zeta^2\!\left(\frac{G - \zeta}{G - 1}\right) - 1\right] = \frac{\varepsilon_0}{(1 - \varepsilon_0)}\left[\left(1 + \frac{t}{\tau_c}\right)^2\!\left(\frac{G - 1 - (t/\tau_c)}{G - 1}\right) - 1\right]$$
$$(4.2.18)$$

which provides an explicit relationship between X and t. We should stress that the validity of Eq. (4.2.18) is restricted to conditions where the overall rate is controlled by chemical kinetics and hence the concentration field is uniform within the particle. When the reaction rate is so fast that the diffusive transport is no longer rapid enough to maintain a uniform reactant concentration field within the solid, the effect of pore diffusion will also become important. Under these conditions the governing equations will have to include an expression describing the diffusive transport through the solid. This problem has been treated by Petersen [4] and is also the subject of a recent article by Hashimoto and Silveston [42].

One of the major assumptions in Petersen's model is that the pores are cylindrical and uniform in size. In reality, most porous pellets are made up of particles tightly bound together, whose pores vary in size and shape. A further deficiency of the model is that it ignores the coalescence of growing adjacent pores that might reasonably be expected to occur as reaction proceeds. Hence Eq. (4.2.8) may not be valid from $\varepsilon = 0$ to $\varepsilon = 1$.

Applying the model to the reaction of porous graphite rods with carbon dioxide, however, Petersen [4] found reasonable agreement. Furthermore, this model has the useful feature of describing the fact that the surface area per unit volume may increase, go through a maximum, and then decrease. Few other models can represent such behavior.

Example 4.2.2   Calculate the time required to bring about the complete gasification of a graphite particle under conditions where external mass transport and pore diffusion present a negligible resistance to the progress of reaction.

DATA

$$\varepsilon_0 = 0.05, \qquad\qquad\qquad b = 2$$
$$C_A = 2 \times 10^{-5} \text{ g-mole/cm}^3, \qquad \rho_s = 0.19 \text{ g-mole/cm}^3$$
$$r_0 = 5 \times 10^{-4} \text{ cm}, \qquad\qquad k = 10^{-4} \text{ cm/sec}, \qquad n = 1$$

SOLUTION   Equation (4.2.13) becomes

$$0.0074 G^3 - G + 1 = 0$$

which may be solved to give (as the only positive root)

$$G = 11.089$$

Then

$$\tau_c = r_0 \rho_s / bk C_A^n = 5 \times 10^{-4} \times 0.19/(2 \times 10^{-4} \times 2 \times 10^{-5}) = 2.38 \times 10^4 \text{ sec}$$

Using Eq. (4.2.18) for $X = 1$, we obtain

$$1 = \frac{0.05}{(1 - 0.05)} \left[\left(1 + \frac{t}{\tau_c} \right)^2 \left(\frac{G - 1 - t/\tau_c}{G - 1} \right) - 1 \right]$$

which is a cubic in t/τ_c that can be solved to give, as the only positive root,

$$t/\tau_c = 6.54$$

Hence t, the time required for complete gasification, is

$$t = 6.54 \times 2.38 \times 10^4 \text{ sec} = 1.55 \times 10^5 \text{ sec}$$

SHRINKING PARTICLES WITH REACTION OCCURRING NEAR THE EXTERNAL SURFACE (REGIMES II AND III)

As the reaction rate becomes very high, the gaseous reactant cannot penetrate deeply into the solid. Consequently, the reaction occurs mainly in a thin layer near the external surface, and the overall size will shrink while the interior of the solid remains largely unreacted. The system is illustrated schematically in Fig. 4.3.

We shall now develop a quantitative description of this behavior using the analogy of these systems to those encountered in heterogeneous catalysis. The following assumptions are made:

(1)   A constant average effective diffusivity and specific surface area may be used within the thin reaction zone.

(2)   The structure of the reaction zone is maintained constant as it moves toward the interior.

A mass balance within the pellet then gives the following equation for an irreversible reaction:

$$D_e \nabla^2 C_A - k S_v C_A^n = 0 \qquad\qquad (4.2.19)$$

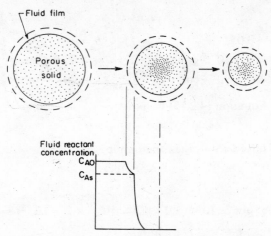

FIG. 4.3. Gasification of a porous solid under conditions where the external dimensions of the particle shrink.

where D_e is the effective diffusivity of the gaseous reactant in the porous solid.

In the case of a pellet where the reaction occurs mainly near the external surface, the overall geometry of the pellet is unimportant and the reaction zone may be considered a flat plate except in the final stages. Therefore, the following analysis applies to a pellet of any geometry. Equation (4.2.19) then may be written as

$$D_e \, d^2C_A/dx^2 - kS_v C_A^n = 0 \qquad (4.2.20)$$

where x is the distance normal to the external surface, and the boundary conditions are given as

$$C_A = C_{As} \quad \text{at} \quad x = 0 \quad \text{(outer surface of the solid)} \qquad (4.2.21)$$

and

$$C_A = dC_A/dx = 0 \quad \text{as} \quad x \to \infty \qquad (4.2.22)$$

The second boundary condition implies that the gaseous-reactant concentration becomes zero at some distance toward the interior of the pellet, and the pellet size is large enough to be considered as infinity compared with the thickness of the reaction zone. This condition is met [3] when the Thiele parameter, based on the volume V_p and external surface area A_p of the pellet at a given time, satisfies the relationship

$$N_{Th} \equiv \frac{V_p}{A_p} \sqrt{\left(\frac{n+1}{2}\right) \frac{kS_v C_{As}^{n-1}}{D_e}} > 3 \qquad (4.2.23)$$

The solution for Eq. (4.2.20) then gives†

$$\frac{dC_A}{dx} = -\left(\frac{2}{n+1}\frac{kS_v}{D_e}C_A^{n+1}\right)^{1/2} \tag{4.2.24}$$

The overall rate of reaction per unit area of external surface is given by

$$\mathscr{R}_s = -D_e\left(\frac{dC_A}{dx}\right)_{at\ ext\ sur} = \left(\frac{2}{n+1}kS_vD_e\right)^{1/2}C_{As}^{(n+1)/2} \tag{4.2.25}$$

Similar results have been obtained from the exact solutions for an infinite slab, a long cylinder, and a sphere [3].

It is noted that in general S_v, the surface area per unit volume in the reaction zone, is different from the value for the initial solid due to the consumption of the solid reactant during the reaction. Its value is expected, however, to remain constant as the overall size shrinks.

We also note from Eq. (4.2.25) that the rate is proportional to $\sqrt{kD_e}$ with the result that the apparent activation energy for the reaction is one-half the intrinsic activation energy, assuming that effective diffusivity is weakly dependent on temperature. The apparent reaction order is also changed from n to $(n+1)/2$.

We note that in region II, where the diffusional resistance is high, the gasification of a porous solid may be described by an expression that is formally equivalent to a relationship developed for nonporous solids. Let us recall that in Chapter 3 we developed the following expression for the gasification of a nonporous solid in the absence of external mass transfer effects:

$$(bkC_{As}^n/\rho_B)(A_p/F_p V_p)t = 1 - (1-X)^{1/F_p} \tag{3.2.9}$$

An identical expression would hold for the gasification of a porous solid, provided k is replaced by k' defined as

$$k' \equiv \left(\frac{2}{n+1}kS_vD_e\right)^{1/2} \tag{4.2.25a}$$

and C_{As}^n is replaced by $C_{As}^{(n+1)/2}$.

In many earlier studies of gas–solid reactions the investigators drew no distinction whether they were dealing with porous or nonporous solids. It is readily seen from the discussion presented above that under certain circumstances it is possible to represent rate measurements made on porous solid reactants with expressions developed for nonporous systems. It must be stressed, however, that while the data may appear to be consistent with the model, the intrinsic kinetic parameters cannot be deduced from such studies.

† Substitute $p \equiv dC_A/dx$; then $d^2C_A/dx^2 = p\,dp/dC_A$. Solve for p with C_A as the independent variable.

As seen from Eq. (4.2.25a), rather than obtaining the "real" reaction rate constant one would obtain the quantity appearing on the r.h.s. of Eq. (4.2.25). Moreover, the apparent reaction order may also be falsified for reactions that are not first order. A readable discussion of this problem has been presented by Turkdogan [61].

In the above derivation a power-law kinetic expression was assumed for the intrinsic chemical reaction. For the systems where large concentration differences occur between the bulk fluid and the interior of the porous solid, the Langmuir–Hinshelwood type rate expression should be used because this provides a better description of the rate of heterogeneous reactions.

The rates of many gas–solid reactions may be expressed by

$$\mathscr{R}_A = kC_A/(1 + KC_A) \qquad (4.2.26)$$

where $\mathscr{R}_A$ is the rate of reaction in moles of gaseous reactant per unit area of solid surface per unit time.

It can be shown [5] using an appropriate transformation of variables, that Eq. (4.2.26) also represents a class of more complex mechanisms with the general rate expression

$$\mathscr{R}_A = \frac{k(p_A - p_x/K_e)}{1 + K_A p_A + \sum_i K_i p_i + \sum_j K_{ij} p_{ij}} \qquad (4.2.27)$$

which takes into account a reversible surface chemical reaction and the adsorption of species that participate in the reaction as well as the adsorption of inert species.

It may be shown that, for systems where diffusional effects play a major role and the chemical step obeys the rate law given by Eq. (4.2.26), the overall rate of reaction per unit area of external surface is

$$\mathscr{R}_s = (2kS_v D_e)^{1/2}\{[KC_{As} - \ln(1 + KC_{As})]^{1/2}/|KC_{As}|\}C_{As} \qquad (4.2.28)$$

In the derivation of Eqs. (4.2.25) and (4.2.28) it was assumed that mass transport within the pores of the solid occurs by equimolar counterdiffusion. This should be a reasonable approximation for systems where there is a volume change on reaction, provided there is an excess of inert constituents present. A net generation of gas molecules (e.g., $C + O_2 \rightarrow 2CO$) results in an outward flow of gaseous products and reduces the rate of diffusion of the reactant into the solid. A net consumption of gas molecules (e.g., $2H_2 + C \rightarrow CH_4$) results in the opposite effect.

Weekman and Gorring [6] reported calculated results on the effect of net generation or consumption of gas at constant pressure, extending earlier work by Thiele [7].

In the regime where diffusion is important. most reaction occurs near the external surface. The mass balance within the porous solid undergoing a

reaction with a net generation or consumption of gas may be represented by
the equation

$$D_e\left[\frac{d^2C_A}{dx^2} - \frac{\theta/C_{As}}{1 + (\theta C_A/C_{As})}\left(\frac{dC_A}{dx}\right)^2\right] - kS_v C_A{}^n\left(1 + \theta\frac{C_A}{C_{As}}\right) = 0 \quad (4.2.29)$$

where

$$\theta \equiv (v - 1)y_{As} \quad (4.2.30)$$

Here v is the number of moles of gaseous species produced per unit mole of
gaseous reactant, and y_{As} is the mole fraction of reactants in the bulk. In
Eq. (4.2.29) a power-law rate expression was used for the chemical reaction
step. The boundary conditions are the same as those given by Eqs. (4.2.21)
and (4.2.22). The solution of Eq. (4.2.29) then gives the overall rate of reaction
per unit area of external surface.

For a zeroth-order reaction ($n = 0$),

$$\mathscr{R}_s = (2kS_v D_e)^{1/2}[\theta^{-1} \ln(1 + \theta)]^{1/2}C_{As}^{1/2} \quad (4.2.31)$$

For a first-order reaction ($n = 1$),

$$\mathscr{R}_s = (2kS_v D_e)^{1/2}[\theta^{-1} - \theta^{-2} \ln(1 + \theta)]^{1/2}C_{As} \quad (4.2.32)$$

For a second-order reaction ($n = 2$),

$$\mathscr{R}_s = (2kS_v D_e)^{1/2}[\tfrac{1}{2}\theta^{-1} - \theta^{-2} + \theta^{-3} \ln(1 + \theta)]^{1/2}C_{As}^{3/2} \quad (4.2.33)$$

For the case of extremely low porosity or very fast reaction, the reaction
on the external surface may become significant [2]. Then the overall rate per
unit area of external surface using Eq. (4.2.25) is written

$$\mathscr{R}_s = \left(\frac{2}{n + 1} kS_v D_e\right)^{1/2} C_{As}^{(n+1)/2} + kfC_{As}^n \quad (4.2.34)$$

where f is the roughness factor for the external surface, which is defined as the
ratio of the true external surface area to the apparent projected external
surface area. We may use this expression instead of Eq. (3.2.1) in order to
calculate the overall conversion.

For a first-order irreversible reaction, Eq. (3.2.25) now becomes

$$-\frac{\rho_B}{b}\frac{dr_c}{dt} = \frac{C_{Ab}}{\left[\left(\dfrac{2}{n + 1} kS_v D_e\right)^{1/2} + kf\right]^{-1} + \dfrac{1}{h_D}} \quad (4.2.35)$$

and the subsequent derivations in Section 3.2 may be carried out in a similar
way.

Equation (4.2.35) describes both regimes II and III. When $h_D \gg k$, that is, the external mass transport is very fast, Eq. (4.2.35) reduces to

$$-(\rho_B/b)\, dr_c/dt = (\{[2/(n+1)]kS_v D_e\}^{1/2} + kf)C_{Ab} \qquad (4.2.36)$$

Unless the solid has very low porosity the second term in Eq. (4.2.36) is small, and the following equation applies:

$$-(\rho_B/b)\, dr_c/dt = \{[2/(n+1)]kS_v\, D_e\}^{1/2}C_{Ab} \qquad (4.2.37)$$

which corresponds to regime II. The case where the second term is much more important would correspond to the system of nonporous solid discussed in Section 3.2.1.

Where $k \gg h_D$, Eq. (4.2.35) reduces to

$$-(\rho_B/b)\, dr_c/dt = h_D\, C_{Ab} \qquad (4.2.38)$$

which is equivalent to Eq. (3.2.12) for $K_E \to \infty$ and corresponds to regime III. Let us illustrate in the following example how the numerical values of the system parameters determine the regime to which the system would correspond.

Example 4.2.3   Consider the gasification of graphite with carbon dioxide, which may be assumed to be first order with respect to carbon dioxide concentration. At 1100°C pertinent parameter values for kinetics and mass transport were given in Example 4.2.1. It is further assumed that $f = 1$ and $D_e = 0.1$ cm^2/sec. Suppose h_D is 2 cm/sec and determine the relative importance of resistances of various steps.

SOLUTION

$$kf = 10^{-3}\ \text{cm/sec}, \qquad \{[2/(n+1)]kS_v D_e\}^{1/2} = 0.5\ \text{cm/sec}$$

Thus it is seen that the reaction at the very external surface contributes little to the overall reaction rate, and the external mass transport offers much less resistance than does the reaction at the internal-pore walls.

Now suppose temperature is raised such that $k = 0.1$ cm/sec. Assuming diffusivity remains about the same, we obtain

$$kf = 0.1\ \text{cm/sec}, \qquad \{[2/(n+1)]kS_v\, D_e\}^{1/2} = 5\ \text{cm/sec}$$

We see that the kf term is still negligible compared with the internal reaction term. However, now the external mass transport offers a greater resistance than the chemical reaction step. Further, if the solid has internal surface area S_v of only 250 cm^2/cm^3, then

$$kf = 0.1\ \text{cm/sec}, \qquad \{[2/(n+1)]kS_v\, D_e\}^{1/2} = 0.5\ \text{cm/sec}$$

In this case the reaction at the external surface contributes significantly to the overall rate.

NONISOTHERMAL BEHAVIOR

In,systems that fall in regime II, it is possible that temperature gradient due to the heat effect of the reaction may play a role in determining the overall rate. The analysis of the general nonisothermal systems involves the heat balance equation in addition to the mass balance, and usually requires complicated and time-consuming numerical solution. However, under certain circumstances, simple approximate solutions are possible. We shall limit our discussion to these cases.

Let us assume that in regime II, that is, when the movement of the external surface toward the center of the solid is very slow, the concentration profile corresponds to the steady state profile at that overall size. Under the pseudo–steady state conditions, the rate of diffusion of reactants across a boundary normal to the diffusion path is equal to the rate of reaction within the boundary. Furthermore, let us make a pseudo–steady state assumption for the heat conduction in the solid, i.e., that the rate of heat accumulation within the boundary is negligible compared with the heat generated by reaction or conducted out of the boundary. Consequently the heat generated or consumed by the reaction is all transferred across the boundary. Therefore,

$$(-\Delta H)D_e \, dC_A/dx = -k_e \, dT/dx \qquad (4.2.39)$$

where k_e is the effective thermal conductivity of the porous solid. If D_e is constant in the reaction zone and the temperature variation of ΔH and k_e is negligible, Eq. (4.2.39) may be integrated (assuming k_e and D_e independent of x) to give

$$T - T_s = [(-\Delta H)D_e/k_e](C_{As} - C_A) \qquad (4.2.40)$$

where T_s is the temperature at the external surface of the porous solid.

Thus, the temperature at any position within the solid is related to the concentration at the same position. Petersen [3] gives a more complete and rigorous derivation of Eq. (4.2.40) with the same result. This relationship was first derived by Damköhler [8] and shown by Prater [9] to be valid for all kinetics and geometries of the solid.

An important consequence of Eq. (4.2.40) is that it may be used to calculate the maximum temperature in the porous solid undergoing exothermic reaction, where the reactant concentration becomes zero in the interior of the solid. Setting $C_A = 0$ in Eq. (4.2.40) we obtain

$$T_{max} - T_s = (-\Delta H)D_e C_{As}/k_e \qquad (4.2.41)$$

In order to obtain the overall rate of reaction, the concentration profile in the pellet must be determined as in the isothermal case. The specific reaction rate constant, k in Eq. (4.2.20) is expressed as an Arrhenius function of temperature, and Eq. (4.2.40) is substituted to give

$$D_e \frac{d^2 C_A}{dx^2} - k_s S_v C_A{}^n \exp\left[\frac{\gamma\beta(C_{As} - C_A)}{\beta(C_{As} - C_A) + C_{As}}\right] = 0 \qquad (4.2.42)$$

where

$$k_s \equiv k(T_s) \qquad \text{(the reaction rate constant at temperature } T_s\text{)} \qquad (4.2.43)$$

$$\gamma \equiv E/RT_s \qquad (4.2.44)$$

$$\beta \equiv (-\Delta H) D_e C_{As}/k_e T_s = [(T - T_s)/T_s])_{max} \qquad (4.2.45)$$

and E is the activation energy.

There have been a number of methods used for obtaining the solution to Eq. (4.2.42) or its variation. Beek [10] applied an approximate method using linearization, and Schilson and Amundson [11] used an iterative method. Tinkler and Pigford [12] suggested perturbation techniques for the case when temperature gradients are not large. Others [13–15] used machine computation.

Petersen [3, 16] developed a method for obtaining an asymptotic solution for regime II. We will summarize Petersen's development here. In regime II the penetration of reactants within the porous solid is small, and the boundary conditions given by Eqs. (4.2.21) and (4.2.22) are valid. Then the solution to Eq. (4.2.42) is given by

$$\frac{dC_A}{dx} = -\left\{\frac{2k_s S_v}{D_e} \int_0^{C_A} C_A{}^n \exp\left[\frac{\gamma\beta(C_{As} - C_A)}{\beta(C_{As} - C_A) + C_{As}}\right] dC_A\right\}^{1/2} \qquad (4.2.46)$$

The overall rate of reaction per unit external surface area is then given by

$$\mathscr{R}_s = -D_e \left(\frac{dC_A}{dx}\right)_{\text{at ext sur}}$$

$$= (2k_s S_v D_e)^{1/2} \left\{\int_0^{C_{As}} C_A{}^n \exp\left[\frac{\gamma\beta(C_{As} - C_A)}{\beta(C_{As} - C_A) + C_{As}}\right] dC_A\right\}^{1/2} \qquad (4.2.47)$$

We note that when $\gamma\beta = 0$, Eq. (4.2.47) reduces to the solution for the isothermal case given by Eq. (4.2.25).

A further approximation may be made by considering the fact that for many systems, the difference between the maximum temperature in the solid and the surface temperature is small compared with the absolute surface temperature, that is,

$$[(T - T_s)/T_s]_{max} = \beta \ll 1$$

Then, using Eq. (4.2.47) the overall rate of reaction per unit area of external surface becomes

$$\mathscr{R}_s = (2k_s S_v D_e)^{1/2} \left\{\int_0^{C_{As}} C_A{}^n \exp\left[\delta\left(1 - \frac{C_A}{C_{As}}\right)\right] dC_A\right\}^{1/2} \qquad (4.2.48)$$

where

$$\delta \equiv \gamma\beta = E(-\Delta H)D_e C_{As}/R\mathcal{k}_e T_s^2 \tag{4.2.49}$$

We see that one now needs only one parameter δ instead of γ and β to account for nonisothermality.

For a first-order reaction ($n = 1$), we obtain

$$\mathcal{R}_s = (2k_s S_v D_e)^{1/2}[(e^\delta - 1 - \delta)^{1/2}/|\delta|]C_{As} \tag{4.2.50}$$

The corresponding solution for a second-order reaction ($n = 2$) is

$$\mathcal{R}_s = (2k_s S_v D_e)^{1/2}\{[2(e^\delta - 1 - \delta) - \delta^2]/\delta^3\}^{1/2}C_{As}^{3/2} \tag{4.2.51}$$

Petersen [3, 16] showed that for the case where reaction occurs mainly near the external surface, Eqs. (4.2.50) and (4.2.51) agree almost exactly with the exact solution, for the values of $|\delta|$ from zero up to at least 5. The criterion for the reaction occurring sufficiently close to the external surface is given by the condition [3, 16]

$$N_{Th} = \frac{V_p}{A_p}\sqrt{\left(\frac{n+1}{2}\right)\frac{S_v k_s C_{As}^{n-1}}{D_e}} > 2$$

When the reaction on the external surface is appreicable, in addition to the reaction on pore wall, one may add the term, $k_s f C_{As}^n$ to the r.h.s. of Eqs. (4.2.50) and (4.2.51) as was done in Eq. (4.2.34) for isothermal systems. The resulting equation may be used in lieu of Eq. (3.2.1) for calculating the overall conversion of a solid pellet.

In the transition region between regimes I and II where the chemical reaction and diffusion present a comparable resistance to the overall progress of reaction, multiple solutions may occur and the possibility of instability arises when the reaction is exothermic [15]. The criteria for the existence of multiple steady state for chemical reactions in porous catalyst pellets have been studied extensively [17–21]. The effect of net gas generation or consumption on nonisothermal reaction in a porous solid was analyzed by Weekman [22].

Let us illustrate the possible effects of nonisothermal behavior by the following example.

Example 4.2.4   Again consider the reaction of porous graphite with carbon dioxide as in Examples 4.2.1 and 4.2.3. Determine the nonisothermal effect on the rate.

ADDITIONAL DATA

$(-\Delta H) = -41,000$ cal/g-mole, $\qquad E = 86,000$ cal/g-mole

$D_e = 0.1$ cm^2/sec, $\qquad\qquad C_{As} = 10^{-5}$ g-mole/cm^3

$\mathcal{k}_e = 0.1$ cal/sec cm °K, $\qquad\quad T_s = 1100$°K

SOLUTION Thus, $\delta = -0.01$. The ratio between the nonisothermal and isothermal rates from Eqs. (4.2.50) and (4.2.25) is

$$(\mathscr{R}_s)_{\text{noniso}}/(\mathscr{R}_s)_{\text{iso}} = 2^{1/2}(e^\delta - 1 - \delta)^{1/2}/|\delta| \simeq 1$$

We see that there is very little nonisothermal effect. Porous carbon has a much smaller k_e (≈ 0.005). Assuming that other parameters remain the same, we have $\delta = -0.2$. The ratio now becomes

$$(\mathscr{R}_s)_{\text{noniso}}/(\mathscr{R}_s)_{\text{iso}} = 0.97$$

The effect is still quite small.

CONCLUDING REMARKS

In this section we considered the gasification of porous solids; by definition this means systems where there is no solid reaction product, so that (in the absence of inerts) upon completion of the reaction all the solid will disappear. As illustrated in Fig. 4.1 the gasification process may be classified into the following three regimes.

In region I the overall rate is controlled by chemical kinetics and the concentration of the gaseous reactant is spatially uniform within the porous solid. Under these conditions all the surface area is available for reaction and the particle reacts at a spatially uniform rate. Provided an assumption can be made for relating the total surface area to the extent of reaction (e.g., as done by Petersen) a simple analytical expression may be developed for describing the course of gasification in this region.

In region II both diffusion and chemical kinetics play an important role and here most of the reaction will occur in a finite zone close to the outer surface of the particle; it follows that the particle will shrink in the course of the reaction. By making certain simplifying assumptions, analytical expressions may be derived for representing such systems. Formally there is a parallel between these expressions and those previously given for nonporous solids, although the physical nature of these two systems is quite different: for nonporous solids reaction occurs at a macroscopic sharp boundary, whereas in the gasification of porous solids in regime II, the reaction zone is diffuse. Because of this major discrepancy in the physical nature of these two systems, great care must be taken in the use of these equations, as discussed in the text.

In region III the overall rate is controlled by external mass transfer and gasification takes place at the outer surface of the shrinking solid. Under these conditions the solid structure is unimportant and one may use the expressions previously developed for nonporous systems for describing the progress of the reaction. Figure 4.1 and Example 4.2.3 provide simple illustrations of how the parameters of the system determine the reaction regime.

In general, one may say that at low temperatures and for small particles the system tends to region I, while at high temperatures mass transfer and pore diffusion become important, causing the system to tend to regime II or III.

The above classification was developed for isothermal systems. However, nonisothermal behavior would not alter the basic nature of these three mechanisms, although the existence of temperature gradients within the solid could cause additional complications. A discussion of some of these effects is also given in the text.

4.3   Porous Solids of Unchanging Overall Sizes

In many gas–solid reactions encountered in chemical and metallurgical processes, solid products are formed. Such reactions may be described by the general reaction equation

$$A(g) + bB(s) = cC(g) + dD(s) \qquad (4.3.1)$$

The overall volume of solid may increase or decrease depending on the relative density of the solid product compared with that of the solid reactant. In most cases, however, the change is rather small so that the overall size may be regarded as constant.

Examples of such a system are the regeneration of coked catalysts, the reduction of some metal oxides, and the roasting of sulfide ores.

When the solid reactant is nonporous, there is a sharp boundary between the unreacted core and the completely reacted layer. In the general case of a porous solid there is a gradual change in the degree of conversion throughout the particle. The external layer will be completely reacted after a certain time, and the thickness of this completely reacted layer will increase toward the interior of the particle, as shown in Fig. 4.4. Under these conditions, in contrast to nonporous solids, the reaction within the partially reacted zone occurs simultaneously with diffusion of fluid reactants in this zone.

When chemical reaction presents the major resistance to the overall progress of reaction, the concentration of the fluid reactant will be constant everywhere and the reaction will occur uniformly throughout the volume of the solid. If, on the other hand, pore diffusion presents the major resistance, the reaction will occur in a narrow boundary between the unreacted and the completely reacted zones, where the gaseous reactant concentration becomes zero (or takes on its equilibrium value in the case of reversible reaction). This latter case is identical to the diffusion-controlled shrinking unreacted-core system of a nonporous solid, which was discussed in detail in Chapter 3.

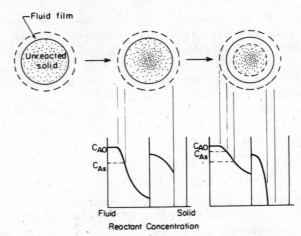

FIG. 4.4. The progress of a reaction of a porous solid reactant with a gas to form a porous solid product.

In the intermediate region where the resistances presented by chemical reaction and pore diffusion are of comparable magnitude, both processes must be considered simultaneously.

In the subsequent sections the limiting cases of pure chemical reaction and diffusion-controlled regimes will be discussed first. Then the general case of mixed control will be studied, showing how these limiting cases are the particular extremes of the general solution and under what conditions the asymptotic situation is reached.

The model to be presented here is based on the work of Sohn and Szekely [23–26], which may be regarded as a generalization of the numerous models [27, 28, 31, 32, 34, 35] that have been proposed to represent the diffuse reaction zone of reacting porous solids. Henceforth we shall refer to the model as the "grain model."

Before we start the analysis of the reaction of a porous solid, it is necessary to characterize the system. A pellet is usually formed by compacting fine grains of particle with or without some binding agents. The overall shape may be approximated by that of a slab, a long cylinder, or a sphere. It is, however, not so easy to characterize the shape of the individual grains that make up the pellet. Difficulties will arise because the grains are usually of irregular shape and are distributed in size; moreover, the shape of the grains may change in the course of the reaction.

To allow a convenient statement of the problem let us consider a pellet that may have the shape of a sphere, a long cylinder, or a flat plate, made up of individual grains of equal size, which again could be spheres, long cylinders,

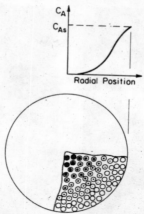

FIG. 4.5. Representation of the grain model for the reaction of a porous solid with a gas.

or flat plates. We will see subsequently that the assumption of this idealized structure is not a necessary part of the model to be presented. However, it does facilitate the presentation of the model and we shall use it here. Thus, we may have spherical pellets made up of spherical, cylindrical, or flat platelike grains, cylindrical pellets made up of spherical grains, etc.; nine combinations in all. A schematic diagram of a spherical pellet made up of spherical grains is shown in Fig. 4.5.

Figure 4.6 is a scanning electron micrograph of a nickel oxide pellet before and after reduction by hydrogen. In considering such a pellet to be made up of large numbers of discrete grains, we are not far from the truth, although in general the assumptions of a uniform grain size and regular shape are not accurate.

The special case of a spherical pellet made up of flat plates† was studied by Ausman and Watson [33] and Ishida and Wen [30, 34]. Szekely and Evans [27, 28, 35, 36] considered spherical pellets made up of spherical grains and have applied their model to the reduction of nickel oxide pellets with hydrogen [28, 36]. Sohn and Szekely [23] made the generalization to allow three basic geometries described above for both the grains and the pellet. As will be discussed later, this model was found satisfactory for the interpretation of experimental results and for the planning of experimental programs for systems where structural changes on reaction may be neglected [24]. Before

---

† The shape of the grains is an interpretation supplied by the authors of this book; in their original articles these workers merely postulated that the local rate of reaction within the pellet is independent of the local extent of reaction of the solid.

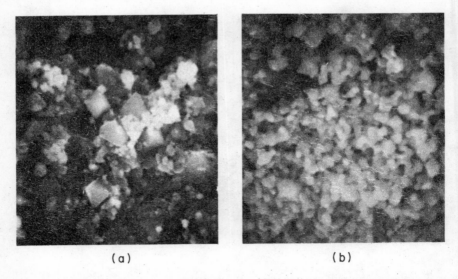

(a)   (b)

FIG. 4.6. Scanning electron microscope photographs of a nickel oxide compact (a) before and (b) after reduction to nickel by hydrogen.

stating the full governing equations, let us consider the two asymptotic regimes, the existence of which is readily postulated on the basis of physical reasoning, namely:

(a) Region of chemical control: the pellets react uniformly throughout;
(b) Region of diffusion control: sharp reaction interface.

4.3.1   Uniformly Reacting Porous Pellets

As noted above, this situation occurs when the rate of diffusion through the interstices among the grains presents a negligible resistance to the progress of reaction and thus chemical kinetics controls the process.

Such a system may be considered as an agglomerate of individual grains reacting in the absence of mass transport resistance. Thus the gaseous reactant concentration is uniform throughout the solid, as shown in Fig. 4.7. The criteria for the validity of assuming this condition will be developed subsequently in the discussion of simultaneous chemical reaction with diffusion. The grains are ordinarily nonporous and we assume that within each grain the reaction front retains its original geometrical shape (i.e., concentric spheres, coaxial cylinders, or parallel planes) as reaction proceeds. Then the results of the analysis obtained for the reaction of nonporous particles in Chapter 3 apply directly to the individual grains.

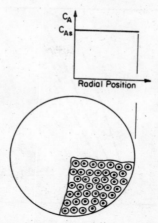

FIG. 4.7. The grain model for the case of a solid undergoing reaction under conditions of chemical control.

Thus, from Eq. (3.3.33) for a first-order reaction we may write with no net generation of gas ($c = 1$)

$$t^* = g_{F_g}(X) + \hat{\sigma}_g^2 p_{F_g}(X) \qquad (4.3.2)$$

The dimensionless parameters and conversion functions are now based on the size and the shape of the grains, i.e.,

$$t^* = \frac{bk(C_{As} - C_{Cs}/K_E)}{\rho_s}\left(\frac{A_g}{F_g V_g}\right)t \qquad (4.3.3)$$

$$\hat{\sigma}_g^2 = \frac{k}{2D_g}\frac{V_g}{A_g}\left(1 + \frac{1}{K_E}\right) \qquad (4.3.4)$$

where C_{As} is the gaseous reactant concentration at the surface of the grains, which in the absence of external mass transfer effects assumed here is the same as C_{A0}, the reactant concentration in the bulk (assumed to contain negligible quantities of gaseous product). A_g, V_g, and F_g are the surface area, volume, and shape factor for the grain, respectively; ρ_s is the molar density of the solid reactant, K_E the equilibrium constant for the reaction; and D_g the effective diffusivity of the gaseous reactant in the product layer around each individual grain. Irreversible reactions ($K_E \to \infty$) may be treated simply by dropping the last factor in parentheses from the r.h.s. of Eq. (4.3.4).

If the product forms a nonporous layer around the grain then the reactant must be transported by solid state diffusion. Then $\hat{\sigma}_g^2$ may become very large and the overall process may be controlled by the slow solid state diffusion. This has been observed experimentally in the reduction of iron oxides at temperatures where iron forms a dense layer around the wüstite grains [43].

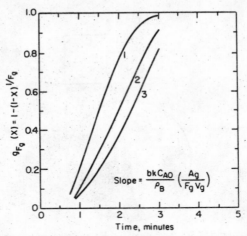

FIG. 4.8. Determination of the grain shape factor (F_g) and reaction rate constant (k) by curve fitting. 1, 2, 3 are assumed values of F_g.

When the product layer has any degree of porosity in the form of whiskers or wormholes, σ_g^2 becomes very small. Under these conditions, we may write, for an nth-order reaction,

$$(bkC_{As}^n/\rho_s)(A_g/F_g V_g)t = g_{F_g}(X) \equiv 1 - (1 - X)^{1/F_g} \tag{4.3.5}$$

It is noted that the time required to attain a certain conversion is proportional to the grain size in this case, whereas for nonporous solids it is proportional to the overall size of the particle [see Eq. (3.2.9)].

This relationship enables one to determine from experimental conversion data the intrinsic reaction rate constant and the grain shape factor [23, 24]. In Fig. 4.8 the conversion function $g_{F_g}(X)$ is plotted against time assuming a value of F_g. The grain shape factor is determined as the value of F_g that gives the best straight line. The reaction rate constant may then be obtained from the slope of such a plot. This procedure is illustrated in Fig. 4.8. Szekely $et\ al.$ [24] and Evans $et\ al.$ [37] have applied this method to the study of the reduction of nickel oxide by hydrogen. Note that the experiment for this purpose must be performed under conditions where the pellet offers no resistance to the diffusion of the reacting species.

4.3.2   Reaction of Porous Pellets Controlled by Diffusion through the Product Layer

If the chemical reaction step presents a negligible resistance to the progress of reaction, compared with the resistance due to pore diffusion, the overall rate is controlled by diffusion. The reaction occurs in a narrow zone separating the unreacted core and the completely reacted layer, where the

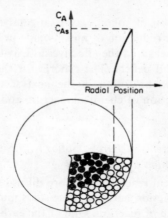

FIG. 4.9. The grain model for a solid undergoing reaction under conditions of pore diffusion control.

reactant concentration drops to a very small value. This situation (shown in Fig. 4.9) is identical to the diffusion-controlled shrinking unreacted-core system of a nonporous solid, discussed previously. The analyses presented in Section 3.3.2 apply directly to this case, and the relationships between conversion and time derived there may be used in identical form provided t^+ is defined in the following manner †

$$t^+ \equiv [2bF_p D_e (C_{A0} - C_{C0}/K_E)(1 - \varepsilon)\rho_s](A_p/F_p V_p)^2([K_E/(1 + K_E)])t \quad (4.3.6)$$

Then we have from Eq. (3.3.17) for a reaction with no net generation of gas (or where the gaseous reactant exists at low concentrations in an inert component)

$$t^+ = p_{F_p}(X) \quad (4.3.7)$$

where

$$p_{F_p}(X) \equiv X^2 \quad \text{for} \quad F_p = 1 \quad (4.3.8)$$

$$\equiv X + (1 - X) \ln(1 - X) \quad \text{for} \quad F_p = 2 \quad (4.3.9)$$

$$\equiv 1 - 3(1 - X)^{2/3} + 2(1 - X) \quad \text{for} \quad F_p = 3 \quad (4.3.10)$$

In this regime the time required to attain a certain conversion is proportional to the square of the characteristic dimension of the pellet.

The effective diffusivity in the product layer may be determined by plotting, according to Eq. (4.3.7), the conversion data obtained under conditions of diffusion control. This method was applied to spherical systems by Kawasaki et al. [38] for the reduction of iron oxide pellets, by Weisz and Goodwin

† When the reactant solid matrix contains inerts the quantity ε appearing in Eq. (4.3.6) has to include the volume occupied by the inerts in addition to the void space present in the matrix.

[39] for the combustion of coke deposit on catalyst pellets, and by Szekely and Sohn [26] for the reduction of nickel oxide pellets by hydrogen. Szekely and coworkers [24] measured the effective diffusivity by using Eq. (4.3.7) for slablike and cylindrical pellets. The criteria for the appropriateness of this asymptotic region will be found in the next section, which discusses the analysis of the interaction of chemical reaction and diffusion within a porous solid.

4.3.3   Isothermal Systems of Porous Solids with Simultaneous First-Order Chemical Reaction and Diffusion

When both chemical reaction and pore diffusion control the overall rate of reaction, the interaction between these steps must be considered in determining the overall rate of reaction. One of the earliest analyses was made by Ausman and Watson [33], who considered the regeneration of coked catalyst pellet, assuming that the reaction is first order with respect to the gaseous reactant and independent of local conversion of solid so long as there remains some unreacted solid. By further assuming that the solid structure remained unchanged during reaction, they obtained a solution for the conversion of solid in a spherical pellet. Ishida and Wen [34] extended the analysis to incorporate the structural change by assigning different diffusivities for the completely reacted layer and the partially reacted inner core. Similar work is described in a paper by Lahiri and Seshadri [44].

In general, the rate of reaction of a solid particle with a gas depends on the conversion of the solid reactant due to the changing surface area available for reaction. Let us therefore develop the mathematical relationships describing the reaction of a porous solid in terms of the generalized grain model described above [23].

The simplifying assumptions are as follows:

(1)   The pseudosteady state approximation is appropriate for describing the concentration of the gaseous reactants within the pellet [40, 41].

(2)   The resistance due to external mass transport is negligible. (This simplification will be relaxed in a later discussion where its effect is studied.)

(3)   The solid structure is macroscopically uniform and is unaffected by the reaction.

(4)   The system is isothermal.

(5)   Diffusion within the pellet is either equimolar counterdiffusion or at low concentration of the diffusing species, and the effective diffusivities of gaseous reactant and product are equal and uniform throughout the pellet.

(6)   Diffusion of the gaseous reactants through the product layer of the individual grain does not affect the rate.

(7)   The viscous flow contribution to mass transport in the pores may be neglected.

Within the framework of these assumptions the problem may be stated by combining equations for the conservation of the gaseous reactant and product with a mass balance for the solid reactant. The former may be written

$$D_e \nabla^2 C_A - v_A = 0 \tag{4.3.11}$$

$$D_e \nabla^2 C_C + v_A = 0 \tag{4.3.12}$$

where v_A is the local rate of consumption of the gaseous reactant A, in moles per unit volume of porous solid per unit time.

Regarding the conservation of the solid reactant, we shall assume that within each grain the reaction front retains its original geometrical shape (i.e., concentric spheres, coaxial cylinders, or parallel planes) for successive times. Then the local rate of reaction at a solid surface may be expressed for a reaction that is first order with respect to gaseous species:

$$-\rho_s \, \partial r_c / \partial t = bk(C_A - C_C/K_E) \tag{4.3.13}$$

where r_c is the coordinate of the shrinking core surface within the grain.

An expression may now be obtained for the local rate of reaction v_A. For platelike, cylindrical, and spherical grains, v_A is given as†

$$v_A = (1 - \varepsilon)k(A_g/V_g)(A_g r_c/F_g V_g)^{F_g - 1}(C_A - C_C/K_E) \tag{4.3.14}$$

The local (position-dependent) rate of consumption of gaseous reactant in a unit volume of pellet is equal to the local rate of consumption of solid reactant per unit volume of pellet.

We may now proceed by dividing Eq. (4.3.12) by K_E and subtracting from Eq. (4.3.11) to yield

$$D_e \nabla^2 (C_A - C_C/K_E) - v_A(1 + 1/K_E) = 0 \tag{4.3.15}$$

Substituting for v_A from Eq. (4.3.14) and introducing the dimensionless variables

$$\psi \equiv \left(C_A - \frac{C_C}{K_E}\right) \bigg/ \left(C_{As} - \frac{C_{Cs}}{K_E}\right) \tag{4.3.16}$$

$$\xi \equiv A_g r_c / F_g V_g \tag{4.3.17}$$

$$t^* \equiv \frac{bk}{\rho_s}\left(\frac{A_g}{F_g V_g}\right)\left(C_{As} - \frac{C_{Cs}}{K_E}\right)t \tag{4.3.18}$$

$$\sigma \equiv \frac{F_p V_p}{A_p}\sqrt{\frac{(1 - \varepsilon)k}{D_e}\frac{A_g}{V_g}\left(1 + \frac{1}{K_E}\right)} \tag{4.3.19}$$

---

† When the reactant solid matrix contains inerts the quantity ε appearing in Eq. (4.3.14) must include the volume occupied by the inerts in addition to the void space present in the matrix.

into Eqs. (4.3.13) and (4.3.15), we obtain

$$\nabla^{*2}\psi - \sigma^2\psi\xi^{F_g - 1} = 0 \tag{4.3.20}$$

and

$$\partial\xi/\partial t^* = -\psi \tag{4.3.21}$$

Note that in the above equations C_{Cs} is the gaseous product concentration in the bulk gas stream, which would normally be negligibly small in a laboratory experiments, and ∇^{*2} is the Laplacian operator with η as the position coordinate, defined as

$$\eta \equiv A_p R/F_p V_p \tag{4.2.22}$$

where R is the spatial coordinate measured from the center of the pellet.

The initial and boundary conditions for Eqs. (4.3.20) and (4.3.21) are

$$\text{at} \quad t^* = 0, \qquad \xi = 1 \qquad \text{for all} \quad \eta \tag{4.3.23}$$

$$\text{at} \quad \eta = 1, \qquad \psi = 1 \qquad \text{for all} \quad t^* \geq 0 \tag{4.3.24}$$

$$\text{at} \quad \eta = 0, \quad \partial\psi/\partial\eta = 0 \quad \text{for all} \quad t^* \geq 0 \tag{4.3.25}$$

The dimensionless representation of the governing equations shows that the concentration driving force for reaction ψ and the local extent of reaction of the solid reactant ξ are related to the position η and time t^* through a single parameter σ. This parameter σ, which is analogous to the Thiele modulus in heterogeneous catalysis, incorporates both kinetic and structural properties and is the measure of the relative magnitude of chemical reaction and diffusion rates.

In general, the governing equations must be solved numerically. For most practical purposes the results sought are the overall extent and rate of reaction. In terms of the parameters used in the formulation these quantities are defined as

$$X = \int_0^1 \eta^{F_p - 1}(1 - \xi^{F_g}) \, d\eta \bigg/ \int_0^1 \eta^{F_p - 1} \, d\eta \tag{4.3.26}$$

$$\frac{dX}{dt^*} = \frac{\int_0^1 \eta^{F_p - 1}(-F_g \xi^{F_g - 1} \, \partial\xi/\partial t^*) \, d\eta}{\int_0^1 \eta^{F_p - 1} \, d\eta} \tag{4.3.27}$$

which, on combining with Eq. (4.3.21), can also be written

$$dX/dt^* = F_g F_p \int_0^1 \eta^{F_p - 1} \xi^{f_g - 1} \psi \, d\eta \tag{4.3.28}$$

A simpler expression for the overall rate of reaction may be obtained, however, by recognizing that at steady state the overall rate is equal to the rate of diffusion into the pellet:

$$\frac{(1 - \varepsilon)\rho_B V_p}{b} \frac{dX}{dt} = D_e A_p \frac{dC_A}{dR}\bigg|_{R = F_p V_p / A_p} \qquad (4.3.29)$$

from which we obtain

$$\frac{dX}{dt^*} = \frac{F_g F_p}{\sigma^2} \frac{d\psi}{d\eta}\bigg|_{\eta = 1} \qquad (4.3.30)$$

ASYMPTOTIC SOLUTIONS

Before presenting the complete solution to Eqs. (4.3.20) and (4.3.21), it is worthwhile to examine the asymptotic behaviors that have been already discussed in previous sections.

As σ approaches zero, pore diffusion presents a negligible resistance to the progress of reaction. Under this condition the reactant concentration is uniform throughout the pellet and is equal to that in the bulk ($\psi = 1$). Therefore, ξ is independent of η. From the result obtained in Section 4.3.1, we thus have

$$t^* = g_{F_g}(X) \equiv 1 - (1 - X)^{1/F_g} \qquad (4.3.31)$$

As σ approaches infinity, reaction occurs mostly in a narrow zone between the unreacted core and the completely reacted layer. The concentration driving force for reaction ($C_A - C_C/K_E$) drops to nearly zero at this reaction zone, and the diffusion through the product layer controls the overall rate. From the analysis discussed in Section 4.3.2, we have

$$t^+ = (2F_g F_p / \sigma^2) t^* = p_{F_p}(X) \qquad (4.3.32)$$

The form of Eq. (4.3.32) suggests the following generalized modulus:

$$\hat{\sigma} \equiv \frac{\sigma}{\sqrt{2F_g F_p}} = \frac{V_p}{A_p} \sqrt{\frac{(1 - \varepsilon)k F_p}{2D_e} \left(\frac{A_g}{F_g V_g}\right)(1 + 1/K_E)} \qquad (4.3.33)$$

In terms of this modulus Eq. (4.3.32) may now be rewritten as

$$t^* / \hat{\sigma}^2 = p_{F_p}(X) \qquad (4.3.34)$$

Equations (4.3.31) and (4.3.34) correspond to asymptotic solutions for pure chemical reaction control and pure diffusion control, respectively.

For the intermediate values of $\hat{\sigma}$, Eqs. (4.3.20) and (4.3.21) must be solved with the initial and boundary conditions given by Eqs. (4.3.23)–(4.3.25). The complete solution will also indicate the range of values of $\hat{\sigma}$ for which these asymptotic regimes are approached.

ANALYTICAL SOLUTIONS

Analytical solutions may be obtained for systems where $F_g = 1$ [23, 26, 34]. Let us consider an infinite slab of a porous solid made up of flat grains, that is, $F_g = F_p = 1$. Equations (4.3.20) and (4.3.21) may then be written

$$d^2\psi/d\eta^2 - 2\hat{\sigma}^2\psi = 0 \qquad \text{for} \quad \xi > 0 \qquad (4.3.35)$$

$$\partial\xi/\partial t^* = -\psi \qquad \text{for} \quad \xi > 0 \qquad (4.3.36)$$

$$d^2\psi/d\eta^2 = 0 \qquad \text{for} \quad \xi = 0 \qquad (4.3.37)$$

with the same initial and boundary conditions given by Eqs. (4.3.23)–(4.3.25). It is apparent on physical grounds that ξ does not become negative.

Due to the concentration gradient within the pellet, the reaction is faster near the external surface than in the interior. After a certain time a completely reacted layer will be formed in the outer region of the pellet. The solution is obtained by considering separately the periods before and after the complete conversion of the outermost layer.

PERIOD BEFORE THE COMPLETE CONVERSION OF THE EXTERNAL LAYER

In this period the concentration profile described by Eq. (4.3.35) remains unchanged with time and is given by

$$\psi = \cosh(\sqrt{2}\hat{\sigma}\eta)/\cosh(\sqrt{2}\hat{\sigma}) \qquad (4.3.38)$$

Substituting ψ into Eq. (4.3.36) and integrating, we get

$$\xi = 1 - [\cosh(\sqrt{2}\hat{\sigma}\eta)/\cosh(\sqrt{2}\hat{\sigma})]t^* \qquad (4.3.39)$$

The time required to react completely the external layer of the pellet, t_1^*, may be obtained by substituting $\eta = 1$ and $\xi = 0$ in Eq. (4.3.39). Thus we have

$$t_1^* = 1 \qquad (4.3.40)$$

The overall conversion of the solid, X, during this period is determined by using Eq. (4.3.26):

$$X = t^* \int_0^1 [\cosh(\sqrt{2}\,\hat{\sigma}\eta)/\cosh(\sqrt{2}\,\hat{\sigma})]\,d\eta = [\tanh(\sqrt{2}\,\hat{\sigma})/\sqrt{2}\,\hat{\sigma}]t^* \quad (4.3.41)$$

At the end of the period, the overall conversion and the position of the reaction interfaces within the grains are obtained from Eqs. (4.3.41) and (4.3.39):

$$X_1 = \tanh(\sqrt{2}\hat{\sigma})/\sqrt{2}\hat{\sigma} \qquad (4.3.42)$$

$$\xi_1 = 1 - [\cosh(\sqrt{2}\hat{\sigma}\eta)/\cosh(\sqrt{2}\hat{\sigma})] \qquad (4.3.43)$$

PERIOD AFTER THE COMPLETE CONVERSION OF THE
EXTERNAL LAYER

After the external layer of the solid is completely reacted, there are two zones within the pellet, that is, the completely reacted outer layer and the partially reacted inner core, as shown in Fig. 4.4. In the outer layer only the diffusion of reactant gas occurs without reaction since the solid reactant has been completely consumed, whereas in the inner core chemical reaction continues to occur simultaneously with diffusion. Let η_m be the position of the boundary between the partially and completely reacted zones. Let I and II represent the inner core and the outer layer, respectively. Then the governing equations are given by the following:

Within the inner core (zone I)

$$d^2\psi^I/d\eta^2 - 2\hat{\sigma}^2\psi^I = 0 \qquad (4.3.44)$$

$$\partial\xi^I/\partial t^* = -\psi^I \qquad (4.3.45)$$

with the boundary conditions

$$\psi^I = \psi_m^{\;I} \qquad \text{at} \quad \eta = \eta_m \qquad (4.3.46)$$

$$d\psi^I/d\eta = 0 \qquad \text{at} \quad \eta = 0 \qquad (4.3.47)$$

For the outer layer (zone II)

$$d^2\psi^{II}/d\eta^2 = 0 \qquad (4.3.48)$$

$$\xi^{II} = 0 \qquad (4.3.49)$$

with the boundary conditions

$$\psi^{II} = 1 \qquad \text{at} \quad \eta = 1 \qquad (4.3.50)$$

$$\psi^{II} = \psi_m^{II} \qquad \text{at} \quad \eta = \eta_m \qquad (4.3.51)$$

We also have the following matching conditions at $\eta = \eta_m$:

$$\psi_m^{\;I} = \psi_m^{II} = \psi_m \qquad (4.3.52)$$

$$(d\psi^I/d\eta)_{\eta=\eta_m} = (d\psi^{II}/d\eta)_{\eta=\eta_m} \qquad (4.3.53)$$

The initial condition for ξ in this period is, from Eq. (4.3.43),

$$\xi^I = 1 - \cosh(\sqrt{2}\hat{\sigma}\eta)/\cosh(\sqrt{2}\hat{\sigma}) \qquad \text{and} \qquad \eta_m = 1 \quad \text{at} \quad t^* = 1 \qquad (4.3.54)$$

The detailed solution of Eqs. (4.3.44)–(4.3.54) is presented in the literature [23]. The positions of the reaction fronts within the grains of zone I are given by

$$\xi^I = 1 - \cosh(\sqrt{2}\hat{\sigma}\eta)/\cosh(\sqrt{2}\hat{\sigma}\eta_m) \qquad (4.3.55)$$

The overall conversion during the period after the complete conversion of the external layer is

$$X = 1 - \eta_m + \tanh(\sqrt{2}\,\hat{\sigma}\eta_m)/\sqrt{2}\,\hat{\sigma} \qquad (4.3.56)$$

where η_m, the dimensionless inner radius of the completely reacted zone, is given by

$$t^* = 1 + \hat{\sigma}^2(1 - \eta_m)^2 + \sqrt{2}\,\hat{\sigma}(1 - \eta_m)\tanh(\sqrt{2}\,\hat{\sigma}\eta_m) \qquad (4.3.57)$$

The required relationship between t^* and X may then be obtained from Eqs. (4.3.56) and (4.3.57).

From the results obtained by Ishida and Wen [34], we may write the solution for the case of a spherical pellet made up of flat grains ($F_p = 3$ and $F_g = 1$) as follows:

When $0 \leq t^* \leq 1$

$$\psi = \sinh(\sqrt{6}\,\hat{\sigma}\eta)/\eta \sinh(\sqrt{6}\,\hat{\sigma}) \qquad (4.3.58)$$

$$\xi = 1 - [\sinh(\sqrt{6}\,\hat{\sigma}\eta)/\eta \sinh(\sqrt{6}\,\hat{\sigma})]t^* \qquad (4.3.59)$$

and

$$X = (1/2\hat{\sigma}^2)[\sqrt{6}\,\hat{\sigma}\coth(\sqrt{6}\,\hat{\sigma}) - 1]t^* \qquad (4.3.60)$$

When $t^* \geq 1$

$$\psi^{\mathrm{I}} = [\eta_m \sinh(\sqrt{6}\,\hat{\sigma}\eta)/\eta \sinh(\sqrt{6}\,\hat{\sigma}\eta_m)]\psi_m \qquad (4.3.61)$$

$$\psi^{\mathrm{II}} = \frac{\eta_m(1 - \eta)}{\eta(1 - \eta_m)}\psi_m + \frac{1 - \eta_m/\eta}{1 - \eta_m} \qquad (4.3.62)$$

where

$$\psi_m = \{1 + (1 - \eta_m)[\sqrt{6}\,\hat{\sigma}\eta_m \coth(\sqrt{6}\,\hat{\sigma}\eta) - 1]\}^{-1} \qquad (4.3.63)$$

and

$$t^* = 1 + \hat{\sigma}^2(1 - \eta_m)^2(1 + 2\eta_m) + (1 - \eta_m)[\sqrt{6}\,\hat{\sigma}\eta_m \coth(\sqrt{6}\,\hat{\sigma}\eta_m) - 1] \quad (4.3.64)$$

$$\xi^{\mathrm{I}} = 1 - \eta_m \sinh(\sqrt{6}\,\hat{\sigma}\eta)/\eta \sinh(\sqrt{6}\,\hat{\sigma}\eta_m) \qquad (4.3.65)$$

$$X = 1 - \eta_m{}^3 + (\eta_m/2\hat{\sigma}^2)[6\,\hat{\sigma}\eta_m \coth(\sqrt{6}\,\hat{\sigma}\eta_m) - 1] \qquad (4.3.66)$$

Similarly, an analytical solution may be developed for the case of cylindrical pellets made up of flat grains ($F_p = 2$ and $F_g = 1$).

As $\hat{\sigma}$ approaches 0 or ∞, the analytical solutions for the overall conversion, Eqs. (4.3.41), (4.3.56), and (4.3.57) for $F_g = F_p = 1$, and Eqs. (4.3.60), (4.3.64), and (4.3.66), reduce to the corresponding asymptotic solutions given by Eqs. (4.3.31) and (4.3.34), respectively.

THE GENERAL CASE—NUMERICAL SOLUTION

In the general case involving nonflat grains ($F_g \neq 1$), the system of Eqs. (4.3.20) and (4.3.21) with the boundary conditions (4.3.23)–(4.3.25) must be solved numerically. Details of a computer program for carrying out a numerical solution may be found in a thesis by Evans [45].

In this development of the grain model it has been assumed that the reaction is first order with respect to gaseous reactant. In a subsequent section we shall discuss the case where the rate of reaction is a nonlinear function of the gaseous reactant concentration. In this latter case the numerical solution is more difficult and usually involves an iterative approach [25].

Since we have analytical solutions for a first-order reaction in systems with $F_g = 1$, we will present here the numerical solutions for a first-order reaction in systems with $F_g \neq 1$ [23].

Figures 4.10 and 4.11 show computed results for small and large values of $\hat{\sigma}$, respectively. Also shown in these figures are the analytical solutions for $F_g = 1$. The coordinates are so chosen as to allow the convenient presentation of the appropriate asymptotic solutions and to show the deviation of the curves for the intermediate values of $\hat{\sigma}$ from the two asymptotes.

Thus, Fig. 4.10 shows a plot of $g_{F_g}(X)$ against t^* with $\hat{\sigma}$ as a parameter, for various values of the shape factors F_p and F_g. Also shown is the line for $\hat{\sigma} = 0$, that is, $g_{F_g}(X) = t^*$, which is valid for all geometries. It is readily seen that the $\hat{\sigma} = 0$ asymptote is appropriate for $\hat{\sigma} < 0.3$ ($\hat{\sigma}^2 < 0.1$) regardless of the geometry of the porous solid. This information is useful, since it obviates the need for a detailed computation within this region. Moreover we can neglect the effect of diffusion, and the rate data obtained are those of intrinsic chemical reaction.

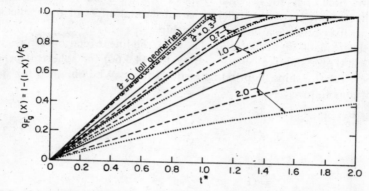

FIG. 4.10. Plot of the conversion function $g_{F_g}(X)$ against dimensionless time for several values of the reaction modulus $\hat{\sigma}$ and for various geometries [23]. - - -, $F_g = 1$, $F_p = 1$; ———, $F_g = 1$, $F_p = 3$; · · ·, $F_g = 3$, $F_p = 3$.

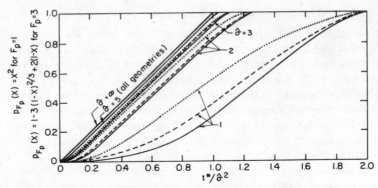

FIG. 4.11. Plot of the conversion function $p_{F_p}(X)$ against $t^*/\hat{\sigma}^2$ for several values of the reaction modulus and for various geometries [23]. - - -, $F_g = 1$, $F_p = 1$; ———, $F_g = 1$, $F_p = 3$; $\cdots$, $F_g = 3$, $F_p = 1$.

Another important feature of Fig. 4.10 is that a linear relationship between $g_{F_g}(X)$ and time is observed in all cases for $t^* \leq 1$, that is, before the external layer is completely reacted. For the systems with $F_g = 1$, this is expected since the rate of reaction per unit volume of the pellet remains constant during the reaction until the solid reactant is completely consumed. The analytical solution indicates this feature, as can be seen in Eqs. (4.3.41) and (4.3.60). For the systems with $F_g > 1$, the reason for the linear relationship is not so obvious, although a qualitative explanation can be put forward. The linear relationship persists up to a high conversion value, even for values of $\hat{\sigma}$ as large as 1, that is, when chemical reaction and diffusion are of comparable importance. This indicates that the existence of a linear relationship between $g_{F_g}(X)$ and time does not necessarily mean that the system is controlled by chemical reaction; even when there is a substantial effect of intrapellet diffusion, one may observe a linear relationship.

For pellets made up of flat grains ($F_g = 1$), the linear relationship is given by Eq. (4.3.41) for slablike pellets, and by Eq. (4.3.60) for spherical pellets. From these relationships, approximate equations may be obtained for small values of $\hat{\sigma}$ as follows:

For $F_g = F_p = 1$

$$X = (1 - \tfrac{2}{3}\hat{\sigma}^2 + \tfrac{8}{15}\hat{\sigma}^4 - \cdots)t^* \tag{4.3.67}$$

For $F_g = 1$ and $F_p = 3$

$$X = (1 - \tfrac{2}{5}\hat{\sigma}^2 + \tfrac{8}{35}\hat{\sigma}^4 - \cdots)t^* \tag{4.3.68}$$

When diffusion is significant ($\hat{\sigma} > 0.3$) the increase in the slope of the straight line with increasing temperature is smaller than it would be in the absence of the diffusional effect. This would result in a smaller apparent activation energy than the intrinsic value.

Figure 4.11 shows a plot of $p_{F_p}(X)$ against $t^*/\hat{\sigma}^2$ with $\hat{\sigma}$ as a parameter, for large values of $\hat{\sigma}$. As in the case of small $\hat{\sigma}$ values, the choice of the coordinates was dictated by the form of the asymptotic solution.

We see that the $\hat{\sigma} \to \infty$ asymptote is approached when $\hat{\sigma} > 3.0$ ($\hat{\sigma}^2 > 10$) for all geometries considered. This finding provides a definition for the range of validity of the asymptotic solution obtained for diffusion control.

It is of interest again to note that the linear relationship between $p_{F_p}(X)$ and time persists even when the value of $\hat{\sigma}$ does not correspond to infinity (say, $\hat{\sigma} > 2$). When $2 < \hat{\sigma} < \infty$, a straight line is obtained that does not pass through the origin upon extrapolation. The data on the reduction of iron oxide obtained by Kawasaki et al. [38] and those on the reduction of nickel oxide pellets by Szekely et al. [24] show this behavior, indicating the effect of chemical kinetics at least at the outset of the reaction.

In view of the fact that the curves for the systems of different geometries lie close to each other in this range of $\hat{\sigma}$ values, the equation for the linear relationship may be obtained from the analytic solution. Thus from Eqs. (4.3.56) and (4.3.57), or Eqs. (4.3.64) and (4.3.66), we obtain the following for large $\hat{\sigma}$ and $F_g = 1$:

$$t^*/\hat{\sigma}^2 = p_{F_p}(x) + 1/2\hat{\sigma}^2 \qquad (4.3.69)$$

This equation will hold true for other values of F_g since as we see from Fig. 4.11 at high $\hat{\sigma}$ the conversion time relationship is nearly independent of F_g.

Equation (4.3.69) suggests that the effective diffusivity in the porous solid may be determined from the slope of the straight line by plotting according to Eq. (4.3.69) the data obtained under the condition of diffusion control. It is also seen that the extrapolation of the straight line to the zero conversion $[p_{F_p}(X) = 0]$ will give the chemical rate constant. It must be borne in mind, however, that in many gas–solid reactions an induction period exists in the beginning of the reaction [24], which complicates the determination of the reaction rate constant in this manner. Therefore, the much more satisfactory method of determining the rate constant discussed in Section 4.3.1. is recommended wherever possible.

Example 4.3.1 It is proposed to reduce FeO particles with hydrogen at 900°C. The reaction is assumed to be first order. Experimental measurements showed that with $K_E = 0.5$

$$A_g/F_g V_g = 5 \times 10^4 \text{ cm}^{-1}, \qquad k = 10^{-4} \text{ cm/sec}$$

If the reaction is to be carried out with spherical pellets, made up by agglomerating the powder, estimate the largest allowable pellet size for which diffusional resistance will still be negligible. Neglect external mass transfer effects and assume that $D_e = 0.8$ cm^2/sec and $\varepsilon = 0.3$.

SOLUTION   The effect of diffusion is negligible, when $\hat{\sigma} < 0.3$, i.e.,

$$0.3 > \frac{V_p}{A_p} \sqrt{\frac{0.7 \times 10^{-4} \times 5 \times 10^4 \times 3}{1.6}}$$

Thus $(R_p \times 2.56)/3 < 0.3$, so that $R_p < 0.35$ cm should ensure the absence of diffusional limitations.

The solutions for intermediate values of $\hat{\sigma}$ depend on the geometry of the system, as can be seen in Figs. 4.10 and 4.11. These cannot be expressed analytically except for the particular case of $F_g = 1$. Even the solutions for $F_g = 1$ give the relationship between conversion and time in an indirect form through the parameter η_m. A simple and direct relationship between conversion and time may be obtained by an appropriate manipulation of the numerical solutions shown in Figs. 4.10 and 4.11 [23].

If we plot t^* for the complete conversion against $\hat{\sigma}$ for various geometries, it is found that $t^*_{x=1}$ is a function only of $\hat{\sigma}$ regardless of the geometry of the sample. From the analytical solution for the case of $F_g = 1$, the relationship between $t^*_{x=1}$ and $\hat{\sigma}$ may be obtained from Eq. (4.3.57) or (4.3.64). Thus we have

$$t^*_{x=1} = 1 + \hat{\sigma}^2 \qquad (4.3.70)$$

For other values of conversion the relationship between t^* and $\hat{\sigma}$ varies with conversion values and geometries. In order to show the effect of $\hat{\sigma}$ on t^*, a plot of

$$t^*_{\hat{\sigma}=0}/t^* \qquad \text{against} \qquad \hat{\sigma}\sqrt{p_{F_p}(X)/g_{F_g}(X)}$$

is shown in Fig. 4.12. The solution for $X = 1$ given by Eq. (4.3.70) is also included as a particular case. The coordinates were so chosen as to bring the asymptotes for various conversion values to the same positions. Thus, regardless of the conversion value, the function on the ordinate approaches unity as $\hat{\sigma} \to 0$, and the relationship approaches the asymptote shown by the dashed line as $\hat{\sigma} \to \infty$ according to Eq. (4.3.34).

As mentioned above, the relationship for conversion values other than unity varies with geometry of the pellet. When such coordinates were used, however, the results for various geometries are brought quite close. The lines in Fig. 4.12 for $X \neq 1$ have been obtained by averaging out the small variations for different geometries. Then, by noting that the curves lie in close proximity to that for $X = 1$, the relationship may be written approximately, using Eq. (4.3.70), as

$$\text{(reciprocal of ordinate)} = 1 + \text{(abscissa)}^2 \qquad (4.3.71)$$

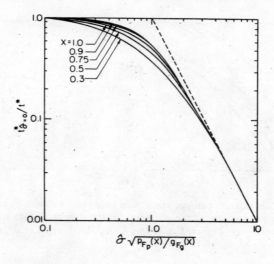

FIG. 4.12. Generalized plot for reaction time versus conversion [23].

from which we obtain

$$t^*/t^*_{\hat{\sigma}=0} = t^*/g_{F_g}(X) \cong 1 + \hat{\sigma}^2 p_{F_p}(X)/g_{F_g}(X) \qquad (4.3.72)$$

or

$$t^* \cong g_{F_g}(X) + \hat{\sigma}^2 p_{F_p}(X) \qquad (4.3.73)$$

In Eq. (4.3.72) the relationship given by Eq. (4.3.31), $t^*_{\hat{\sigma}=0} = g_{F_g}(X)$, was used.

Figure 4.13 shows the comparison between the exact numerical solution and Eq. (4.3.73) for systems with $F_g = 1$ and $F_p = 3$, and with $F_g = F_p = 3$. It is seen that the approximate solution gives a satisfactory representation of the exact solution. The comparison has been made for the intermediate values of $\hat{\sigma}$, for which the difference is largest. For smaller and larger values of $\hat{\sigma}$, the agreement is better than shown in the figure. Furthermore, the approximate solution is exact at $X = 1$ and is asymptotically correct as $\hat{\sigma} \to 0$ or $\hat{\sigma} \to \infty$.

Figures 4.14 and 4.15 show the comparison between the rates of overall conversion obtained by exact numerical method and by differentiating the approximate solution. The latter give

$$dX/dt^* = [g'_{F_g}(X) + \hat{\sigma}^2 p'_{F_p}(X)]^{-1} \qquad (4.3.74)$$

The comparison is again made for the intermediate values of $\hat{\sigma}$, for which the approximate solution shows the least satisfactory agreement.

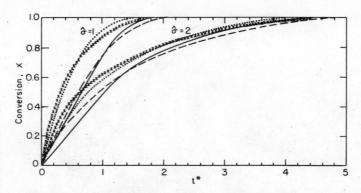

FIG. 4.13. A comparison between conversion time plots obtained by exact solution of the grain model equations and those obtained from the approximate equation (4.3.73) [23]. Exact solution: ———, $F_g=1$, $F_p=3$; $\cdots$, $F_g=3$, $F_p=3$. $g_{F_g}(X)+\hat{\sigma}^2 p_{F_p})X)= t^*$: ---, $F_g=1$, $F_p=3$; xxx, $F_g=3$, $F_p=3$.

It is seen that, as expected, there is only approximate agreement; the error is particularly large in the early stage of the reaction. The early error, however, is compensated in the later stage of reaction in such a way that the times required for the complete conversion becomes identical. Furthermore, it is expected that the agreement is better for the limiting cases of large or small values of $\hat{\sigma}$; thus the agreement is better for $\hat{\sigma} = 2$ than for $\hat{\sigma} = 1$.

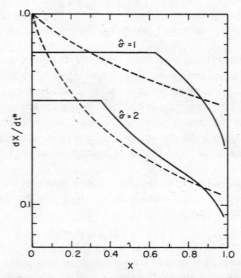

FIG. 4.14. A comparison between reaction rates obtained from the exact solution and from Eq. (4.3.74). $F_g = F_p = 1$; ———, exact solution; ---, $dX/dt^* = [g'_{F_g}(X) + \hat{\sigma}^2 p'_{F_g}(X)]^{-1}$.

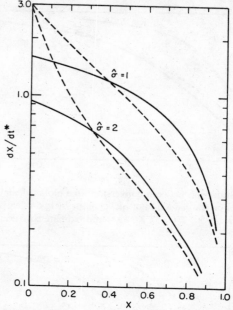

FIG. 4.15. A comparison between reaction rates obtained from the exact solution and from Eq. (4.3.74). $F_g = F_p = 3$; ———, exact solution; - - -, $dX/dt^* = [g'_{r_g}(X) + \hat{\sigma}^2 p'_{F_g}(X)]^{-1}$.

The results of these analyses were applied to the reduction of nickel oxide pellets with hydrogen by systematically designed experiments [24]. Information was first obtained on the intrinsic chemical kinetics and on effective diffusivity by making experimental runs in the appropriate asymptotic regimes, and by making use of Eqs. (4.3.31) and (4.3.34). Then the parameters were combined to calculate $\hat{\sigma}$, and the predicted relationship between conversion and time was obtained from Eq. (4.3.73). Figures 4.16 and 4.17 show examples of the comparison between the predicted results and the experimental data in the intermediate regime. (The starting time of the predicted curve for one of the runs shown in Fig. 4.17 has been shifted to take care of the induction period. For a discussion on induction period the reader is referred to the original article [24].)

We note that there is general agreement over most of the conversion range. From the beginning of the reaction up to about 50% conversion, the approximate equation predicts a higher conversion than the experimental value. It also gives a higher value than that predicted by the exact solution, as shown in Fig. 4.13. This suggests that the exact solution would have given much better agreement. Nevertheless, at the expense of the somewhat greater accuracy, the approximate solution provides a closed-form relation-

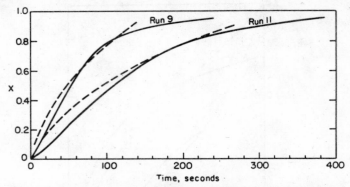

FIG. 4.16. A comparison between conversion time plots calculated from Eq. (4 3.73) and experimental data on the reduction of nickel oxide by hydrogen [24]. Run 9, $\hat{\sigma} = 1.07$, $T = 683°K$; run 11, $\hat{\sigma} = 1.50$, $T = 657°K$; - - -, approximate equation.

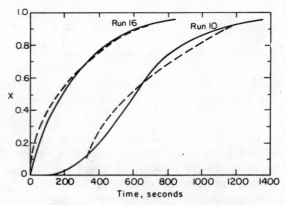

FIG. 4.17. A comparison between conversion time plots calculated from Eq. (4.3.73) and experimental data on the reduction of nickel oxide by hydrogen [24]. Run 16, $\hat{\sigma} = 3.29$, $T = 683°K$; run 10, $\hat{\sigma} = 1.32$, $T = 520°K$; - - -, approximate equation.

ship between conversion and reaction time for all values of $\hat{\sigma}$ and all geometries considered. The clear advantage is that the closed-form equation obviates the necessity for the numerical solution of the differential equation for each value of $\hat{\sigma}$. If one considers the problem of applying the results of a single-pellet study to multiparticle systems, the usefulness of the closed-form solution becomes much greater.

Another important feature of the approximate solution is that it provides greater insight into the problem by showing clearly the relative importance of chemical reaction and diffusion through the parameter $\hat{\sigma}$. Thus, even a crude estimate of $\hat{\sigma}$ might provide enough information as to whether one needs data on chemical kinetics or diffusion, or both, in order to predict the relationship between conversion and time: If $\hat{\sigma}$ is much smaller than unity, one needs kinetic parameters only; if it is much larger than unity, only diffusional parameters. If $\hat{\sigma}$ is around unity, both would be required.

Furthermore, Eq. (4.3.73) shows that the time required to reach a certain conversion is the sum of the time to reach the same conversion under the chemical reaction control and that under the diffusion control. Thus

$$t^*(X) \cong t^*(X)\bigg|_{\hat{\sigma}=0} + t^*(X)\bigg|_{\hat{\sigma}\to\infty} \qquad (4.3.75)$$

where, with reference to Eqs. (4.3.31) and (4.3.34),

$$t^*(X)\bigg|_{\hat{\sigma}=0} = g_{F_g}(X) \qquad (4.3.76)$$

and

$$t^*(X)\bigg|_{\hat{\sigma}\to\infty} = \hat{\sigma}^2 p_{F_p}(X) \qquad (4.3.77)$$

The applications of Eq. (4.3.73) to actual experiment and optimization of gas–solid-reaction systems have been discussed [23, 24].

4.3.4 Comparison and Similarity between the Solutions for Porous and Nonporous Solids

It is of interest to note that Eq. (4.3.73) is of the same form as the relationship between conversion and time for the shrinking-core system of initially *nonporous* particles, where the overall rate is controlled by both diffusion and chemical kinetics. For a first-order reaction this relationship, given by Eq. (3.3.33), may be put in the following form, which is valid for spheres, cylinders, and flat plates:

$$\frac{bk(C_{As} - C_{Cs}/K_E)}{\rho_s}\left(\frac{A_p}{F_p V_p}\right)t = g_{F_p}(X) + \frac{k}{2D_e}\left(1 + \frac{1}{K_E}\right)\frac{V_p}{A_p}p_{F_p}(X) \quad (4.3.78)$$

where D_e is the effective diffusivity in the product layer.

On examining Eq. (4.3.78) it is seen that the time required to attain a certain conversion for a nonporous system is the sum of two terms: the time to reach the same conversion under chemical reaction control and that under pure diffusion control. This is an important result that is generally true for systems consisting of first-order rate processes in series. It is to be noted, moreover, that even in the case of a porous system where diffusion and reaction occur in parallel, this additivity holds, if approximately, as seen in Eq. (4.3.75). The first term on the r.h.s. corresponds to t^* in the absence of diffusional resistance and the second term to t^* under diffusion control.

Equation (4.3.78) becomes identical to Eq. (4.3.73) upon replacing F_g by F_p, A_g by A_p, and V_g by V_p and setting $\varepsilon = 0$. Furthermore, the examination of these two equations may explain why the topochemical model with chemical reaction control is approached more readily for nonporous solid reactants, and why the diffusion-controlled shrinking-core model becomes applicable for porous solids. This behavior is readily apparent on considering that, in general, $(V_p/A_p) \gg (V_g/A_g)$.

This explains the fact that, for the reduction of iron oxide, two extremely different mechanisms have been proposed: purely chemical reaction-controlled [46–48] and entirely diffusion-controlled mechanisms [38]. On examining these systems it is found that the former applied to high-density solids (less than 10% porosity) and the latter to relatively porous pellets (about 30% porosity).

Wilhelm and St. Pierre [49] found in the reduction of hematite and magnetite that the chemical reaction-controlled mechanism does not describe their experimental data with porous pellets. The same mechanism adequately described the data with dense samples [46, 50]. Furthermore, Hansen et al. [51] reported that for the reduction of wüstite the assumption of chemical reaction control is reasonable for high-density particles, giving an activation energy of 27 kcal/g-mole; however, when they forced this assumption on porous pellets, they obtained an activation energy of 6 kcal/g-mole, which indicates a large influence of diffusional processes.

4.3.5   The Effect of External Mass Transport

In the preceding discussion we assumed that external mass transport does not influence the overall rate. In laboratory scale experimental studies such conditions are readily realized by using sufficiently high gas velocities. In practical systems involving packed or fluidized beds of solid particles, however, external mass transport may be fully or partially rate controlling.

When the effect of external mass transport cannot be neglected, the governing Eqs. (4.3.20) and (4.3.21) and the conditions (4.3.23) and (4.3.25) still apply, but the boundary condition (4.3.24) must be replaced by the following:

$$\text{at} \quad \eta = 1, \quad d\psi/d\eta = N_{Sh}^*(1 - \psi) \tag{4.3.79}$$

where

$$N_{Sh}^* \equiv (h_D/D_e)(F_p V_p/A_p) = N_{Sh}(D_M/D_e) \tag{4.3.80}$$

As in the case of negligible resistance to the progress of reaction by external mass transport ($N_{Sh}^* \to \infty$) discussed in the previous sections, analytical solutions including the effect of N_{Sh}^* can be obtained for systems of $F_g = 1$ [26, 34].

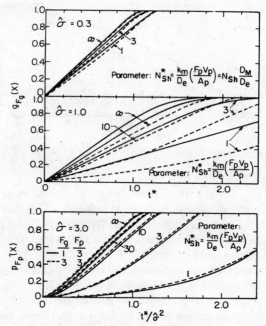

FIG. 4.18. The effect of external mass transfer on the relationship between conversion and time [23]. ——, $F_g = 1$, $F_p = 3$; - - -, $F_g = 3$, $F_p = 3$.

In the general case of $F_p = 3$, Sohn and Szekely [23] obtained numerical solutions, which are shown in Fig. 4.18.

It is seen in Fig. 4.18 that when $\hat{\sigma}$ is small, variations in N_{Sh}^* play only a minor role, which is to be expected, because in this case the system is chemically controlled.

While N_{Sh}^* may play an important role in determining the overall rate for larger values of $\hat{\sigma}$, the effect of external mass transport becomes unimportant when $N_{Sh}^* > 30$.

A relationship analogous to Eq. (4.3.73) may be derived for the case where the external mass transport has an influence on the overall rate of reaction. This relationship is

$$t^* = g_{F_g}(X) + \hat{\sigma}^2[p_{F_p}(X) + (2X/N_{Sh}^*)] \qquad (4.3.81)$$

The reader is referred to the literature [23] for the derivation of Eq. (4.3.81).

It is to be stressed that for use in conjunction with Eq. (4.3.81) t^* is defined as

$$t^* = \frac{bk}{\rho_s}(A_g/F_g V_g)t[C_{A0} - C_{C0}/K_E]$$

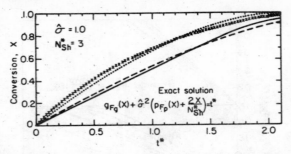

FIG. 4.19. Comparison of calculations made using Eq. (4.3.81) with the exact solution [23].

where C_{A0} and C_{C0} denote the concentrations of the gaseous reactants and products in the bulk of the gas stream, respectively. This designation is not inconsistent with the previous definition of t^* [in Eq. (4.3.18)] because in that case external mass transfer was absent so that the surface concentration and the bulk concentration were the same.

Figure 4.19 shows the comparison between the exact solution and Eq. (4.3.81) for $\hat{\sigma} = 1.0$ and $N_{Sh}^* = 3$. It is seen that Eq. (4.3.81) gives a satisfactory representation of the exact solution. For smaller and larger values of $\hat{\sigma}$ the agreement is better than shown in this figure. Furthermore, Eq. (4.3.81) converges to the exact solution as $\hat{\sigma} \to 0$ or $\hat{\sigma} \to \infty$. For $X = 1$, Eq. (4.3.81) again becomes identical to the exact solution regardless of the $\hat{\sigma}$ value and geometry.

Example 4.3.2 It is proposed to reduce a FeO pellet with hydrogen at 900°C. The following information is available:

$$F_g = 3, \qquad\qquad F_g V_g / A_g = 2 \times 10^{-4} \text{ cm}$$
$$D_e = 0.8 \text{ cm}^2/\text{sec}, \qquad\qquad \varepsilon = 0.35$$
$$k = 5 \times 10^{-4} \text{ cm/sec}, \qquad\qquad \rho_s = 0.0794 \text{ g-mole/cm}^3$$
$$N_{Sh}^* = 10, \qquad\qquad K_E = 0.5$$

Assume that the chemical reaction is a first-order irreversible reaction and calculate the time required to achieve 90% reaction for spherical pellets that are 0.2, 2, and 4 cm in diameter.

The equation that should be employed for calculation is

$$t^* = g_{F_g}(X) + \hat{\sigma}^2[p_{F_p}(X) + 2X/N_{Sh}^*] \qquad (4.3.81)$$

where

$$g_{F_g}(X) = 1 - (1 - X)^{1/F_g}$$

For $X = 0.9$, $g_{F_g}(X) = 0.5358$. Since $F_p = 3$

$$p_{F_p}(X) = 1 - 3(1 - X)^{2/3} + 2(1 - X)$$

For $X = 0.9$, $p_{F_p}(X) = 0.5538$. From Eqs. (4.3.33) we can calculate $\hat{\sigma}$ for these cases:

for a 0.2 cm pellet, $\hat{\sigma}^2 = 0.0102$;
for a 2 cm pellet, $\hat{\sigma}^2 = 1.02$;
for a 4 cm pellet, $\hat{\sigma}^2 = 4.08$.

Put all the values into Eq. (4.3.81):

for a 0.2 cm pellet, $t^* = 0.5358 + 0.0102\,(0.5537 + 0.18) = 0.5433$;
for a 2 cm pellet, $t^* = 0.5358 + 1.02\,(0.5537 + 0.18) = 1.2842$;
for a 4 cm pellet, $t^* = 0.5358 + 4.08\,(0.5537 + 0.18) = 3.5293$.

By using the equation that defines t^*,

$$t^* = \frac{bk}{\rho_s} \frac{A_g}{F_g V_g}\, t \left(C_{A0} - \frac{C_{CO}}{K_E} \right)$$

one can translate t^* to t:

for a 0.2 cm pellet, $t = 1662$ sec;
for a 2 cm pellet, $t = 3929$ sec;
for a 4 cm pellet, $t = 10{,}799$ sec.

4.3.6 The Effect of Diffusion on Reactions Obeying Langmuir–Hinshelwood Kinetics

In the previous section we developed a generalized model for the reaction between a porous solid and a reactant gas, for systems with first-order kinetics with respect to the reactant gas. The model provided general criteria for relating structural effects to the reactivity of porous solids, and the reaction modulus defined therein allowed the assessment of the relative importance of pure diffusion and chemical kinetics.

While almost all the mathematical models for gas–solid reaction systems are based on the assumption of first-order kinetics, in many instances the Langmuir–Hinshelwood type rate expression provides a more realistic description of the system, especially over a wide range of reactant concentrations. Examples of gas–solid reactions that have been found to follow a Langmuir–Hinshelwood type kinetics include the reduction of iron oxides by hydrogen [52, 53], the reduction of nickel oxide by hydrogen [54], the oxidation of uranium–carbon alloys [55], and the reaction of carbon with various gases [2].

FORMULATION

In this section we will outline the incorporation of the Langmuir–Hinshelwood type kinetics in the reactions between a porous solid and a gas [25].

While the use of nonlinear kinetics in gas–solid reactions (involving porous solids) is a rather unexplored field, a great deal of work has been done on the effect of Langmuir–Hinshelwood kinetics on the effectiveness factor in heterogeneous catalysis [5, 56–59].

Let us consider the following rate expression for an irreversible surface reaction:

$$\mathscr{R}_A = kC_A/(1 + KC_A) \tag{4.3.82}$$

where $\mathscr{R}_A$ is the local reaction rate in moles per unit surface per unit time, C_A the concentration of reactant A at the reaction surface, and k and K the constants in the Langmuir–Hinshelwood rate expression.

Many gas–solid reactions, especially those involving metal compounds, obey this rate law. It can be shown, furthermore, that by an appropriate transformation [5] the results obtained for rate expressions of the type given by Eq. (4.3.82) are valid for the description of systems obeying more complex rate laws, e.g.,

$$\mathscr{R}_A = \frac{k(C_A - C_x/K_{eq})}{1 + K_A C_A + \sum_i K_i C_i + \sum_j K_{1j} C_{1j}} \tag{4.3.83}$$

This rate expression takes into account a reversible chemical reaction on the solid surface and the adsorption of species that participate in the reaction as well as the adsorption of inert species.

Within a narrow reactant gas concentration range Eq. (4.3.82) may be approximated by the empirical relationship

$$\mathscr{R}_A \propto C_A^{1/n} \qquad (n > 1) \tag{4.3.84}$$

which has been used frequently to described the dependence of the rate of a gas–solid reaction on gaseous reactant concentration.

Note, however, that in the reaction of porous solids the diffusional resistance may lead to the existence of a gradient in the partial pressure of the reactant gas within the pellet. Under these conditions the apparent "reaction order" may change from a low value at the outside surface to a value approaching unity in the interior of the pellet. This behavior is the major disadvantage of Eq. (4.3.84) for the interpretation of experimental data when both diffusional and kinetic effects are important.

When the rate expression given by Eq. (4.3.82) is used instead of the first order kinetics, the conservation of the gaseous reactant shown by Eq. (4.3.20) becomes

$$\nabla^{*2}\psi - 2F_g F_p \hat{\sigma}^2 \xi^{F_\Psi - 1}(1 + \kappa)\psi/(1 + \kappa\psi) = 0 \tag{4.3.85}$$

and the mass balance on the solid grains is given as

$$\partial\xi/\partial t^* = -(1 + \kappa)\psi/(1 + \kappa\psi) \tag{4.3.86}$$

where

$$\hat{\sigma} \equiv \frac{V_p}{A_p} \sqrt{\frac{(1-\varepsilon)F_p}{2D_e} \left(\frac{k}{1+\kappa}\right) \left(\frac{A_g}{F_g V_g}\right)} \tag{4.3.87}$$

$$\kappa \equiv KC_{A0} \tag{4.3.88}$$

and

$$t^* \equiv \frac{bkC_{A0}}{\rho_s(1+\kappa)} \frac{A_g^{\prime\prime}}{F_g V_g} t \tag{4.3.89}$$

In the absence of external mass transfer resistance, the initial and boundary conditions are the same as those given by Eqs. (4.3.23)–(4.3.25).

The solution of Eqs. (4.3.85) and (4.3.86) will again yield ξ as a function of η and t^*. The overall conversion can then be calculated using Eq. (4.3.26).

The forms of the above dimensionless variables were chosen such that, as κ becomes ∞, the equations reduce directly to those for a zeroth-order reaction. When κ is zero, the governing equations reduce to those previously given for a first-order irreversible reaction.

The generalized gas–solid reaction modulus $\hat{\sigma}$, which is defined by taking into consideration the shapes of both the pellet and the individual grains, has proved quite useful in the previous discussion in presenting and generalizing the conversion versus time relationship for various geometries. The parameter $\hat{\sigma}$ also provides a convenient criterion for the assessment of the relative importance of chemical and diffusion control without regard to geometry. As will be shown subsequently, through the definition of $\hat{\sigma}$ in Eq. (4.3.87), similar considerations will apply to Langmuir–Hinshelwood kinetics.

ASYMPTOTIC BEHAVIOR

In general, Eqs. (4.3.85) and (4.3.86) must be solved numerically. However, as in the case of a first-order reaction, simple asymptotic solutions may be obtained. Thus, as $\hat{\sigma}$ approaches zero, the following relationship holds:

$$t^* = g_{F_g}(X) \tag{4.3.90}$$

This expression, which is identical to Eq. (4.3.31) previously given for first-order kinetics, holds regardless of the value of κ in the Langmuir–Hinshelwood expression.

Information on F_g and on the intrinsic reaction parameters can be readily obtained within this region by using Eq. (4.3.90) as described in Section 4.3.1.

As $\hat{\sigma}$ approaches infinity, it is readily apparent that the following relationship will hold:

$$t^*/\hat{\sigma}^2 = p_{F_p}(X) \tag{4.3.91}$$

ANALYTICAL SOLUTIONS—SPECIAL CASES

The case of $\kappa \to \infty$ (i.e., zeroth-order reaction) has an analytical solution; the reader is referred to the literature [25] for further details.

THE COMPLETE SOLUTION

In general Eqs. (4.3.85) and (4.3.86) must be solved numerically. To obtain such solutions the technique of linearization was used [25] about a trial solution and subsequent iteration, as described by Newman [60].

Figures 4.20a and 4.20b show the effect of κ on the relationship between the conversion and time for different values of $\hat{\sigma}$. The geometries considered are a slablike pellet made up of flat grains ($F_g = F_p = 1$) and a spherical pellet made up of spherical grains ($F_g = F_p = 3$). These two systems are thought to represent the two extremes of the nine possible geometries.

The effect of κ is largest for $\hat{\sigma}$ around unity, i.e., when chemical reaction and diffusion are of comparable importance. When $\hat{\sigma}$ approaches zero, Eq. (4.3.90) holds and is valid for all values of κ. On the other hand, when $\hat{\sigma}$ approaches ∞, Eq. (4.3.91) becomes the solution regardless of the reaction kinetics.

The curves for $\hat{\sigma} = 1$ show that, as far as the conversion versus time relationship is concerned, first-order kinetics ($\kappa = 0$) may be assumed for $\kappa < 0.2$ and zeroth-order kinetics ($\kappa = \infty$) may be assumed for $\kappa > 100$.

In the previous discussion of first-order kinetics a graphical procedure

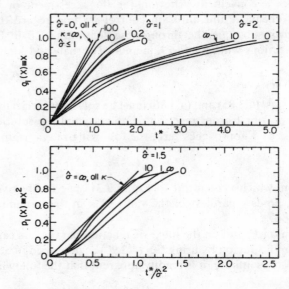

FIG. 4.20a.

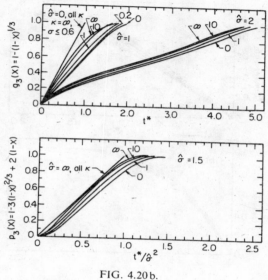

FIG. 4.20 b.

FIG. 4.20. Effect of the parameter κ on conversion versus time relationship for different values of $\hat{\sigma}$ [25] (a) $F_g = F_p = 1$, (b) $F_g = F_p = 3$. Parameter κ.

was used to arrive at the approximate relationship (4.3.73). If plots similar to Fig. 4.12 are made for various values of κ it is found that the curves for $\kappa < 1$ lie close to those for $\kappa = 0$ (first-order kinetics) for which Eq. (4.3.73) is valid. Figure 4.21 shows explicitly the comparison between Eq. (4.3.73) and the exact solutions for different values of κ. It can be said that Eq. (4.3.73) approximates the exact solution for $\kappa < 1$. The comparison is made for $\hat{\sigma} = 1$ for which the effect of κ is largest. For smaller and larger values of $\hat{\sigma}$ the agreement is better than shown in this figure. Equation (4.3.73) becomes asymptotically correct as $\hat{\sigma} \to 0$ or ∞ regardless of the value of κ.

We see that, even if the intrinsic reaction is one to which Langmuir-Hinshelwood kinetics apply, the results previously calculated assuming first-order kinetics are approximately true provided $\kappa < 1$.

4.3.7 The Effect of Structural Changes during Reaction

Perhaps the most critical assumption made in the derivation of the generalized model hitherto discussed is that structural changes do not occur during reaction.

In many gas–solid-reaction systems, particularly in the majority of oxide reduction processes a change in the structure and the porosity is inevitable. Structural changes may also occur due to the sintering of solid under the high

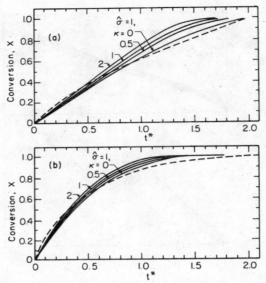

FIG. 4.21. Comparison of Eq. (4.3.73) with the exact solution for different values of κ [25]. (a) $F_g = F_p = 1$; ———, exact solution; - - -, $t^* = g_{F_g}(X) + \hat{\sigma}^2 p_{F_p}(X)$. (b) $F_g = F_p = 3$; ———, exact solution; - - -, $t^* = g_{F_g}(X) + \hat{\sigma}^2 p_{F_p}(X)$.

temperatures encountered in many gas–solid reactions. Structural changes can either increase the diffusional resistance through sintering or lead to a more open pore structure, which would increase the effective diffusivity in the pores.

A SIMPLIFIED (TWO-DIFFUSIVITY) MODEL

We shall present in this section an approximate model for gas–solid reactions with structural changes. The diffusivity will, in general, change with the degree of local conversion of the solid reactant. As a first approximation, however, let us assume that we may account for the structural changes by assigning different diffusivities to the completely reacted region on the one hand and to the unreacted or partially reacted zones on the other.

Let us denote the diffusivities of the reacted and the unreacted (or partially reacted) regions by D_e and D_e', respectively, and let us further assume that these quantities do not change in the course of the reaction: Thus, we neglect sintering of the completely reacted layer.

Intuitively, at low values of $\hat{\sigma}$ (nearly in the chemically controlled regime) D_e' is the appropriate diffusivity for describing the progress of reaction. At high values of $\hat{\sigma}$ (diffusion-controlled regime) D_e is appropriate. In the intermediate regime both diffusivities will be important.

Ishida and Wen [34] used such a two-diffusivity model for a spherical pellet with the internal structure corresponding to $F_g = 1$, and obtained an analytical solution. For first-order reactions, analytical solutions are possible, by following a similar procedure for slabs and cylinders made up of flat grains ($F_g = 1$) [26].

For the general case ($F_g \neq 1$) a numerical procedure must be employed to solve the governing equations. For a description of the results of such calculations the reader is referred to the literature [26].

In developing various mathematical models in this chapter we have assumed that the solid structure remains unchanged, or at worst, changes only due to reaction. At elevated temperatures, however, structural changes due to sintering may occur. Such changes depend not only on the temperature to which the solid is exposed but also on the duration of the exposure. As discussed in an earlier chapter, such dependence is usually very complicated and at present not very well understood. Therefore, one must attack each system separately and thus a generalization would not be very meaningful, even if possible.

Figure 4.22 is a set of scanning electron microscope pictures (taken by Turkdogan and Vinters [61]) of iron produced by the reduction of hematite

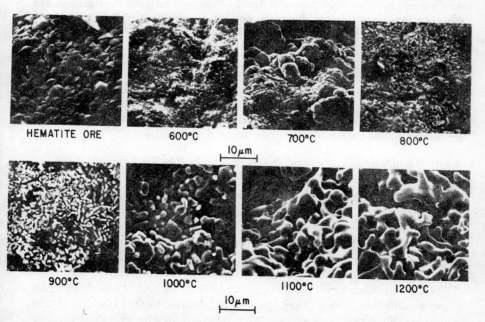

FIG. 4.22. Fracture surfaces of hematite ore and porous iron reduced in hydrogen at 600 to 1200°C as viewed in the scanning electron microscope [61].

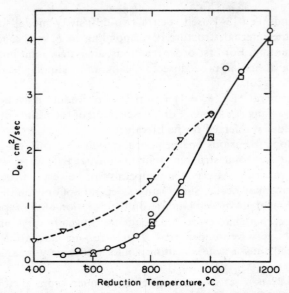

FIG. 4.23. Hydrogen–water vapor effective diffusivity derived from data for the reduction of hematite to iron compared with that obtained by direct measurements and calculated from pore structure. From the work of Turkdogan *et al.* [62]. ▽, previous work; △, present work (both direct measurements); □, calculated from pore structure;○, derived from reduction data.

by hydrogen at various temperatures. Clearly the structure of the iron depends markedly on the reduction temperature and this is reflected in the measurements of effective diffusivities carried out by Turkdogan and coworkers [62] the results of which are given in Fig. 4.23. This is illustrative of the difficulties encountered when structural changes accompany reaction.

4.3.8   The Effect of Diffusional Resistance of the Product Layer around Individual Grains

The discussion up to this point has been directed at studying the effect of the interaction between intrinsic chemical kinetics and the diffusion of matter through the pores of the solid. We have treated the systems consisting of very fine grains (or grains that although initially dense develop porosity on reaction) and hence neglected the effect of diffusional resistance in the product layer around the individual grain, as discussed in Section 4.3.1.

When this assumption is not valid, one has to include the effect of intragranular diffusion in describing the overall process. In this case the rate of regression of the reaction interface within a grain may be written, for a first-order reaction, as follows:

$$\partial \xi / \partial t^* = -\psi [1 - \hat{\sigma}_g^2 q'_{F_g}(\xi)]^{-1} \qquad (4.3.92)$$

where $\hat{\sigma}_g{}^2$ has been defined by Eq. (4.3.3) and $q'_{F_g}(\xi)$ is the derivative with respect to ξ of $q_{F_g}(\xi)$, which is defined as

$$q_{F_g}(\xi) \equiv (1 - \xi)^2 \qquad \text{for} \quad F_g = 1 \tag{4.3.93}$$

$$\equiv 1 - \xi^2 + \xi^2 \ln \xi^2 \qquad \text{for} \quad F_g = 2 \tag{4.3.94}$$

$$\equiv 1 - 3\xi^2 + 2\xi^3 \qquad \text{for} \quad F_g = 3 \tag{4.3.95}$$

The function $q_{F_g}(\xi)$ has the same significance as $p_{F_p}(X)$, that is, it describes the progress of reaction of a grain under diffusional influence. It is noted, however, that $q_{F_g}(\xi)$ is defined with ξ as the independent variable, whereas $p_{F_p}(X)$ is defined in terms of X, the overall conversion. Equation (4.3.92) can be easily verified on combining Eq. (3.3.32) with Eqs. (4.3.13), (3.2.8), and (4.3.4).

The conservation of the gaseous reactant within the pellet now gives

$$\nabla^{*2}\psi - 2F_g F_p \hat{\sigma}^2 \{\xi^{F_g - 1}/[1 - \hat{\sigma}_g{}^2 q'_{F_g}(\xi)]\}\psi = 0 \tag{4.3.96}$$

Equations (4.3.92) and (4.3.96) may be solved with the initial and boundary conditions given by Eqs. (4.3.23)–(4.3.25).

The overall conversion of the solid may then be determined by using Eq. (4.3.26). Calvelo and Smith [63] obtained numerical solution for a system with $F_g = F_p = 3$. The effect of intragrain diffusion is shown in Fig. 4.24. It is seen that as $\hat{\sigma}_g$ decreases, the solution approaches the $\hat{\sigma}_g{}^2 = 0$ asymptote.

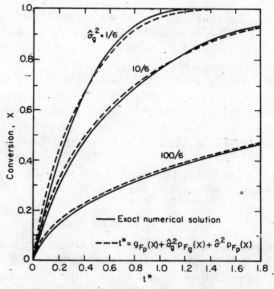

FIG. 4.24. Comparison of the approximate Eq. (4.3.97) with exact solution (Calvelo and Smith [63]) for $\hat{\sigma} = \sqrt{2}/3$ and $F_g = F_p = 3$.

From the results of Section 3.3.3, it may be expected that the $\hat{\sigma}_g{}^2 = 0$ asymptote is valid for $\hat{\sigma}_g{}^2 < 0.1$. In this regime the system becomes identical to the grain model presented earlier. On the other hand, when $\hat{\sigma}_g{}^2 > 10$ the reaction of an individual grain is expected to be controlled by the intragrain diffusion. For this condition Pigford and Sliger [29] have proposed an approximate method of solution that is valid for low conversion of the solid.

Example 4.3.3   It is assumed that the reduction of FeO with hydrogen at 900°C is a first-order reversible reaction with $K_E = 0.5$. Suppose that the grains are spherical, i.e., $F_g = 3$.

The product formed is nonporous, with the result that the reactant must be transported by solid state diffusion.

If the following data are available,

$$F_g V_g / A_g = 2 \times 10^{-4} \text{ cm}, \qquad k = 5 \times 10^{-3} \text{ cm/sec}$$

calculate the values of $\hat{\sigma}_g$ corresponding to the following solid state diffusivities:

$$\text{(i)} \quad D_g = 1.0 \times 10^{-5} \text{ cm}^2/\text{sec}$$
$$\text{(ii)} \quad D_g = 1.0 \times 10^{-7} \text{ cm}^2/\text{sec}$$

SOLUTION

(i) $\hat{\sigma}_g{}^2 = (k/2D_g)(V_g/A_g)(1 + 1/K_E)$
$= (5 \times 10^{-3}/2 \times 10^{-5})(2 \times 10^{-4}/3)3 = 5.0 \times 10^{-2} = 0.05$

(ii) $\hat{\sigma}_g{}^2 = (5 \times 10^{-3}/2 \times 10^{-7})(2 \times 10^{-4}/3)3 = 5$

In the first case one can simply neglect the intragrain diffusion ($\hat{\sigma}_g{}^2 < 0.1$). But when $D_g = 1.0 \times 10^{-7}$ cm²/sec, the intragrain diffusion should be taken care of, although it is not the controlling factor (when $\hat{\sigma}_g{}^2 > 10$, intragrain diffusion is the controlling step).

Figure 4.25 shows the effect of intrapellet diffusion by varying $\hat{\sigma}$ for a constant value of $\hat{\sigma}_g{}^2$. As $\hat{\sigma}$ approaches zero, the concentration in the interstices of the pellet becomes uniform at $\psi = 1$. Then the solution becomes identical to that for a single nonporous particle given by Eq. (3.3.33). In the other extreme case of $\hat{\sigma} \to \infty$, the solution approaches the asymptote given by Eq. (4.3.34).

Also shown in Figs. 4.24 and 4.25 is the aproximate solution analogous to Eq. (4.3.73), obtained by assuming Eq. (4.3.75) also applies to this system [66]:

$$t^* \cong g_{F_g}(X) + \hat{\sigma}_g{}^2 p_{F_g}(X) + \hat{\sigma}^2 p_{F_p}(X) \qquad (4.3.97)$$

The approximate solution is seen to give a satisfactory representation of the exact solution that must be obtained numerically for different geometries and for each combination of $\hat{\sigma}_g$ and $\hat{\sigma}$. From the discussion presented in Section 4.3.3, it is expected that Eq. (4.3.97) will also apply to systems with

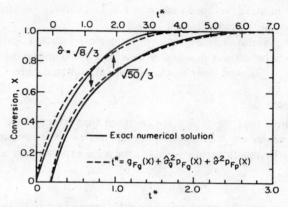

FIG. 4.25. Comparison of the approximate Eq. (4.3.97) with exact solutions (Calvelo and Smith [63]) for $\hat{\sigma}_g^2 = \frac{1}{8}$ and $F_g = F_p = 3$.

geometries other than $F_g = F_p = 3$. The approximate solution also enables one to incorporate easily the effect of external mass transfer into the solution. Thus, we get

$$t^* = g_{F_g}(X) + \hat{\sigma}_g{}^2 p_{F_g}(X) + \hat{\sigma}^2[p_{F_p}(X) + (2X/N_{Sh}^*)] \qquad (4.3.98)$$

4.3.9   Nonisothermal Systems

The nonisothermal reaction between a porous solid and a gas to give porous solid product is, in general, very complex, although a few simple asymptotic cases render themselves amenable to systematic analysis. The most important aspects of nonisothermal behavior in this group of reactions are instability and multiplicity. The upper and lower stable steady states usually correspond to the diffusion-controlled and the chemical reaction-controlled regimes, respectively, and the rate in these regimes can be described rather simply. The temperature rise in a diffusion-controlled system has been discussed in Section 3.3.6, the results of which apply directly to the reaction of a porous solid as well as that of a nonporous solid. Thus, the main problems of nonisothermal systems to be discussed here reduce to the prediction of and criteria for the multiple states and the transitional instability. The first involves determining whether multiplicity may exist and under what initial condition will the system start reacting at the upper or the lower state. The second concerns whether a transition from the lower to the upper steady state, or vice versa, may occur during the reaction of a given system, and what are the criteria for such transition.

Most of the studies on the nonisothermal effects in gas–solid reactions have been based on the shrinking unreacted-core model. Ishida et al. [64]

have applied a model developed earlier by Ishida and Wen [30] for isothermal reactions to nonisothermal gas–solid reactions.

Calvelo and Smith [63] obtained a numerical solution for the nonisothermal reaction between a gas and a spherical pellet made up of spherical grains ($F_p = F_g = 3$), and their work will be summarized here.

FORMULATION

Let us assume that the grains are so small that temperature gradients in them can be neglected. Then Eqs. (4.3.92) and (4.3.96) remain applicable except that k, the reaction rate constant must be expressed as a function of temperature:

$$k = k_0 \exp(-E/RT) \tag{4.3.99}$$

The energy balance in the pellet, assuming the pseudosteady state approximation for heat conduction, is given as

$$\ell_e \nabla^{*2} T + (-\Delta H) C_{As} D_e \{2 F_g F_p \hat\sigma^2 \xi^{F_g - 1} / [1 - \hat\sigma_g^2 q'_{F_g}(\xi)] \} \psi = 0 \tag{4.3.100}$$

On combining Eqs. (4.3.96) and (4.3.100) and integrating, we obtain

$$T - T_s = [(-\Delta H) D_e C_{As} / \ell_e](1 - \psi) \tag{4.3.101}$$

where T is the temperature at the external surface of the pellet. This relationship is identical to that obtained for reactions in porous catalysts [3, 8, 9] and for the gasification of porous solids discussed in Section 4.2. Equation (4.3.101) makes it possible to eliminate the energy balance.

Substitution of Eq. (4.3.101) in Eq. (4.3.99) gives

$$k = k_s \exp\{\gamma\beta(1 - \psi) / [1 + \beta(1 - \psi)]\} \tag{4.3.102}$$

where k_s, γ, and β have been defined previously by Eqs. (4.2.43)–(4.2.45), respectively.

When the above expression for k is substituted in $\hat\sigma$ and $\hat\sigma_g$ in Eqs. (4.3.92) and (4.3.96), the following differential equations result†:

$$\nabla^{*2}\psi - \sigma^2(T_s) \frac{\xi^{F_g - 1}\psi \exp[f(\psi)]}{1 - \hat\sigma_g^2(T_s) q'_{F_g}(\xi) \exp[f(\psi)]} = 0 \tag{4.3.103}$$

$$\frac{\partial \xi}{\partial t_s^*} = -\frac{\psi \exp[f(\psi)]}{1 - \hat\sigma_g^2(T_s) q'_{F_g}(\xi) \exp[f(\psi)]} \tag{4.3.104}$$

where

$$f(\psi) \equiv \gamma\beta(1 - \psi) / [1 + \beta(1 - \psi)] \tag{4.3.105}$$

and $\sigma(T_s)$, $\hat\sigma_g(T_s)$, and t_s^* are defined with k_s, that is, k evaluated at T_s.

---

† Since Calvelo and Smith use a parameter which corresponds to σ rather than to $\hat\sigma$, we will also use σ rather than $\hat\sigma$, although $\hat\sigma$ is a more general definition with respect to various geometries. Also note that $\sigma = 2F_g F_p \hat\sigma$.

The simultaneous nonlinear differential Eqs. (4.3.103) and (4.3.104), with the initial and boundary conditions Eqs. (4.3.23)–(4.3.25) may be solved using the technique of linearization about a trial solution and subsequent iteration [25, 60, 65].

THE EFFECT ON CONVERSION

Figure 4.26 shows the profile of ξ for a moderately exothermic reaction ($\beta = 0.2$) at different times. The most interesting feature is that complete local conversion ($\xi = 0$) can occur for grains located inside the external layer when the rest of the pellet is only partially reacted. This phenomenon occurs because the gaseous reactant concentration decreases toward the center of the pellet, while temperature increases, thus leading to a maximum local rate of reaction in an intermediate region. In general, time required to attain a certain conversion is shorter with a smaller value of $\hat{\sigma}$. For highly exothermic reactions, however, this rule may not apply, as illustrated in Fig. 4.27. It is seen that conversion for a smaller $\hat{\sigma}$ may become lower than that for a larger $\hat{\sigma}$ at the same time. This behavior may be best understood by studying the change in the rate of reaction with time, as will be discussed below.

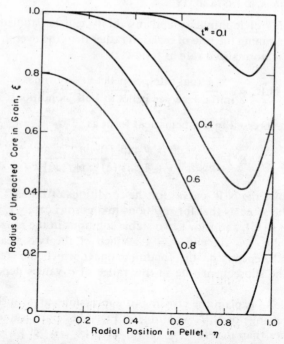

FIG. 4.26. Radius of unreacted core in grain versus radial position in pellet for $\sigma = 4$, $\hat{\sigma}_g{}^2 = \frac{1}{6}$, $\beta = 20$, $\gamma = 0.2$ [62].

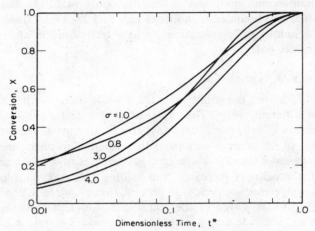

FIG. 4.27. Conversion versus time for highly exothermic reaction for large $\hat{\sigma}_g^2$, $\beta = 0.6$, $\sigma_g^2 = \frac{1}{6}$ [62].

MULTIPLICITY AND INSTABILITY

The problems of multiplicity and instability can be studied more conveniently by examining the rate of reaction rather than the overall conversion. Let us define the normalized rate as follows:

$$\text{N.R.} = \frac{\text{actual rate per pellet at any time}}{\text{initial rate per pellet at bulk conditions}} \qquad (4.3.106)$$

which may be expressed in mathematical form as

$$\text{N.R.} = F_p \int_0^1 \frac{\xi^{F_g - 1} \psi \, \exp[f(\psi)] \eta^{F_p - 1} \, d\eta}{1 + \hat{\sigma}_g^2 q_{F_g}'(\xi) \exp[f(\psi)]} \qquad (4.3.107)$$

Figure 4.28 shows the N.R. curves for the conditions of Fig. 4.27. The curve for $t_s^* = 0$ is the same as that for reactions in a porous catalyst, obtained by Weisz and Hicks [15], and shows two stable solutions in the range of $0.3 \leq \sigma \leq 0.6$ for $\beta = 0.6$, $\gamma = 20$, and $\hat{\sigma}_g = \frac{1}{6}$. Which of the two steady states is actually reached depends on the condition under which the steady state is approached. The subsequent rate in this range of σ values depends on the rate at $t_s^* = 0$.

It is possible to explain the crossing of conversion curves in Fig. 4.27 by examining the change of N.R. with time in Fig. 4.28. For instance, N.R. for $\sigma = 0.8$ is greater than that for $\sigma = 1.0$ or 3.0 at $t_s^* = 0$. At $t_s^* = 0.01$, however, N.R. for $\sigma = 0.8$ has become smaller than that for $\sigma = 1.0$, and later it

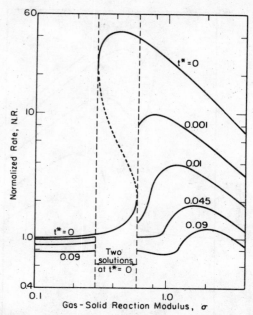

FIG. 4.28. Variation of the normalized reaction rate with time and reaction modulus for $\hat{\sigma}_g^2 = \frac{1}{4}$, $\gamma = 20$, $\beta = 0.6$ [62].

becomes smaller than that for $\sigma = 3.0$. As a result, the conversion curve for $\sigma = 0.8$ falls below the curves for $\sigma = 1.0$ and 3.0 as time progresses (Fig. 4.27).

Figure 4.29 shows a similar plot for a much smaller value of $\hat{\sigma}_g^2 \ (=1/600)$ for which the multiplicity of the solution occurs in a wider range of reaction time, and hence the transitional instability can be illustrated more readily. For $t_s^* > 0$, only curves for upper solutions are given. The two dashed lines are the loci for the transition from a high-temperature to a low-temperature state, the upper being a locus for N.R. from which the transition occurs and the lower being that for the solution after the transition. This transition appears as a discontinuous change in slope in the conversion curves as shown in Fig. 4.30, where the dashed line represents the locus of the points at which the transition occurs.

The multiplicity of solutions at finite time values may be analyzed more explicitly by applying the following treatment of Amundson and Raymond [17] to noncatalytic gas–solid reactions in a slab. Equation (4.3.100) may be written for an infinite slab as follows:

$$\lambda_e \, d^2T/d\eta^2 = -(\Delta H)C_{As}D_e\{\sigma^2\xi^{F_g-1}/[1 - \hat{\sigma}_g^2 q'_{F_g}(\xi)]\}\psi \qquad (4.3.108)$$

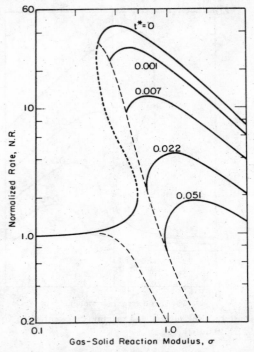

FIG. 4.29. Variation of the normalized reaction rate with time and reaction modulus for $\hat{\sigma}_g^2 = 1/600$, $\gamma = 20$, $\beta = 0.6$ [62]. – –, locus of jump from high to low solution.

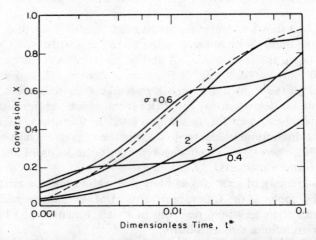

FIG. 4.30. Conversion versus time for highly exothermic reaction and small $\hat{\sigma}_g^2$, $\beta = 0.6$, $\hat{\sigma}_g^2 = 1/600$ [62]. – –, locus of jump from high to low solution.

When ψ is substituted in terms of temperature using Eq. (4.3.101) and the resultant equation is rearranged in dimensionless form, Eq. (4.3.108) becomes

$$d^2\Theta/d\eta^2 = -\Phi(\Theta, \xi) \qquad (4.3.109)$$

where

$$\Theta \equiv T/T_s \qquad (4.3.110)$$

$$\Phi(\Theta, \xi) \equiv \frac{\sigma^2(T_s)\xi^{F_g-1}[\beta - (\Theta - 1)]\exp[\gamma(1 - 1/\Theta)]}{1 - \hat\sigma_g{}^2(T_s)q'_{F_g}(\xi)\exp[\gamma(1 - 1/\Theta)]} \qquad (4.3.111)$$

Similarly, Eq. (4.3.104) may be rewritten in terms of Θ instead of ψ to give

$$\frac{\partial \xi}{\partial t_s{}^*} = -\frac{[1 - (\Theta - 1)/\beta]\exp[\gamma(1 - 1/\Theta)]}{1 - \hat\sigma_g{}^2(T_s)q'_{F_g}(\xi)\exp[\gamma(1 - 1/\Theta)]} \qquad (4.3.112)$$

Integrating Eq. (4.3.109),† and applying the boundary condition that $d\Theta/d\eta = 0$ at the center of symmetry, we obtain

$$d\Theta/d\eta = -\left[2\int_\Theta^{\Theta(0)} \Phi(\Theta, \xi)\, d\Theta\right]^{1/2} \qquad (4.3.113)$$

Further integration from the external surface to the center yields

$$\int_0^1 d\eta = 1 = \int_1^{\Theta(0)}\left[2\int_\Theta^{\Theta(0)} \Phi(\Theta, \xi)\, d\Theta\right]^{-1/2} d\Theta = S[\Theta(0), \xi] \quad (4.3.114)$$

With the aid of Eq. (4.3.112), the integral in Eq. (4.3.114) can be evaluated at any $t_s{}^*$. Figure 4.31 shows this integral evaluated for the values of parameters used in Figs. 4.29 and 4.30 along with $\sigma = 0.16$. The solution of Eq. (4.3.114) is given by the points where the curves intersect the horizontal line of $S = 1$. The curve for $t_s{}^* = 0$ corresponds to the solution for a catalytic reaction in a porous pellet. It is seen that in the early stage of reaction there are three solutions. Therefore, if the upper state is achieved initially, the system will jump to the lower state at some time $0 < t_s{}^* < 0.002$. Note that at $t_s{}^* = 0$, $\sigma = 0.16$ for an infinite slab would correspond approximately to $\sigma = 3 \times 0.16$ for a sphere if temperatures were uniform. This is within the range of multiple solutions in Fig. 4.29 although the analogy may be crude.

Thus we have seen that the reaction between a porous solid and a gas can undergo a sharp transition while the ambient conditions remain unchanged. Another important type of instability is that under the changing bulk conditions, as discussed in Chapter 3. Such an instability would have a significant effect on the operation of a packed bed in which exothermic gas–solid reactions occur.

† Substitute $p = d\Theta/d\eta$; then $d^2\Theta/d\eta^2 = p\, dp/d\Theta$. Solve for p with Θ as the independent variable.

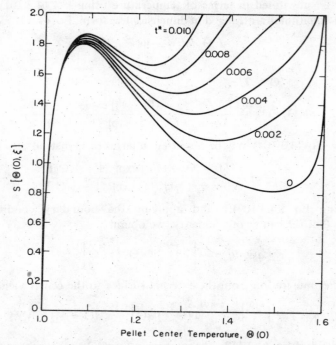

FIG. 4.31. $S[\Theta(0), \xi]$ as a function of center temperature for different times. Slablike geometry $\sigma = 0.16$, $\sigma_g^2 = 1/600$, $\gamma = 20$, $\beta = 0.6$ [62].

4.4   Concluding Remarks

In this section we developed the *grain model* for representing the reaction between porous solids and a gas. We note that this grain model is one of the distributed models that have become available in recent years. The principal feature of all these models is that they allow the reaction to occur over a diffuse or distributed zone within the porous pellet.

The grain model is thought to be attractive because the structural parameters appearing in it may be related to microscopically or macroscopically observable quantities, even though this may involve some idealization.

A further important feature of the grain model is that it allows one to define quantitatively the asymptotic regimes of chemical reaction control and diffusion control and thus enables the rational planning of experimental programs. A full discussion of this problem is presented in Chapter 6.

Perhaps the major objection that might be raised to the grain model, as it has been presented here, is that the envisioned solid structure and topology of the reaction process are unrealistically idealized. In fact, an alternative approach may be used that avoids these idealizations. Using, for the moment, the approach described above we see that the *local extent of reaction* within a small region of the porous pellet is given by

$$\chi = 1 - \xi^{F_g} = 1 - C_B/C_{BI} \qquad (4.3.115)$$

where C_B is the number of moles of solid reactant per unit volume of porous solid and C_{BI} the initial number of moles of solid reactant per unit volume of porous solid. The local rate of reaction (in moles per unit volume of porous solid) is

$$v_A = k(1 - \varepsilon) \frac{A_g}{V_g} \left(C_A - \frac{C_C}{K_E} \right) \xi^{F_g - 1} \qquad (4.3.116)$$

Consequently the local rate of reaction is related to the local extent of reaction by

$$v_A = kS_v(C_A - C_C/K_E)(1 - \chi)^m = kS_v(C_A - C_C/K_E)(C_B/C_{BI})^m \quad (4.3.117)$$

where S_v is the surface area per unit volume of porous solid (a quantity readily measured, e.g., by the BET method) and

$$m = (F_g - 1)/F_g \qquad (4.3.118)$$

It is possible to arrive at all the results of the grain model by postulating a relationship between the local rate of reaction and the local extent of reaction (or solid reactant "concentration") of the form of Eq. (4.3.117). This is then substituted in Eqs. (4.3.11) and (4.3.12) and Eq. (4.3.13) is replaced by

$$\partial C_B/\partial t = -bv_A = -bkS_v(C_A - C_C/K_E)(1 - \chi)^m$$
$$= -bkS_v(C_A - C_C/K_E)(C_B/C_{BI})^m \qquad (4.3.119)$$

with initial condition $C_B = C_{BI}$ at $t = 0$. The solution of these equations is then obtained as before with dimensionless variables:

$$\xi = (C_B/C_{BI})^{1-m} \qquad (4.3.120)$$

$$t^* = \frac{bkS_v(1 - m)}{C_{BI}} (C_{A0} - C_{C0}/K_E)t \qquad (4.3.121)$$

$$\sigma = \frac{V_p}{A_p} \sqrt{\frac{kF_p(1 - m)S_v}{2D_e} \left(1 + \frac{1}{K_E} \right)} \qquad (4.3.122)$$

Equations (4.3.20) and (4.3.21) together with all subsequent equations in the development of the grain model are then unchanged and the same results as before are generated.

This alternative approach is the one used by Ishida and Wen [34], Ausman and Watson [33], Lahiri and Seshadri [44], and Tien and Turkdogan [31], all of whom took the value of m to be zero, i.e., the reaction was zeroth order with respect to solid "concentration." As mentioned above, this is equivalent to the grain model with $F_g = 1/(1 - m)$ equal to unity, i.e., slablike grains. The grain model with long cylindrical grains would be equivalent to a model developed from the postulate of Eq. (4.3.117) with $m = \frac{1}{2}$, while the grain model with spherical grains is equivalent to a model developed from Eq. (4.3.117) with $m = \frac{2}{3}$.

The construction of a structural model by the postulation of Eq. (4.3.117) has the advantage of avoiding the assumption of an idealized solid structure. The structural parameters are now S_v and ε rather than A_g/V_g and ε, and the values of m ($0 < m < 1$) would have to be determined experimentally.

It is thus seen that these two approaches can be shown to lead to identical relationships; in the body of the text we used the grain model for the development of the governing equations because it was thought that in this way the connection between the equations and the physical system could be more readily grasped.

In reality, the dependence of the local rate of reaction on the local extent of reaction of the solid may be far more complex than the simple power law relationship of Eq. (4.3.117). In many cases the rate is low during the initial stages of reaction (mention of these "induction periods" is found elsewhere in the text), reaches a maximum after reaction has proceeded to some extent, and declines thereafter [67]. Equation (4.3.117) describes a reaction in which the rate declines monotonically as reaction proceeds for constant gas concentration and $m < 1$.

Bartlett et al. [68] have considered a grain model in which there is a distribution of grain sizes within the pellet. For certain grain size distributions the behavior becomes equivalent to that which would be predicted from Eq. (4.3.117) with $m = 1$. In other words the local rate of reaction is first order with respect to solid "concentration."

Some comment is appropriate on the extent to which diffuse interface models have been tested against experiment. The grain model was found to be consistent with experimental measurements of the rate of reduction of nickel oxide by hydrogen below the temperature at which serious structural changes occurred due to sintering [24, 36, 37]. Figure 4.32 is taken from the work of Tien and Turkdogan [31] and shows reasonable agreement between the predictions of their diffuse interface model (equivalent to a grain model with $F_g = 1$) and experimental results on the reduction of hematite by hydrogen. Ishida et al. [64] have been able to show reasonable agreement between their nonisothermal grain model and experimental measurements on the combustion of carbon particles in a porous ceramic matrix. Papanastassiou and Bitsianes [69] explained the results of their experiments on the oxidation of porous magnetite pellets in terms of a grain model, although no quantitative comparison was attempted.

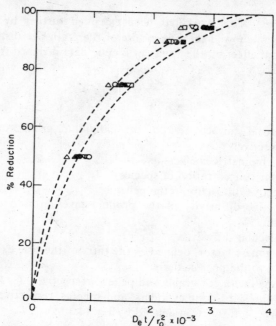

FIG. 4.32. Plot of the extent of reaction against dimensionless time showing a comparison between measurements and predictions based on a diffuse model for the reduction of hematite with hydrogen [31]. Temperature (°C): ▲, 600; ●, 700; ■, 800; ▽, 900; △, 1000; □, 1100; ○, 1200. Calculated curves, – – –.

Finally, it should be stressed that in many instances the classical shrinking-core model described in Chapter 3 is applicable to the reaction of porous pellets. This would be the case at elevated temperatures where diffusion through a porous product layer is the rate-controlling step. A large number of studies, in which experimental data on the reaction of a porous pellet have been interpreted in terms of a diffusion-controlled shrinking-core model, are therefore sound. Under these circumstances the measurement of effective diffusivities suffices to enable the prediction of the rate of progress of reaction and measurement of the structural parameter A_g/V_g (or S_v) is unnecessary. It should be appreciated, however, that any attempt to measure intrinsic chemical rate constants by studying the reaction of a porous pellet and interpreting the results in terms of a shrinking-core model is doomed to failure since the model is valid for porous pellets only under conditions of diffusion control. The major advantages of the grain model are that it allows interpretation and correlation of experimental data for porous pellets under conditions of chemical or mixed control, and enables the prediction of overall reaction rates for various solid structures (different porosities, specific surface area, grain shape, etc.). This work has been extended further by Szekely and Propster (70) who showed that the nature of the grain size distribution may have a significant effect on the apparent kinetic data deduced from measurements.

Notation

A_g, A_p	external surface area of an individual grain and the pellet, respectively
b, c, d	stoichiometric coefficients
C_i	molar concentration of species i
D_e	effective diffusivity in the pellet
D_g	effective diffusivity in the product layer around an individual grain
D_M	molecular diffusivity
f	roughness factor, defined as the ratio of the true external surface area to the projected area
F_g, F_p	shape factor for grains and pellets, respectively ($=1$, 2, and 3 for infinite slabs, long cylinders, and spheres, respectively)
$g_{F_g}(X)$	conversion function defined by Eq. (4.3.5)
G	parameter defined by Eq. (4.2.10)
h_D	mass transfer coefficient
$(-\Delta H)$	heat of reaction
k	reaction rate constant
K	parameter in Langmuir–Hinshelwood expression, Eq. (4.2.26)
K_E	equilibrium constant
k_e	effective thermal conductivity of the pellet
L	parameter defined by Eq. (4.2.4)
m, n	reaction orders
N_{Sh}, N_{Sh}^*	Sherwood and modified Sherwood numbers, respectively, defined in Eq. (4.3.80)
$p(X)$	conversion function defined by Eqs. (4.3.8)–(4.3.10)
$q(\xi)$	conversion function defined by Eqs. (4.3.93)–(4.3.95)
r	radius of the pore
r_c	position of moving reaction front in the individual grain
R	distance from the center of symmetry in the pellet
$\mathcal{R}_s$	rate of reaction per unit surface area
S_v	surface area per unit volume of the pellet
t	time
t^*	dimensionless time defined by Eq. (4.3.3) or (4.3.89)
T	temperature
v	reaction rate per unit volume of the pellet

V_g, V_p	volume of grain and pellet, respectively
x	distance coordinate
X	fractional conversion of the solid
β	dimensionless quantity defined by Eq. (4.2.45)
γ	dimensionless group defined by Eq. (4.2.44)
δ	dimensionless quantity defined by Eq. (4.2.49)
ε	initial porosity of the pellet plus fraction of volume occupied by inert solids, if any
ζ	dimensionless distance defined by Eq. (4.2.9)
η	dimensionless distance defined by Eq. (4.3.22)
θ	dimensionless quantity defined by Eq. (4.2.30)
Θ	dimensionless temperature defined by Eq. (4.3.110)
κ	dimensionless quantity defined by Eq. (4.3.88).
ξ	dimensionless position of reaction interface in the grain, defined by Eq. (4.3.17)
ρ_s	molar concentration of solid reactant
σ	dimensionless parameter defined by Eq. (4.3.19)
$\hat{\sigma}$	generalized gas–solid reaction modulus defined by Eq. (4.3.33)
$\hat{\sigma}_g^2$	shrinking-core reaction modulus for the grain defined by Eq. (4.3.3)
τ_c	dimensionless time defined by Eq. (4.2.16)
Φ	dimensionless function defined by Eq. (4.3.111)
ψ	dimensionless concentration defined by Eq. (4.3.16)
∇^{*2}	Laplacian operator with η as the position coordinate
χ	local extent of reaction defined by Eq. (4.3.115)

Subscripts

A, B	gas A and solid B
b	bulk property
m	value at the boundary between the partially and completely reacted zones
0	original value or bulk property
s	property of solid reactant, or value at the external surface

References

1. E. Wicke, *Symp. Combust.*, *5th* p. 245. Van Nostrand-Reinhold, Princeton, New Jersey, 1955.
2. P. L. Walker, Jr., F. Rusinko, Jr., and L. G. Austin, *Advan. Catal.* **11**, 133 (1959).
3. E. E. Petersen, "Chemical Reaction Analysis," Chapter 4. Prentice-Hall, Englewood Cliffs, New Jersey, 1965.
4. E. E. Petersen, *AIChE J.* **3**, 443 (1957).

5. P. Schneider and P. Mitschka, *Chem. Eng. Sci.* **21**, 455 (1966).
6. V. W. Weekman, Jr. and R. L. Gorring, *J. Catal.* **4**, 260 (1965).
7. E. W. Thiele, *Ind. Eng. Chem.* **31**, 916 (1939).
8. G. Damköhler, *Z. Phys. Chem.* **A193**, 16 (1943).
9. C. D. Prater, *Chem. Eng. Sci.* **8**, 284 (1958).
10. J. Beek, *AIChE J.* **7**, 337 (1961).
11. R. E. Schilson and N. R. Amundson, *Chem. Eng. Sci.* **13**, 226, 237 (1961).
12. J. D. Tinkler and R. L. Pigford, *Chem. Eng. Sci.* **15**, 326 (1961).
13. J. J. Carberry, *AIChE J.* **7**, 350 (1961).
14. J. D. Tinkler and A. B. Metzner, *Ind. Eng. Chem.* **53**, 663 (1961).
15. P. B. Weisz and J. S. Hicks, *Chem. Eng. Sci.* **17**, 265 (1962).
16. E. E. Petersen, *Chem. Eng. Sci.* **17**, 987 (1962).
17. N. R. Amundson and L. R. Raymond, *AIChE J.* **11**, 339 (1965).
18. G. R. Gavalas, *Chem. Eng. Sci.* **21**, 477 (1966).
19. J. C. W. Kuo and N. R. Amundson, *Chem. Eng. Sci.* **22**, 1185 (1967).
20. D. Luss and N. R. Amundson, *Chem. Eng. Sci.* **22**, 253 (1967).
21. J. Wei, *Chem. Eng. Sci.* **20**, 729 (1965).
22. V. W. Weekman, Jr., *J. Catal.* **5**, 44 (1966).
23. H. Y. Sohn and J. Szekely, *Chem. Eng. Sci.* **27**, 763 (1972).
24. J. Szekely, C. I. Lin, and H. Y. Sohn, *Chem. Eng. Sci.* **28**, 1975 (1973).
25. H. Y. Sohn and J. Szekely, *Chem. Eng. Sci.* **28**, 1169 (1973).
26. J. Szekely and H. Y. Sohn, *Inst. Mining Met. Trans. Sect. C* **82**, C 92 (1973).
27. J. Szekely and J. W. Evans, *Chem. Eng. Sci.* **25**, 1091 (1970).
28. J. Szekely and J. W. Evans, *Chem. Eng. Sci.* **26**, 1901 (1971).
29. R. L. Pigford and G. Sliger, *Ind. Eng. Chem. Process Des. Develop.* **12**, 85 (1973)
30. M. Ishida and C. Y. Wen, *Chem. Eng. Sci.* **26**, 1031 (1971).
31. R. H. Tien and E. T. Turkdogan, *Met. Trans.* **3**, 2039 (1972).
32. D. Papanastassiou and G. Bitsianes, *Met. Trans.* **4**, 477 (1973).
33. J. M. Ausman and C. C. Watson, *Chem. Eng. Sci.* **17**, 323 (1962).
34. M. Ishida and C. Y. Wen, *AIChE J.* **14**, 311 (1968).
35. J. Szekely and J. W. Evans, *Met. Trans.* **2**, 1691 (1971).
36. J. Szekely and J. W. Evans, *Met. Trans.* **2**, 1699 (1971).
37. J. W. Evans, C. Leon-Sucre, and S. Song, *Met. Trans.* **7B**, 55 (1976).
38. E. Kawasaki, J. Sanscrainte, and T. J. Walsh, *AIChE J.* **8**, 48 (1962).
39. P. B. Weisz and R. D. Goodwin, *J. Catal.* **2**, 397 (1963).
40. K. B. Bischoff, *Chem. Eng. Sci.* **18**, 711 (1963); **20**, 783 (1965).
41. D. Luss, *Can. J. Chem. Eng.* **46**, 154 (1968).
42. K. Hashimoto and P. L. Silveston, *AIChE J.* **19**, 268 (1973).
43. H. O. Lien, A. E. El-Mehiry, and H. U. Ross, *J. Iron Steel Inst.* **209**, 541 (1971).
44. A. K. Lahiri and V. Seshadri, *J. Iron Steel Inst.* **206**, 1118 (1968).
45. J. W. Evans, Ph.D. Dissertation, SUNY at Buffalo (1970).
46. W. M. McKewan, *Trans. Met. Soc. AIME* **212**, 791 (1958).
47. N. J. Themelis and W. H. Gauvin, *Trans. Met. Soc. AIME* **227**, 290 (1963).
48. F. Habashi, "Extractive Metallurgy," Vol. 1, General Principles, Chapter 8. Gordon and Breach, New York, 1969.
49. A. J. Wilhelm and F. R. St. Pierre, *Trans. Met. Soc. AIME* **221**, 1267 (1961).
50. J. M. Quets, M. E. Wadsworth, and J. R. Lewis, *Trans. Met. Soc. AIME* **218**, 545 (1960).
51. J. P. Hansen, T. N. Rushton, and S. E. Khalafalla, U.S. Bur. Mines Rep. Invest. 6712 (1966).

52. J. M. Quets, M. E. Wadsworth, and J. R. Lewis, *Trans. Met. Soc. AIME* **221**, 1186 (1961).
53. W. M. McKewan, *Trans. Met. Soc. AIME* **224**, 387 (1962).
54. T. Kurosawa, R. Hasegawa, and T. Yagihashi, *Nippon Kinzoku Gakkaishi* **34**, 481 (1970).
55. C. Moreau and J. Philippot, *Proc. Int. Symp. Reactivity Solids, 6th* p. 377 (1969).
56. A. Ya. Rozovski and V. V. Shchekin, *Kinet. Katal.* **1**, 313 (1960).
57. C. Chu and O. A. Hougen, *Chem. Eng. Sci.* **17**, 167 (1962).
58. G. W. Roberts and C. N. Satterfield, *Ind. Eng. Chem. Fundamentals* **4**, 288 (1965).
59. G. W. Roberts and C. N. Satterfield, *Ind. Eng. Chem. Fundamentals* **5**, 317 (1966).
60. J. Newman, *Ind. Eng. Chem. Fundamentals* **7**, 514 (1968).
61. E. T. Turkdogan and J. V. Vinters, *Met. Trans.* **2**, 3175 (1971).
62. E. T. Turkdogan, R. G. Olsson, and J. V. Vinters, *Met. Trans.* **2**, 3189 (1971).
63. A. Calvelo and J. M. Smith, *Proc. Chemeca* (1970).
64. M. Ishida, C. Y. Wen, and T. Shirai, *Chem. Eng. Sci.* **26**, 1043 (1971).
65. E. S. Lee, "Quasilinearization and Invariant Imbedding." Academic Press, New York, 1968.
66. H. Y. Sohn and J. Szekely, *Chem. Eng. Sci.* **29**, 630 (1974).
67. H. Charcosset, R. Frety, Y. Trambouze, and M. Prettre, *Reactivity Solids, Proc. Int. Symp., 6th, 1968* (J. W. Mitchell, ed.). Wiley, New York, 1969.
68. R. W. Bartlett, N. G. Krishnan, and M. C. Van Hecke, *Chem. Eng. Sci.* **28**, 2179 (1973).
69. O. Papanastassiou and G. Bitsianes, *Met. Trans.* **4**, 487 (1973).
70. J. Szekely and M. Propster, *Chem. Eng. Sci.* **30**, 1049 (1975).

Chapter 5 | **Reactions between Solids Proceeding through Gaseous Intermediates**

5.1 Introduction

In Chapters 3 and 4 we developed a general statement of the problem involved in the reaction between a porous or a nonporous solid and a reactant gas. Through the models proposed it was possible to relate the structural parameters of the solid (grain size, shape, porosity, tortuosity, total surface area, etc.) to its reaction characteristics. The treatment presented in these earlier chapters not only should be attractive for the study of gas–solid reactions, but also could form the basis for tackling more complex reaction systems, of which gas–solid reactions form a component. The discussion of one group of these problems, namely solid–solid reactions proceeding through gaseous intermediates, is the subject of this chapter.

Reactions between solids are of considerable importance in the technology of materials processing. For the purpose of classification these reaction systems may be divided into the following two main groups:

(a)   True solid–solid reactions, which take place in the solid state between two species in contact with each other, or through the migration of species in the solid state; and

(b)   Reactions between solid reactants, which take place through gaseous intermediates.

Perhaps the most important example of true solid–solid reactions is the formation of *metal carbides* through the reaction between metal oxides and carbon:

$$SiO_2 + 3C \rightleftharpoons SiC + 2CO \tag{5.1.1}$$

$$TiO_2 + 3C \rightleftharpoons TiC + 2CO \tag{5.1.2}$$

$$2CaO + 3C \rightleftharpoons Ca_2C + 2CO \tag{5.1.3}$$

The reduction of metal oxides with solid carbon could also be a true solid–solid reaction, provided it is carried out at very low absolute pressures:

$$MeO_n + nC \rightleftharpoons Me + nCO \qquad (5.1.4)$$

It will be shown subsequently that metal oxide reduction with carbon, when carried out at atmospheric pressure, is usually not a true solid–solid reaction but proceeds through gaseous intermediates.

Under these conditions the reaction scheme described in Eq. (5.1.4) must be written,

$$MeO_n + nCO \rightleftharpoons Me + nCO_2 \qquad (5.1.5)$$
$$CO_2 + C \rightleftharpoons 2CO \qquad (5.1.6)$$

where it is seen that the overall reaction involves two coupled gas–solid reaction systems, namely, metal oxide reduction with carbon monoxide and the reaction of CO_2 with solid carbon.

The reduction of iron oxide, lead oxide, nickel oxide, and copper oxide with solid carbon would constitute typical practical examples for the scheme shown in Eqs. (5.1.5) and (5.1.6); of these iron oxide reduction is the only one of current industrial significance.

Other examples of solid–solid reactions proceeding through gaseous intermediates are *segregation roasting* and the *lime pellet concentrate roasting process.*

In segregation roasting a mixed metal oxide is made to react with carbon and an alkaline earth chloride in the presence of water vapor:

$$(FeO)_m NiO + CaCl_2 = (FeO)_m + NiCl_2 \uparrow + CaO$$
$$H_2O + C + NiCl_2 = Ni + 2HCl + CO \qquad (5.1.7)$$
$$(FeO)_m NiO + 2HCl = (FeO)_m + NiCl_2 \uparrow + H_2O$$

The nickel formed on the surface of the carbon particles is then separated by standard mineral-processing techniques.

In the lime concentrate pellet roasting process a sulfide ore is mixed with burned lime, before being made to react with oxygen to yield

$$CuFeS_2 + \tfrac{13}{2}O_2 \longrightarrow CuO + \tfrac{1}{2}Fe_2O_3 + 2SO_2 \qquad (5.1.8)$$
$$2CaO + 2SO_2 + O_2 \longrightarrow 2CaSO_4 \qquad (5.1.9)$$

The reaction scheme designated by (5.1.8) and (5.1.9) involves gas–solid reactions in series, very much like the metal oxide reduction represented by Eqs. (5.1.5) and (5.1.6).

In keeping with the scope of this book, in the following discussion we shall restrict our attention to solid–solid reactions that proceed through intermediates, viz., group (b).

Even within this restriction the treatment will have to be much less comprehensive than that developed in the preceeding chapters because the complex reaction systems have been much less extensively studied; moreover, the theoretical basis available at present for the interpretation of measurements is also much less advanced. Nonetheless, it was thought worthwhile to include at least a cursory treatment of these systems because much more work is warranted in this area. Regarding the organization of this chapter, Section 5.2 will be devoted to the discussion of metal oxide reduction with carbon, with emphasis on the earlier theoretical treatment of these systems; a description of some key experimental studies will also be given here. A brief treatment of roasting operations, including segregation roasting, will be given in Section 5.3. A preliminary theoretical treatment of solid–solid reactions will be proposed in Section 5.4. Finally, concluding remarks will be contained in Section 5.5.

5.2   The Reduction of Metal Oxides with Carbon

As noted earlier, when carried out at atmospheric pressures, the "direct" reduction of metal oxides with solid carbon,

$$MeO_n + nC \rightleftharpoons Me + nCO_2 \qquad (5.2.1)$$

usually takes place through gaseous intermediates and may be written

$$MeO_n + nCO \rightleftharpoons Me + nCO_2 \qquad (5.2.2)$$
$$CO_2 + C \rightleftharpoons 2CO \qquad (5.2.3)$$

Thus the overall reaction involves two gas–solid systems.

The actual stoichiometry is not predetermined and will depend on the relative rates of the reactions designated by Eqs. (5.2.2) and (5.2.3). Numerous investigators have studied such "direct reduction" of metal oxides with carbon. The usual experimental arrangement consisted of suspending either pellets made up of carbon (graphite) and the metal oxides, or mixed powder contained in a crucible, from one arm of a recording balance in a furnace; the progress of the reaction was then followed by monitoring the weight change. Other investigators reacted powders contained in a fixed crucible and monitored the progress of the reaction through continuous gas analysis.

Until recently the majority of the investigators simply reported the kinetic data (e.g., as reaction rates at various temperatures and for different reaction mixtures) without attempting to fit these to any particular model.

Such reaction rates were reported by Baldwin [1], who measured the reduction of solid iron oxide with coke, in an attempt to elucidate the mechanism of "direct reduc ion" in the blast furnace. The reactions between ferric oxide and graphite powders were also studied by Otsuka and Kunii [2], who found that the reaction rates were very sensitive to both the particle size and to the relative proportion of the solid reactants.

The reduction of nickel oxide with various forms of carbon was also studied by Sharma *et al.* [3], who paid particular attention to the role played by the reactivity of the carbon employed.

Kohl and Marincek [4, 5] studied the reaction of iron oxide, nickel oxide, and cobalt oxide with graphite over the temperature range 920–1200°C but used a rather simple rate expression for the interpretation of their data.

Much of this earlier work is largely of historical interest because proper kinetic expressions have not been proposed for representing the reaction rate nor have sufficient data been provided to develop such an interpretation *a posteriori*. In more recent studies of solid–solid reactions, notably by Rao [6, 7], Kondo *et al.* [8, 9], and El-Guindy and Davenport [10], the relationship between solid–solid reactions and the previously discussed gas–solid systems has been recognized and therefore the interpretation of these data may be more readily made. A somewhat more detailed discussion of these results will be made after a brief review of the earlier theoretical work on solid–solid reaction systems.

In contrast to the very extensive modeling studies that have been performed on gas–solid reaction systems, very little theoretical work has been done on solid–solid reactions; as a result no "general" rate expression has been proposed and no clearcut methods have evolved for the interpretation of experimental measurements.

In the early theoretical work use was made of the fact that for special conditions the overall rate of reaction as written by Eq. (5.2.1) would be limited by the rate of only one of the gas–solid reactions, viz., (5.2.2) *or* (5.2.3), that make up the overall sequence. It was thought that for such conditions the system could be represented in terms of a single rate expression, written for the reaction of the solid, which constitutes the rate-controlling step.

In studying the reduction of metal oxides by solid carbon, Jander, [11], Gnistling and Brounshtein [12], and Carter [13] postulated that the reaction of CO_2 with carbon [viz., Eq. (5.2.3)] was the rate-controlling step; the following expressions were proposed by these authors for relating the fractional conversion of the solid reactant X to time t:

$$[1-(1-X)^{1/3}]^2 = \tilde{k}t/r^2 \qquad \text{(Jander)} \qquad (5.2.4)$$

$$1-\tfrac{2}{3}X-(1-X)^{2/3} = \tilde{k}t/r^2 \qquad \text{(Gnistling } et\,al.) \qquad (5.2.5)$$

and

$$\frac{Z-[1+(Z-1)X]^{2/3}-(Z-1)-(1-X)^{2/3}}{2(Z-1)} = \frac{\tilde{k}t}{r^2} \qquad \text{(Carter)} \quad (5.2.6)$$

where r is the initial radius of the oxide particle, X is the fraction of solid reacted, and Z is the ratio (final volume of product)/(initial volume of reactant). The analogy between Eqs. (5.2.4) and (5.2.6) and the corresponding expressions developed in Chapter 3 should be readily apparent. Several investigators have attempted to verify Jander's equation using the Fe_2O_3–C system, but these studies were essentially inconclusive. A good critical review of this earlier work is available in a recent paper by Rao [6], who himself has aone extensive experiments on the "direct" reduction of hematite with graphite. Rao has found that the complexity of the hematite–graphite system (in particular the multistep nature of hematite reduction to iron) precluded the representation of his experimental results by the relatively simple expressions given by Eqs. (5.2.4) and (5.2.6). Nonetheless he found that in the special case where Li_2O catalyst was added, the results conformed to the plot according to chemical reaction control as illustrated in Fig. 5.1. In a

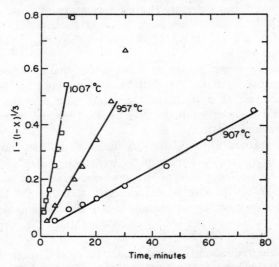

FIG. 5.1. Experimental data on the direct reduction of hematite by graphite in the presence of a lithium oxide catalyst [6].

subsequent paper, Rao [7] developed a more sophisticated interpretation of his earlier published experimental data by making allowance for the complex equilibrium relationship between Fe_2O_3, Fe_3O_4, "FeO" and the $CO + CO_2$, gas mixture. This equilibrium relationship is shown in Fig. 5.2. Rao has considered, furthermore, that the overall rate of reaction is limited by the reaction between CO_2 and carbon, viz.,

$$CO_2 + C \rightleftharpoons 2CO \qquad (5.2.3a)$$

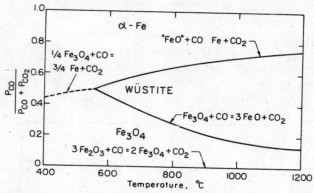

FIG. 5.2. Equilibrium diagram for iron oxides in the presence of carbon monoxide–carbon dioxide mixtures.

and thus the gas mixture within the reacting porous solid was assumed to be in equilibrium with the iron and the iron oxides present.

By allowing for Langmuir–Hinshelwood kinetics for the rate of reaction (5.2.3) and for the diffusive exchange of CO and CO_2 between the porous pellet and the environment, Rao developed a set of modeling equations, which had to be solved numerically.

As seen in Figs. 5.3 and 5.4 there was quite good agreement between Rao's earlier measurements and his subsequent interpretation. El-Guindy and Davenport [10] studied the reaction between ilmenite and solid carbon to obtain metallic iron and titanium dioxide, viz.,

$$FeTiO_3 + CO \rightleftharpoons Fe + TiO_2 + CO_2 \qquad (5.2.7)$$
$$C + CO_2 \rightleftharpoons 2CO \qquad (5.2.3a)$$

In interpreting their results these authors also suggested that reaction (5.2.3a) constituted the rate-controlling step and were able to represent their results adequately by using either Jander's or Gnistling's equation [viz., Eq. (5.2.4) or (5.2.5)].

Finally, in a recent paper Maru and coworkers [9] reported on the reaction between chromiun oxide and chromium carbide, which forms the basis of the Simplex process [14].

The authors proposed that the reaction proceeds according to

$$\tfrac{1}{6}Cr_{23}C_6 + CO_2 \rightleftharpoons \tfrac{23}{6}Cr + 2CO \qquad (5.2.8)$$
$$\tfrac{1}{3}Cr_2O_3 + CO \rightleftharpoons \tfrac{2}{3}Cr + CO_2 \qquad (5.2.9)$$

The authors suggested that the overall rate of the reaction was controlled by the reaction between the CO_2 and the carbon contained in the chromium carbide and, as shown in Fig. 5.5, were indeed able to obtain a more or less adequate representation of their experimental results in terms of chemical reaction control, at least at low conversions.

It should be stressed here that the work described in Rao [6, 7], Maru *et al.* [9], and El-Guindy and Davenport [10] represents a significant advance in the understanding of solid–solid reactions proceeding through gaseous intermediates, because these authors were able to develop a quantitative interpretation for their measurements.

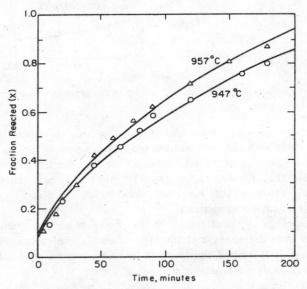

FIG. 5.3. Rao's experimental data and theoretical predictions on the reduction of iron oxide by graphite [7]. ———, theoretical; $\bigcirc$, $\triangle$, experimental.

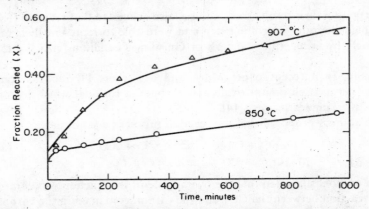

FIG. 5.4. Rao's experimental data and theoretical predictions on the reduction of iron oxide by graphite [7]. ———, theoretical; $\bigcirc$, $\triangle$, experimental.

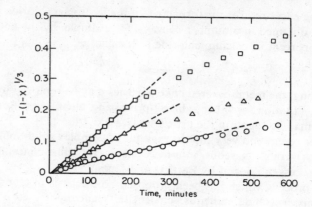

FIG. 5.5.   Conversion function $1 - (1 - X)^{1/3}$ against time for the reaction of chromium oxide with chromium carbide [9]. $\square$, 1100°; $\triangle$, 1075°C; $\bigcirc$, 1050°C.

The equations used for the purpose of interpretation are analogous to those developed in Chapters 3 and 4 for the reaction of a single reactant gas with a single solid. In all this previously cited work the authors assumed that the overall reaction rate was controlled by the reaction of CO_2 with carbon; the proof for the appropriateness of this assumption was indirect, through the fact that the experimental results could be interpreted by the use of Jander's or Gnistling's equation. If we recall the discussion in Chapters 3 and 4 regarding the apparent insensitivity of measurements taken over a limited range of solid conversions to the model used for their interpretation, we would wish for more conclusive proof for the mechanisms suggested. Clearly a great deal of further experimental and analytical work needs to be done in this area.

5.3   Roasting Operations

Some of the more complex roasting operations, viz., *segregation roasting* and *lime pellet* roasting may also be classified as solid–solid reactions proceeding through gaseous intermediates. Very little fundamental kinetic work has been done on these systems so far, and this appears to be a very fruitful area for further investigation.

In segregation roasting the objective is to convert nonferrous metals, e.g., lateritic nickel ores, to volatile metal chlorides, which are then reduced on the surface of solid reductants (such as coke) that are added to the material charged to the roaster [15, 16]. The overall reaction scheme is

$$MeO + 2HCl \longrightarrow MeCl_2 + H_2O \tag{5.3.1}$$
$$H_2O + C + MeCl_2 \longrightarrow Me + 2HCl + CO \tag{5.3.2}$$

The chlorine necessary for the formation of the hydrogen chloride intermediate may be obtained in a number of ways, for example, by the addition of a few weight percent of calcium chloride to the charge

$$CaCl_2 + MeO \longrightarrow MeCl_2 + CaO \qquad (5.3.3)$$

After roasting, the metal-covered coke particles may be removed by magnetic separation, flotation, or other mineral-processing techniques. Some further references may be found in the article by Iwasaki *et al.* [17].

Another recent and very topical example of complex solid–solid reactions proceeding through gaseous intermediates is the lime-concentrate pellet roasting process developed by Bartlett and Huang [18].

In this process, which has been proposed on the basis of laboratory scale studies, copper sulfide minerals, chalcocite, chalcopyrite, etc., are mixed with burned lime, pelletized, and then roasted at a temperature of about 500°C. The principal objective of the operation is to transform the copper

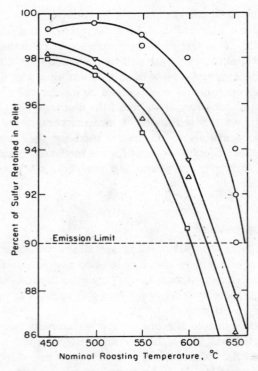

FIG. 5.6.   Bartlett's experimental data on the roasting of copper sulfide concentrates in the presence of lime [18]. CaO stoichiometry; ○, 100%, ▽, 90%; △, 85%; □, 80%

sulfide into leachable copper oxide, but at the same time to ensure that the SO_2 evolved is made to react in situ with the burned lime.

The principal reactions are

$$CuFeS_2 + \tfrac{13}{2}O_2 \longrightarrow CuO + \tfrac{1}{2}Fe_2O_3 + 2SO_2 \qquad (5.3.4)$$
$$2CaO + 2SO_2 + O_2 \longrightarrow 2CaSO_4 \qquad (5.3.5)$$

The authors have shown that by careful temperature control it is possible to retain a very high proportion of the sulfur originally contained in the pellet without having to use burned lime in excess of the stoichiometric requirement. This is illustrated in Fig. 5.6.

While the authors did not perform detailed kinetic studies for the individual reaction steps, viz., Eqs. (5.3.4)–(5.3.5) they seem to have defined adequately the key design parameters for the process, including the way single particle results may be scaled for designed packed-bed systems, which was the principal objective of the work undertaken.

It is thought, however, that some very interesting basic kinetic studies could be made on this system.

5.4   Theoretical Analysis of Solid–Solid Reactions Proceeding through Gaseous Intermediates: Reaction Rates Controlled by Chemical Kinetics

In the preceding sections we cited several practical examples of solid–solid reactions, proceeding through gaseous intermediates. While these reaction systems are of industrial interest and some laboratory scale kinetic measurements have been made, unlike the case of simple gas–solid reactions there are no general modeling equations available for the interpretation of the experimental data.

The absence of comprehensive theoretical treatment is likely to be caused by the complexity of these systems. In this section we shall develop a mathematical representation for solid–solid reactions, for systems where the overall rate is controlled by chemical kinetics [25].† We must acknowledge that this treatment will be applicable only to a limited range of practical situations; nonetheless it represents a first step, upon which subsequent, more comprehensive studies may be based.

Let us consider a porous pellet made up of uniformly mixed, small solid particles as sketched in Fig. 5.7. The individual solid particles (grains) may have the shape of a flat plate, a long cylinder, or a sphere. The relative amounts of solid species may differ and the grains corresponding to the two species may also differ in shape and in size. In the present formulation we

† A more general formulation is available in a recent paper by Szekely *et al.* [19].

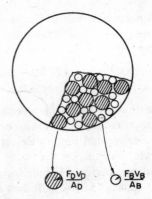

FIG. 5.7. The grain model for a solid–solid reaction proceeding via gaseous intermediates.

will, however, restrict ourselves to systems where any given species is represented by a given uniform grain size rather than a size distribution.

Reactions between solids B and D through the gaseous intermediates A and C with a net generation of gaseous species may be represented by the following:

$$A(g) + bB(s) = cC(g) + eE(s) \qquad (5.4.1a)$$

$$C(g) + dD(s) = aA(g) + fF(s) \qquad (5.4.1b)$$

where $ac > 1$.

The following assumptions are made:

(1)   The overall rate of reaction is controlled by chemical kinetics and the concentration of the gaseous species is uniform throughout the pellet.

(2)   The system is isothermal.

(3)   Diffusion of the gaseous reactants through the product layer of the individual grains is not rate limiting.

Regarding the assumptions made in the derivation of the model, the consideration of uniform particle size and idealized particle shape is clearly an oversimplification. Nonetheless, many systems of practical interest contain particles within a close size range; moreover, the analysis allows the consideration of systems with differently shaped particles.

Assumption (1) is likely to be valid over a wide range of conditions of practical interest; it should hold when the rate of the chemical reaction(s) is much slower than pore diffusion and when the composition of the gaseous environment surrounding the pellet is identical to that in the pores. This would be the case for a system of uniformly mixed, fine particles placed in a closed vessel. In such a system the following analysis is also applicable to reactions with no net generation of gases ($ac = 1$). Assumption (1) should also hold, at least as a good approximation, for fast reactions where the net (convective) outflow of the product gases is sufficiently rapid so as to prevent any diffusive interchange with the gaseous environment. Under these conditions the system is chemically controlled and the reaction proceeds essentially independently of the gaseous environment surrounding the pellet [2]. A criterion for satisfying this latter condition will be developed subsequently.

Assumption (2) will be valid for reactions with a moderate thermal effect. As discussed in Chapter 4, the diffusion of reactants through the product layer of the grains can be neglected when the grain size is small [20, 21].

The net rate at which the gaseous components A and C are being generated may be calculated by subtracting their rate of consumption from the rate of generation. On making allowance for the stoichiometric coefficients we have

$$dn_A/dt = -v_1 + av_2 \tag{5.4.2}$$

$$dn_C/dt = cv_1 - v_2 \tag{5.4.3}$$

where v_1 and v_2 are the net forward rates of reactions (5.4.1a) and (5.4.1b), respectively, per unit volume of the pellet. We shall assume that within each grain the reaction front maintains its original shape (i.e., concentric spheres, coaxial cylinders, or parallel planes) for successive times.

The advancement of the reaction interface within the grain may be expressed as

$$-\rho_B \, dr_B/dt = bf_1(C_A, C_C) \tag{5.4.4}$$

$$-\rho_D \, dr_D/dt = df_2(C_A, C_C) \tag{5.4.5}$$

where $f_1(C_A, C_C)$ and $f_2(C_A, C_C)$ are the kinetic expressions for the net forward gas–solid reactions of (5.4.1a) and (5.4.1b), respectively. The relationships between v_1 and $f_1(C_A, C_C)$ and between v_2 and $f_2(C_A, C_C)$ are

$$v_1 = \alpha_B(A_B/V_B)(A_B r_B/F_B V_B)^{F_B - 1} f_1(C_A, C_C) \tag{5.4.6}$$

$$v_2 = \alpha_D(A_D/V_D)(A_D r_D/F_D V_D)^{F_D - 1} f_2(C_A, C_C) \tag{5.4.7}$$

where α_B and α_D are volumes occupied by the solids B and D, respectively, per unit volume of the pellet. The surface area and volume of the unreacted grain are designated by A and V, respectively, and F is the grain shape factor, which has the value of 1, 2, or 3 for flat plates, long cylinders, or spheres, respectively.

In a constant-pressure system the following relationships hold:

$$dn_A/dt = (C_A/V_P) \, dV/dt \tag{5.4.8}$$

$$dn_C/dt = (C_C/V_P) dV/dt \tag{5.4.9}$$

where dV/dt is the rate of increase in volume of the gaseous mixture, and V_p the volume of the pellet. We also have

$$C_A + C_C = C_T \qquad (5.4.10)$$

where C_T is the total molar concentration of the gaseous species within the pellet.

In deriving Eqs. (5.4.8) and (5.4.9) we made a pseudosteady–state assumption that the gas phase concentrations at any time are at the steady state values corresponding to the amount and sizes of the solid at that time; that is, $C \, dV/dt \gg V \, dC/dt$; thus the concentrations will, in general, change slowly with time.

Equations (5.4.2)–(5.4.10) provide a complete statement of the problem. Note that by combining these equations expressions may be obtained for the concentration of the gaseous species A and C. These values of C_A and C_C may then be used in conjunction with Eqs. (5.4.4) and (5.4.5) to obtain the instantaneous reaction rates of the solid reactants.

The above formulation will hold for any rate expression for the gas–solid reaction designated by $f_1(C_A, C_C)$ and $f_2(C_A, C_C)$. The rate at which a solid reacts with a gas is often related to the gas concentration by a power law type expression; a more generally valid relationship is provided by the Langmuir–Hinshelwood type rate expression [22–24], a simple form of which may be written

$$f_1(C_A, C_C) = k_1(C_A - C_A^*)/(1 + K_1 C_A) \qquad (5.4.11)$$

$$f_2(C_A, C_C) = k_2(C_C - C_C^*)/(1 + K_2 C_C) \qquad (5.4.12)$$

where C_A^* and C_C^* are equilibrium concentrations. Substituting Eqs. (5.4.11) and (5.4.12) into the previously given rate expressions and rearranging the resultant equations in dimensionless forms, we obtain

$$\psi_A + \psi_B = 1 \qquad (5.4.13)$$

$$\frac{(\psi_A - \psi_A^*)(\psi_C + c\psi_A)}{1 + \kappa_1 \psi_A} = \gamma\beta\left(\frac{F_D \xi_D^{F_D - 1}}{F_B \xi_B^{F_B - 1}}\right)\frac{(\psi_C - \psi_C^*)(\psi_A + a\psi_C)}{1 + \kappa_2 \psi_C} \qquad (5.4.14)$$

$$d\xi_B/dt^* = -(1 + \kappa_1)(\psi_A - \psi_A^*)/(1 + \kappa_1 \psi_A) \qquad (5.4.15)$$

$$d\xi_D/dt^* = -\beta(1 + \kappa_2)(\psi_C - \psi_C^*)/(1 + \kappa_2 \psi_C) \qquad (5.4.16)$$

The dimensionless quantities appearing in these equations are defined as

$$\psi_A \equiv C_A/C_C \qquad (5.4.17)$$

$$\psi_C \equiv C_C/C_T \qquad (5.4.18)$$

$$\kappa_1 \equiv K_1 C_T \qquad (5.4.19)$$

$$\kappa_2 \equiv K_2 C_T \tag{5.4.20}$$

$$\xi_B \equiv A_B r_B / F_B V_B \tag{5.4.21}$$

$$\xi_D \equiv A_D r_D / F_D V_D \tag{5.4.22}$$

$$\gamma = (b/d)(\alpha_D/\alpha_B)(\rho_D/\rho_B) \tag{5.4.23}$$

$$\beta \equiv \left(\frac{d}{b}\right)\left(\frac{\rho_B}{\rho_D}\right)\left(\frac{F_A V_B / A_B}{F_D V_D / A_D}\right)\left(\frac{k_2}{1+\kappa_2}\right)\left(\frac{1+\kappa_1}{k_1}\right) \tag{5.4.24}$$

$$t^* \equiv [bk_1 C_T / \rho_B (1+\kappa_1)](A_B / F_B V_B) t \tag{5.4.25}$$

The conversion of the solid reactants (X_B and X_D) may then be calculated by using the expressions

$$X_B = 1 - \xi_B^{F_B} \tag{5.4.26}$$

$$X_D = 1 - \xi_D^{F_D} \tag{5.4.27}$$

Note that the conversion and the rate of reaction are related to the dimensionless time t^* through the two parameters γ and β. The parameter γ represents the relative molar quantity of the two solids in the pellet, and β corresponds to the ratio of reactivities of the two reactant solids. As will be shown later, the quantities γ and β allow us to define asymptotic regimes and to identify the rate-controlling step.

COMPUTED RESULTS

In the following we shall present computed results for systems that obey first-order, irreversible rate expressions, which constitute a special case of the more general rate laws described by Eqs. (5.4.11) and (5.4.12). The consideration of more involved rate expressions would not introduce undue computational difficulties [24], but the interpretation of the results would be rather more involved.

For first-order irreversible kinetics, $\kappa_1 = \kappa_2 = \psi_A^* = \psi_C^* = 0$; and on combining Eqs. (5.4.13) and (5.4.14) with the condition $0 \leq \psi_A \leq 1$, we obtain

$$\psi_A = (v - \sqrt{v^2 - 4uw})/2u \tag{5.4.28}$$

where

$$u \equiv (a-1)\gamma\beta(F_D \xi_D^{F_D-1}/F_B \xi_B^{F_B-1}) - (c-1) \tag{5.4.29}$$

$$v \equiv (2a-1)\gamma\beta(F_D \xi_D^{F_D-1}/F_B \xi_B^{F_B-1}) + 1 \tag{5.4.30}$$

$$w \equiv a\gamma\beta(F_D \xi_D^{F_D-1}/F_B \xi_B^{F_B-1}) \tag{5.4.31}$$

Then Eqs. (5.4.15) and (5.4.16) may be written

$$d\xi_B/dt^* = -\psi_A \tag{5.4.32}$$

$$d\xi_D/dt^* = -\beta(1 - \psi_A) \tag{5.4.33}$$

By substituting for ψ_A in Eqs. (5.4.32) and (5.4.33) and integrating, we obtain the desired relationship between $\xi_B(X_B)$, $\xi_D(X_D)$, and t^*.

ASYMPTOTIC BEHAVIOR

Before proceeding with the presentation of the computed result for the "general" case, it is worthwhile examining certain asymptotic regions, because these will provide a readily improved insight into the behavior of the system.

Small $\gamma\beta$ asymptote When $\gamma\beta$ approaches zero, reaction (5.4.1b) is much slower than reaction (5.4.1a), and hence the former controls the overall rate. The gaseous intermediates consist mainly of the gas C, and the concentration of the gas A will be small. This expected behavior of the system is readily demonstrated by consideration of the governing equations.

A purely physical illustration may be helpful at this stage. If we were to consider, for argument's sake, the reaction system

$$FeO + CO \rightleftharpoons Fe + CO_2 \tag{5.4.34}$$

$$C + CO_2 \rightleftharpoons 2CO \tag{5.3.35}$$

[with $CO : A(g)$, $FeO : B(s)$, $CO_2 : C(g)$ C : D(s)], then the $\gamma\beta \to 0$ asymptote would correspond to a situation where the overall rate is being controlled by reaction (5.4.35) and consequently the gas contained in the pores would be predominantly CO_2. (This argument would also have to contain the postulate that both the abovementioned reactions are irreversible.)

As $\gamma\beta \to 0$, the solution of Eq. (5.4.28) becomes

$$\psi_A \cong a\gamma\beta F_D \zeta_D^{\xi F_D - 1}/F_B \zeta_B^{\xi F_B - 1} \tag{5.4.36}$$

Substituting Eq. (5.4.36) into Eqs (5.4.32) and (5.4.33), we obtain

$$d\xi_B/dt^* = -a\gamma\beta F_D \zeta_D^{\xi F_D - 1}/F_B \zeta_B^{\xi F_B - 1} \tag{5.4.37}$$

$$d\xi_D/dt^* = -\beta \tag{5.4.38}$$

The integration of Eqs. (5.4.38) yields

$$\beta t^* = 1 - \xi_D = 1 - (1 - X_D)^{1/F_D} \equiv g_{F_D}(X_D) \tag{5.4.39}$$

Finally, on combining Eqs. (5.4.39) and (5.4.37) and integrating, we obtain

$$\zeta_B^{F_B} = 1 - a\gamma[1 - (1 - \beta t^*)^{F_D}] \tag{5.4.40}$$

Equations (5.4.39) and (5.4.40) do, of course, correspond to the well known expressions for the "topochemical model" for each grain under kinetic control, for a fixed gas concentration—pure C in the present case.

The existence of this asymptote was obvious on physical grounds, but the mathematical development given here will allow us the definition of the range of parameters for which this asymptote is valid. As will be shown subsequently in the discussion of the general case, the $\gamma\beta \to 0$ asymptote is reasonable when $\gamma\beta < 0.01$.

The conversion X_B or X_D may be obtained by using Eqs. (5.4.26) and (5.4.27). Furthermore, the conversion of the two solids is related by

$$X_B = a\gamma X_D \qquad (5.4.41)$$

The following comments may be made at this stage: From Eq. (5.4.40) or (5.4.41), it is seen that

(1)   When $a\gamma < 1$, solid D will be converted completely while there will remain unreacted solid B; furthermore, the time for complete reaction is given by

$$\beta t_c^* = 1 \qquad (5.4.42a)$$

(2)   When $a\gamma > 1$, solid B will be converted completely and there will remain unreacted solid D; the time for complete reaction is then given as

$$\beta t_c^* = 1 - (1 - 1/a\gamma)^{1/F_D} \qquad (5.4.42b)$$

(3)   Finally, when $a\gamma = 1$, both solids will react completely, and the time for complete reaction is given either by Eq. (5.4.42a) or by Eq. (5.4.42b).

These points may need some additional emphasis. For solid–solid reactions of the type designated by Eqs. (5.4.1a) and (5.4.1b) or (5.4.34) and (5.4.35), the stoichiometry of the system is not immediately defined and recourse must be made to the earlier defined dimensionless quantity γ, which, together with the stoichiometric coefficient, defines the proportion of the solid reactants that will result in the complete consumption of both species.

Within the asymptotic regime, when $\gamma\beta \to 0$, the governing equations for the system reduce to that for a single gas–solid reaction, and thus the treatment of the experimental data may follow a similar pattern to that described in Chapters 3 and 4.

In principle, it would be possible to deduce the kinetic parameters for the rate-controlling gas–solid reaction from measurements conducted with the $\gamma\beta \to 0$ regime. It is suggested, however, that it would be preferable *to conduct independent measurements on the gas–solid reaction system* and then confirm the validity of the postulate by comparing the kinetic parameters deduced from the studies in the $\gamma\beta \to 0$ regime with these independent measurements.

Large $\gamma\beta$ asymptote When $\gamma\beta$ is large, reaction Eq. (5.4.1a) constitutes the rate-controlling step, and the gas phase will consist predominantly of species A [i.e., using Eqs. (5.4.34) and (5.4.35) as a physical example, the gas would consist of CO]. The behavior of the system is readily described as a corollary of the previously discussed case, by replacing species A and B by species C and D, respectively. The mathematical development readily follows along the lines discussed earlier.

Thus, when $\gamma\beta \to \infty$, Eq. (5.4.18) gives

$$1 - \psi_A = \psi_C = (c/\gamma\beta)F_B \xi_B^{F_B - 1}/F_D \xi_D^{F_D - 1} \tag{5.4.43}$$

and

$$d\xi_B/dt^* = -1 \tag{5.4.44}$$

$$d\xi_D/dt^* = -(c/\gamma)F_B \xi_B^{F_B - 1}/F_D \xi_D^{F_D - 1} \tag{5.4.45}$$

from which we obtain

$$t^* = 1 - \xi_B = 1 - (1 - X_B)^{1/F_B} \equiv g_{F_B}(X_B) \tag{5.4.46}$$

$$\xi_D^{f_D} = 1 - (c/\gamma)[1 - (1 - t^*)^{F_B}] \tag{5.4.47}$$

The relationship between X_B and X_D then becomes

$$X_D = (c/\gamma) X_B \tag{5.4.48}$$

It may be readily shown that

(1) When $c/\gamma < 1$, solid B is the component that is completely converted and the time for complete reaction is given by

$$t_c^* = 1 \tag{5.4.49}$$

(2) When $c/\gamma > 1$, solid D is the component that is completely converted, and

$$t_c^* = 1 - [1 - (\gamma/c)]^{1/F_B} \tag{5.4.50}$$

(3) Finally, when $c/\gamma = 1$, both solids will react completely and t_c^* may be obtained from Eq. (5.4.49) or (5.4.50).

In general, for $\gamma\beta \to \infty$ reaction (5.4.1a) controls the overall rate, and information regarding the intrinsic kinetics of the reaction may be obtained with the aid of Eq. (5.4.46).

ANALYTICAL SOLUTION

Analytical solutions are available for the asymptotic regimes of $\gamma\beta \to 0$ and $\gamma\beta \to \infty$. For intermediate values of $\gamma\beta$, analytical solutions are obtained only when both F_B and F_D are unity.

For these latter conditions the interfacial area available for reaction remains constant, and hence both ψ_A and ψ_C are also constant throughout the reaction. Thus, Eqs. (5.4.32) and (5.4.33) are readily integrated giving straight-line relationships between ξ_B (ξ_D) and time. The results for this case will be given below together with those for the general case.

THE GENERAL CASE

In the general case the system of Eqs. (5.4.28), (5.4.32), and (5.4.33) must be solved numerically [25]. The solutions for the cases $F_B = F_D = 3$ and $F_B = F_D = 1$ will be presented subsequently. One of the more important reactions between solids through gaseous intermediates is the reduction of metal oxides with solid carbon, in which case, $a = 2$ and $c = 1$; the following solutions will correspond to these values of a and c.

Figure 5.8 shows the rate of reaction and the concentrations of the gaseous intermediates as functions of γ for various values of β for the case

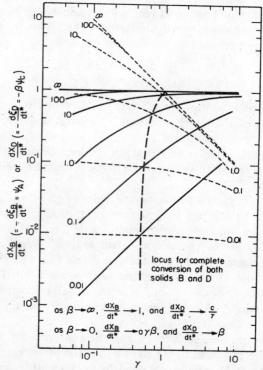

FIG. 5.8. Rate of reaction and concentrations of the gaseous intermediates as a function at the relative molar quantities of the solid reactants [25]. $F_B = F_D = 1$; $a = 2$; $c = 1$; parameter, β; ——, solid B; - - -, solid D.

$F_B = F_D = 1$. Since the rate is constant until either one of the solids is completely reacted, the behavior of a given system would be represented by a point on a curve, as γ is a constant. In general, as γ increases, the rate of reaction of B increases and that of D decreases. This agrees with the experimental results of Rao [6].

We can also see that the small $\gamma\beta$ asymptote is approximated when $\gamma\beta < 10^{-2}$ and the large $\gamma\beta$ asymptote is valid when $\gamma\beta > 10^2$. Note here that the $\gamma\beta \to 0$ and $\gamma\beta \to \infty$ asymptotes correspond to horizontal lines for solids D and B, respectively.

While these data were obtained with a flat-platelike geometry for the grains, subsequent calculations will show that similar $\gamma\beta$ values may be used for defining the asymptotic regimes for the other geometries.

The intersection of the curves depicting the rates of reaction of B and D for identical β values provides the locus for the complete reaction of both B and D in terms of γ and β. Since in essence γ denotes the relative amounts of the solid reactant and β corresponds to a ratio of the kinetic factors, this behavior deserves comment. One might expect that the criteria for the complete conversion of both solid reactants should be set by stoichiometry only without any influence by kinetic factors. The fallacy of this argument is perhaps best demonstrated by the familiar example:

$$FeO + CO \rightleftharpoons CO_2 + Fe$$
$$C + CO_2 \rightleftharpoons 2CO$$

Clearly, the "stoichiometric" proportions of C and FeO will depend on the relative rates of these two reactions, and may range from 0.5 to 1.0. In general, the locus for complete reaction is bounded by $1/a \le \gamma \le c$.

This finding is of considerable practical interest because the criteria for complete reaction are quite important in many practical systems where pure product is desired.

The behavior of systems with spherical grains, i.e., $F_B = F_D = 3$ is shown in Figs. 5.9–5.12. In contrast to flat-platelike grains, in these systems the rate of reaction is a function of conversion.

Figure 5.9a shows the conversion versus reduced time for small values of β in a system where $F_B = F_D = 3$ and $\gamma = 1$, and Fig. 5.9b shows a similar plot for large values of β. It is seen that in this case solid B always reacts faster than solid D, and only when β is infinitely larger do the rates become the same. The infinite $\gamma\beta$ condition is again seen to be approached when $\gamma\beta > 100$, as in the case of $F_B = F_D = 1$.

The rate at which species B reacts and the concentration of the gaseous intermediate A are shown in Fig. 5.10 as a function of the extent of reaction. The reaction rate of D may be obtained from this plot using Eqs. (5.4.32) and (5.4.33). Note that the actual shape of the curve depends on γ.

From Figs. 5.9a and 5.9b we see that for $\gamma = 1$ solid B is completely converted and there remains unreacted solid D. Note that the parameters β and γ provide the criteria as to which of the solid reactants may be completely converted, and perhaps more significantly they define the conditions under which complete conversion of both reactants is possible. This behavior is

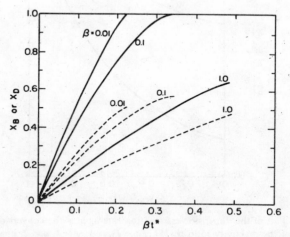

FIG. 5.9a. Extent of reaction versus time for several values of the reactivity ratio β [25]. $F_B = F_D = 3$; $\gamma = 1$, $a = 2$, $c = 1$; ——, solid B; - - -, solid D.

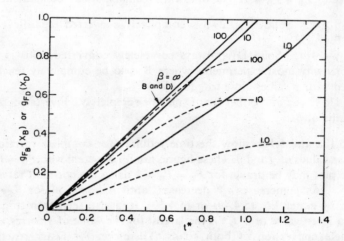

FIG. 5.9b. Plot of the conversion functions $g_{F_B}(X_B)$ and $g_{F_D}(X_D)$ against dimensionless time for several values of the reactivity ratio β [25].

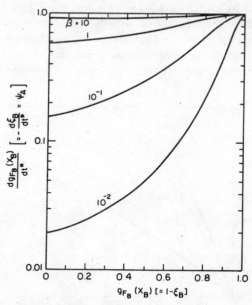

FIG. 5.10. Plot of the dimensionless reaction rate against the conversion function for various values of the reactivity ratio β [25]. $F_B = F_D = 3$, $\gamma = 1$, $a = 2$, $c = 1$.

shown in Fig. 5.11 for $F_B = F_D = 3$. A very similar plot may be obtained in the case of $F_B = F_D = 1$. The following comments may be made here:

(i) $\gamma \geq c$:   Solid B is always completely converted regardless of the value of β.

(ii) $\gamma \leq 1/a$:   Solid D is always completely converted. This is one of the reasons why most experiments where B is to be completely reacted are performed with γ values close to c [2, 6].

(iii) $1/a < \gamma < c$:   Either B or D may be completely converted depending on the value of β.

Figures 5.12a and 5.12b show the time required for complete reaction, t_c^*, for various values of γ and β. These figures are for a system with $\dot{F}_B = F_D = 3$. A similar plot may be drawn for $F_B = F_D = 1$ using the constant rates given in Fig. 5.8. As β increases t_c^* decreases, until it levels off for $\beta \to \infty$, as described by either Eq. (5.4.49) or (5.4.50). It is also seen from Fig. 5.12a that, for a given value of β, t_c^* is the largest for the value of γ corresponding to complete conversion of both solids. Therefore, for faster reaction of given solid materials (i.e., fixed β), γ should be increased if the conversion of solid B is desired. The effect of γ in this case, however, depends on the

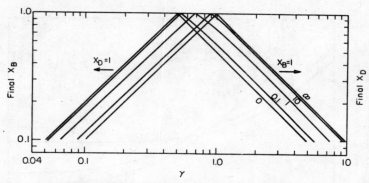

FIG. 5.11. Plot of the extent of reaction of A or D at the end of reaction as a function of the relative molar quantities of solid reactants and the reactivity ratio [25]. $F_B = F_D = 3$, $a = 2$, $c = 1$, parametric $\gamma\beta$.

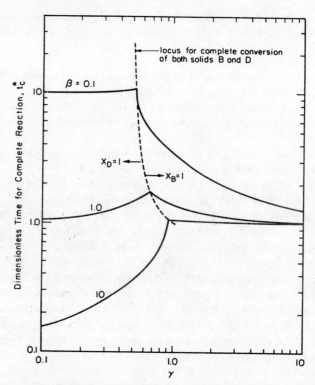

FIG. 5.12a. Plot of dimensionless time for complete reaction against the relative molar quantities of solid reactant for various values of the reactivity ratio β [25]. $F_B = F_D = 3$, $a = 2$, $c = 1$.

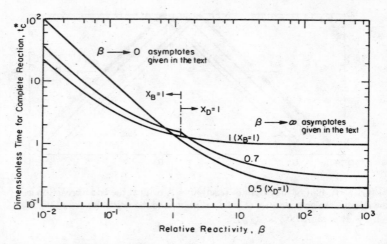

FIG. 5.12b. Plot of dimensionless time for complete reaction against the reactivity ratio β with relative quantities of reactants [25]. $F_B = F_D = 3$, $a = 2$, $c = 1$, parameter γ.

value of β. The smaller the value of β [reaction (5.4.1b) slower than reaction (5.4.1a)] the greater the effect of γ. When β is very large, i.e., when reaction (5.4.1b) is much faster than reaction (5.4.1a), increasing γ (increasing the amount of solid D) will not greatly enhance the overall rate as long as γ is greater than c.

Note also that Fig. 5.12a provides the locus of conditions for complete conversion of both solid species as a function of γ and β.

The effect of grain size or reaction temperature may also be deduced from Fig. 5.4.12b. When β is large, increasing the value of β, and hence decreasing the size of solid D [see the definition of β given by Eq. (5.4.29)], for a fixed size of solid B will not change t_c^*. The overall rate may be increased, however, by reducing the size of solid B, viz., the definition of t^* given by Eq. (5.4.25).

When β is small, increasing its value (hence reducing the size of solid D) will reduce the time for complete conversion. This agrees qualitatively with the experimental results of Rao [6]. Although it is not immediately apparent from Figs. 5.12a and 5.12b, for small values of β the reduction in the size of solid B has little effect on t_c^*. Clearly, these considerations would suggest that the smaller the size of the solids, the faster the overall rate of reaction. It must be noted, however, that for tightly compacted, very fine particles Knudsen diffusion may become an important component of the overall reaction sequence and this situation is not represented by the model given here.

CRITERIA FOR NEGLIGIBLE EFFECT OF DIFFUSION IN A SYSTEM
WITH NET VOLUME GENERATION

In the preceeding analysis, it was assumed that because of the net generation of gases within the porous solid as a result of the reaction, diffusion
from the gas space surrounding the reacting pellet could be neglected. While
this "sweeping effect" of the gas generated in the pores is plausible it is
highly desirable to define the limiting conditions for which this assumption
is valid.

Let us consider a reaction system where a pellet, made up of individual
grains, is surrounded by an inert gas. We shall assume that the reaction
within the pellet results in the formation of a gaseous product at a constant
rate; the objective of the calculation is then to determine the reaction rate
at which the penetration of inert gas is negligible because of the outward
flow of the reaction product.

While these considerations were developed for an inert atmosphere
surrounding the pellet, the criteria should apply, at least as a good approximation for reactive atmospheres.

We define the following symbols: A_p is the external surface of the pellet,
C_1 and C_{1b} the concentrations of ambient gas within the pellet and in the
bulk, respectively, D_e the effective diffusivity, F_p the shape factor for the
pellet ($=1$, 2, and 3 for an infinite slab, a long cylinder, and a sphere, respectively), and R the distance from the center of the pellet. Also, u^* is the
superficial velocity of gas within the pellet, v^* the rate of generation of gases
in g-moles per unit volume per unit time, V_p the volume of the pellet, η the
dimensionless position ($\equiv A_p R/F_p V_p$), ρ the molar density of the gas, and
ψ_4 the dimensionless concentration of ambient gas ($\equiv C_1/C_{1b}$).

By establishing a mass balance we have the following expression for u^*,
the superficial velocity of the product gas stream within the pellet, which is
valid for all the three geometries considered:

$$u^* = (v^*/\rho)(V_p/A_p)\,\eta \qquad (5.4.51)$$

The mass balance on the ambient gas gives

$$D_e \frac{d}{dR}\left(R^{F_p-1}\frac{dC_1}{dR}\right) - \frac{d}{dR}(u^* R^{F_p-1}C_1) = 0 \qquad (5.4.52)$$

Substituting Eq. (5.4.51) in Eq. (5.4.52) and rearranging in dimensionless
form, we obtain

$$\frac{d}{d\eta}\left(\eta^{F_p-1}\frac{d\psi_1}{d\eta}\right) - 2\Phi^2\frac{d}{d\eta}(\eta^{F_p}\psi_1) = 0 \qquad (5.4.53)$$

where

$$\Phi^2 \equiv (v^* F_p/2\rho D_e)(V_p/A_p)^2 \qquad (5.4.54)$$

The boundary conditions are given as

$$d\psi_1/d\eta = 0 \qquad \text{at} \quad \eta = 0 \qquad\qquad (5.4.55)$$

and

$$\psi_1 = 1 \qquad \text{at} \quad \eta = 1 \qquad\qquad (5.4.56)$$

Integration of Eq. (5.4.53) gives

$$\psi_1 = e^{-\Phi^2(1-\eta^2)} \qquad\qquad (5.4.57)$$

The fraction of ambient gas present within the pellet is given by

$$\bar{\psi}_1 = \int_0^1 \psi_1 \eta^{F_p - 1}\, d\eta \Big/ \int_0^1 \eta^{F_p - 1}\, d\eta \qquad\qquad (5.4.58)$$

For $F_p = 1$

$$\bar{\psi}_1 = (1/\Phi) \int_0^\Phi e^{(t^2 - \Phi^2)}\, dt \qquad\qquad (5.4.59)$$

The integral in this equation is the so-called Dawson's integral, the values of which have been tabulated in the literature.†

For $F_p = 2$

$$\bar{\psi}_1 = (1/\Phi^2)(1 - e^{-\Phi^2}) \qquad\qquad (5.4.60)$$

For $F_p = 3$, the integral must be evaluated numerically.

The parameter Φ provides a criterion for the effect of the ambient gas. If it is assumed that the effect of the ambient gas may be neglected when $\bar{\psi}_1$ is less than 5%, it has been shown [25] that the criterion is given as

$$\Phi^2 > \begin{cases} 10 & \text{for a slab} \\ 20 & \text{for a long cylinder} \\ 30 & \text{for a sphere} \end{cases}$$

On defining

$$\hat{\Phi}^2 \equiv (v^*/2\rho D_e)(V_p/A_p)^2 \qquad\qquad (5.4.61)$$

Sohn and Szekely [25] obtained the generalized criterion for any geometry as

$$\hat{\Phi}^2 > 10 \qquad \text{for} \quad \bar{\psi}_1 < 0.05$$

Let us illustrate the use of the $\hat{\Phi}^2$ parameter on a simple numerical calculation.

† For example, Abramowitz and Stegun [26].

Example 5.4.1 Let us consider the reaction of a porous solid matrix of thickness 2.5 cm, in which the following reaction is taking place:

$$FeO + CO = Fe + CO_2$$
$$CO_2 + C = 2CO$$

At 1300°K, 1 atm total pressure, and for an effective diffusivity of 0.5 cm²/sec, a total reaction time of about 100 min would readily meet the criterion for $\Phi^2 < 10$. The reader may verify this statement by substituting into Eq. (5.4.61). The work of Otsuka and Kunii [2] and the majority of Rao's data [6] fall into this category.

We note that this calculation is conservative because, when there is a mass transfer resistance at the surface of the pellet, even a slower reaction rate would be sufficient to suppress diffusional effects.

Finally, let us illustrate the use of the analysis developed here for identifying the rate-controlling mechanism for a reaction system.

Example 5.4.2 Let us consider the reaction system

$$MeO + CO \rightleftharpoons Me + CO_2 \qquad (5.4.62)$$
$$CO_2 + C \rightleftharpoons 2CO \qquad (5.4.63)$$

and identify the rate-controlling step for three sets of values of the process parameters.

(a) When

$$\rho_{MeO} = 5.5 \text{ g/cm}^3, \qquad \rho_C = 2.3 \text{ g/cm}^3$$
$$F_{MeO} = 3, \qquad F_C = 3$$
$$(FV/A)_{MeO} = 8 \times 10^{-5} \text{ cm}, \qquad F_C V_C/A_C = 1 \times 10^{-2} \text{ cm}$$
$$k_1 = 1 \times 10^{-5} \text{ cm/sec (900°C)}, \qquad k_2 = 3 \times 10^{-6} \text{ cm/sec (900°C)}$$

SOLUTION FOR THIS SYSTEM $a = 2$, $b = 1$, $c = 1$, $d = 1$; in this system let $\gamma = 1$; then $\beta = 0.006$.

From Fig. 5.9a we know that this system is controlled largely by reaction (5.4.63). This finding could be confirmed independently by considering that for this case $\gamma\beta < 0.01$.

(b) *The effect of grain size* Let $F_C V_C/A_C = 6 \times 10^{-5}$ cm, rather than the previously given value of 10^{-2} cm, while all the other parameters are kept the same as before; then $\beta = 1$.

From Fig. 5.9a we know that the system is close to the region where both reactions are important.

(c) *The effect of temperature* Finally, let us consider the effect of changing the reaction temperature. Assume that

$$\Delta E_1 = 20,000 \text{ cal/g-mole}, \qquad \Delta E_2 = 80,000 \text{ cal/g-mole}$$

then at 1100°C

$$k_1 = 2 \times 10^{-5} \text{ cm/sec}, \qquad k_2 = 4.4 \times 10^{-4} \text{ cm/sec}$$

Consider, furthermore, that

$$F_C V_C / A_C = 1 \times 10^{-2} \text{ cm}$$

It follows that $\beta = 0.4$.

Inspection of Fig. 5.9a then shows that in this case both reactions are important, and one cannot represent the system in terms of the reaction kinetics of only one of the solid species. The same conclusion could also be reached by forming the $\gamma\beta$ product, which in this case falls outside the asymptotic regions.

5.5 Concluding Remarks

In this chapter we have provided a brief introduction to a very interesting group of reaction systems, namely, solid–solid reactions proceeding through gaseous intermediates. Practical examples of these systems include the reduction of metal oxides with solid carbon, segregation roasting (chloridizing roasting), and the lime concentrate pellet-roasting process.

All these reactions have the common feature that they may be regarded as a coupled system of gas–solid reactions, which is the main reason for including them in this monograph. This field has received very little attention until recently and no comprehensive theoretical developments, paralleling those of gas–solid reactions, have evolved. An asymptotic analysis has been presented in Section 5.4 and shows some promise for planning future experimental studies and for forming a basis of further work.

The whole area of solid–solid reactions proceeding through gaseous intermediates would seem to offer very attractive opportunities for further work by both experimentalists and theoreticians.

Notation

a, b, c, d	stoichiometric coefficients in reactions (5.4.1a) and (5.4.1b)
A_B, A_D	surface area of a grain of solids B and D, respectively
C_A, C_C, C_T	concentration of gases A and C, and total concentration of gases, respectively
F	grain shape factor ($=1$, 2, and 3 for flat plates, long cylinders, and spheres, respectively)
$g(X)$	conversion function defined by Eq. (5.4.39)
k, K	parameters in Langmuir–Hinshelwood rate expression Eqs. (5.4.11) and (5.4.12)
r	position of the moving reaction front in the individual grain

t	time
t^*	dimensionless time defined by Eq. (5.4.25)
t_c^*	dimensionless time for complete reaction
v	net forward reaction rate
V	volume of gas generated by the reaction
V_B, V_D	volume of a grain of solids B and D, respectively
X	fractional conversion of reactant solid
α	fraction of volume occupied by reactant solid per unit volume of pellet
β	ratio of reactivities of solids defined by Eq. (5.4.24)
γ	relative molar quantity of solids defined by Eq. (5.4.23)
κ	dimensionless parameter defined by Eqs. (5.4.19) and (5.4.20)
ξ	dimensionless position of the reaction front in the grain, defined by Eqs. (5.4.21) and (5.4.22)
ρ	molar density of reactant solid
ψ	dimensionless concentration of gaseous intermediate defined by Eqs. (5.4.17) and (5.4.18)

Subscripts

A	gas A
B	solid B
C	gas C
D	solid D
1	reaction (5.4.1a)
2	reaction (5.4.1b)

References

1. B. G. Baldwin, *J. Iron Steel Inst. London* **179**, 30 (1955).
2. K. Otsuka and D. Kunii, *J. Chem. Eng. Japan* **2**(1), 46 (1969).
3. R. A. Sharma, P. P. Bhatnagar, and T. J. Banerjee, *Sci. Ind. Res., New Delhi, India* **16A**, 225 (1957).
4. H. K. Kohl and B. Marinček, *Arch. Eisenhüttenw.* **36**, 851 (1965).
5. H. K. Kohl and B. Marinček, *Arch. Eisenhüttenw.* **38**, 493 (1967).
6. Y. K. Rao, *Met. Trans.* **2**, 1439 (1971).
7. Y. K. Rao, *Chem. Eng. Sci.* **29**, 1435, 1933 (1974).
8. Y. Awakura, Y. Maru, and Y. Kondo, *Proc. Int. Conf. Sci. Technol. Iron Steel, Part 1, Trans. ISIJ* **11**, 483 (1971).
9. Y. Maru, Y. Kuramasu, Y. Awakura, and Y. Kondo, *Met. Trans.* **4**, 2591 (1973).
10. M. I. El-Guindy and W. G. Davenport, *Met. Trans.* **1**, 1729 (1970).
11. W. Jander, *Z. Anorg. Allgem. Chem.* **163**, 1; **166**, 31 (1927).
12. A. M. Gnistling and B. I. Brounshtein, *J. Appl. Chem. USSR* **23**, no. 12, 1327–1338 (1933).
13. R. E. Carter, *J. Chem. Phys.* **34**, 2010–2015 (1961).

14. G. C. Chadwick, *J. Metals* **13**, 455 (1961).
15. T. Rosenqvist, "Principles of Extractive Metallurgy." McGraw-Hill, New York, 1974.
16. F. Habashi, "Principles of Extractive Metallurgy," Vol. 2. Gordon and Breach, New York, 1970.
17. I. Iwasaki, Y. Takohoshi, and H. Kahato, *Trans. Soc. Min. Eng.* 308 (Sept. 1966).
18. R. W. Bartlett and H. H. Huang, *J. Metals* 28 (December, 1973).
19. J. Szekely, C. I. Lin, and Y. A. Liu, *Met. Trans.* (in press).
20. H. Y. Sohn and J. Szekely, *Chem. Eng. Sci.* **27**, 763 (1972).
21. H. Y. Sohn and J. Szekely, *Chem. Eng. Sci.* **29**, 630 (1974).
22. P. L. Walker, Jr., F. Rusinko, Jr., and L. G. Austin, *Advan. Catal.* **11**, 133 (1959).
23. F. Habashi, "Principles of Extractive Metallurgy," Vol. 1, General Principles. Gordon and Breach, New York, 1969.
24. H. Y. Sohn and J. Szekely, *Chem. Eng. Sci.* **28**, 1169 (1973).
25. H. Y. Sohn and J. Szekely, *Chem. Eng. Sci.* **28**, 1789 (1973).
26. M. Abramowitz and I. A. Stegun, "Handbook of Mathematical Functions." Dover, New York, 1965.

| **Experimental Techniques for the Study of Gas–Solid Reactions**

6.1 Introduction

In the preceding chapters we developed a mathematical representation for the progress of gas–solid reactions and showed how the time-dependent conversion or extent of reaction can be related to the various parameters that characterize the system. Many of these parameters, such as the *reaction rate constant* (or more precisely the parameters appearing in the rate expression) and pore diffusion coefficients, are quite specific to both the chemical composition and physical structure of the solid reactants and thus must be determined experimentally.

The modeling equations presented in those chapters would be of very limited use in the absence of reliable information on the kinetic parameters appearing in them. Moreover, the results of even very carefully conducted experimental studies would be of little help in advancing one's understanding of the system unless these data are amenable to interpretation in terms of suitable models.

Unfortunately there appears to be a dichotomy among students of gas–solid reactions dividing them into groups whose primary interest is in mathematical modeling and groups whose main concern is experimental measurement. A better integration of these efforts would be highly desirable and this state of affairs has provided the main motivation for assembling the material in this chapter.

The material to be presented is divided into five sections: preparation of solids, measurement of reaction rates, auxiliary measurements for characterizing the structure of porous pellets, measurement of diffusion coefficients, and the synthesis of this information for the design of experimental studies of gas–solid reaction systems.

6.2   Preparation of Solid Agglomerates

The objective of experimental studies of gas–solid reaction systems is to identify the rate-controlling mechanisms (viz., Chapter 2) for a given set of operating conditions and then to evaluate the numerical values of the various kinetic parameters. Such studies are usually best conducted with single solid pellets or agglomerates, preferably with a regular geometry. The alternatives of using either fine powder or particles assemblies such as packed or fluidized beds are usually less satisfactory. In contacting fine powders with a moving gas stream, care must be taken to avoid elutriation of the particles by the gas; working with packed or fluidized beds of agglomerates may introduce additional complicating factors because of the multiparticle nature of these systems, as will be discussed in Chapter 7.

As discussed in Chapter 2, the rate of the chemical reaction between a gas and a solid surface is frequently dependent on trace impurities, other species absorbed on the solid surface, and even the mechanical history of the surface [1, 2]. For this reason it is necessary to use great care in comparing results of experiments on samples of solid that are apparently chemically identical but have been prepared in different ways. The investigator should therefore aim for consistency in his preparation of solid samples.

Whether the powder, which forms the starting point of the sample preparation, has been prepared in the laboratory (e.g., by the precipitation of hydroxides and their subsequent dehydration or by the oxidation of metal spheres or is purchased from a supplier, for the sake of reproducibility it is desirable to conduct series of measurements with particular batches of material. Substantial batch to batch variations have been found in the chemical reaction parameters of chemically identical (analytical grade) powders provided by the same supplier [3].

The most convenient ways of preparing pellets from fine powder in the laboratory are pressing and pelletizing. Briquetting and extrusion are not attractive techniques for producing the limited numbers of pellets usually required in experiments. Whenever possible, the use of binding agents (or die lubricants, etc.) should be avoided since these may clog pores within the agglomerate or interfere with the chemical reaction between gas and solid. If it is necessary to use a binder, one should be selected that does not react with the solid reactant or product, or better still, that can be driven off by gentle heating. Water or volatile hydrocarbons are likely candidates.

A convential laboratory hydraulic press and suitable punches and dies can be readily used to produce disks, short cylinders, or other shapes of uniform cross section. However, cylinders that have heights approaching or exceeding their diameters will exhibit nonuniform porosity if produced in this way because the compaction pressure is not uniform under these cases.

For spheres or any other shapes, an *isostatic press* may be used. The powder is charged into a thin-walled rubber bag of the same shape as the agglomerate to be produced and the neck of the bag is sealed (a rubber stopper and elastic band suffices) and placed in the isostatic press. The press is really an autoclave connected to a high-pressure pump and filled with a suitable liquid (usually water containing a soluble oil). The press is sealed, brought up to pressure (say about 2000 atm $\simeq$ 30,000 lb/in.^{2}), vented, and opened, and the bag (which is reusable) is peeled back from the compact. If the powder is sufficiently fine, pressing (either conventional or isostatic) frequently yields a pellet that is sufficiently strong to be handled in the laboratory without need for binders. Tableting presses of the sort used by pharmacists can be used if it is necessary to produce a large number of compacts; here the principles are those of the conventional press.

Pelletizing is the production of spherical (or nearly spherical) particles in a pellet "wheel" or drum. Several designs are in use but they all operate by rolling the powder in the presence of some agent that causes the fines to cling together and form balls that gradually grow in size. A suitable machine can be built in the laboratory using a drum open at one end, with its axis inclined a few degrees to the horizontal. The drum is rotated about its axis by means of an electric motor and variable-speed reduction gearing. A small quantity of powder is placed in the drum, which is then run at the speed just below that at which the powder stops tumbling and rotates with the drum. Binding agent is introduced using a spray gun and the drum stopped periodically to discharge the pellets and add more powder. By screening the pellets and returning the undersized ones to the drum a large number of pellets of uniform size are produced. The pelletizing technique has the disadvantages of requiring a binding agent and producing agglomerates that are usually denser on the outside than in the center. The operation is also rather a dusty one, which may present complications in the case of toxic solids.

Sintering, or heat induration as it is sometimes called, can be used to increase the strengths of agglomerates (produced by pressing or pelletizing). As discussed in Chapter 2, it is necessary to bring the pellet to the Tammann temperature (about 0.5 times the melting temperature on the absolute scale) for sintering to occur. Sintering should be used with caution because it may lead to (undesirable) structural changes within the pellet and affect the intrinsic reaction rate between the gas and solid surface. Of course, in certain cases the structural changes involved in sintering may be desired, for instance, if it is wished to study pellets of lower porosity.

The porosity of the pellets can be increased by mixing a volatile inert with the powder prior to agglomeration and then driving off the inert by gentle heating after the agglomerate has been formed. If the inert is a material that sublimes, the method is particularly attractive. Ammonium carbonate and iodine may be used as inerts. In some cases it is possible to adjust the mean pore size of the pellet by altering the particle size of the inert powder. An alternative technique for making small changes in the porosity of the pellet is by screening the powder prior to agglomeration. Powders of a narrow particle size distribution yield higher porosities than pellets of a broad particle distribution, other things being equal.

Thermocouples can be inserted in pellets by drilling, placing the thermocouple in position, and then sealing the hole with refractory cement. However, in the case of pellets made by pressing, it is possible to place the thermocouple in the die (or isostatic pressing bag) along with the powder; the thermocouple leads would then pass through a hole in the punch (or bag stopper). The press is then operated, giving a pellet into which the thermocouple is intimately sealed.

The actual size or shape of the pellet (or the size range of the pellets) that one wishes to use in an experimental investigation depends on one's objectives and on the nature of the system. As a rough preliminary guide, if we are interested in determining the intrinsic chemical rate constants then we must use small pellets so as to minimize the effect of diffusion. In contrast, if we wish to study the regime where diffusional effects are significant then we must use large pellets. A more systematic discussion of this topic will be presented in Section 6.5.

For a detailed discussion of the various techniques employed for the agglomeration of fine powders, the reader is referred to the proceedings of a symposium on this subject [4].

6.3   Measurement of the Reaction Rate

Reaction rates have been usually measured by suspending a solid pellet within a controlled temperature environment (e.g., a furnace) where it is reacted with a moving gas stream. The reaction products may be all gaseous (e.g., in coal gasification) or a solid product may also be produced (e.g., in iron oxide reduction, roasting of sulfides).

The techniques that may be used for following the progress of the reaction may be classified into the following two groups:

(i)   measurement of some change in the properties of the pellet,
(ii)   measurement of some change in the properties of the gas downstream from the pellet.

The most popular techniques in group (i) include the continuous measurement of the weight of the pellet (gravimetric methods), the visual examination of partially reacted pellets [5], and the chemical analysis of (partially) reacted pellets [6]. Other techniques can also be used to advantage, e.g., the measurement of the magnetic susceptibility would seem to offer some promise.

The most frequently used methods in group (ii) involve the analysis of the gas stream for gaseous reaction products. Condensation or absorption of the water vapor produced has been found particularly useful in studying metal oxide reduction with hydrogen.

Gas analysis by infrared absorption or thermal conductivity cells can also be used for the continuous measurement of the composition of the gas exiting the reactor, particularly for simple gas mixtures. Gas chromatography might be used to improve the accuracy of the gas analysis, at the expense of making the analytical procedure intermittent rather than continuous.

The gravimetric method will receive the most attention here since, with equipment available at present, it is the most accurate, and once the equipment has been set up, can be used for many different gas–solid reactions without recalibration. In addition, the method is suitable for reactions in which there is no solid product (gasification reactions) or in which side reactions occur between gas components (e.g., cracking of hydrocarbons in reduction of oxides with natural gas).

THE GRAVIMETRIC METHOD

Perhaps the crudest (and most tedious) gravimetric method is to place several identical pellets, one at a time, into an environment of gaseous reactant at the temperature under study for a given length of time and measure the weight change on a common laboratory balance. Different pellets would, of course, be placed in the reaction apparatus for different lengths of time. However, this procedure relies on being able to maintain exactly the same conditions (of temperature, etc.) for all the pellets, which is a difficult task. A more satisfactory approach is to suspend a pellet within the gaseous reactant environment from one arm of a laboratory beam balance. The weight change can then be followed as the reaction of one pellet proceeds, giving several points on a weight versus time curve from which the whole curve may be interpolated. Moreover, the measurements may be made continuous by using a recording balance.

Several workers have used a quartz spring instead of a beam balance. The extension of the spring is followed through a window by a cathetometer and depends on the suspended mass. Examples of such work are Yannopoulos and Themelis [7] and Landler and Komarek [8]. At least one investigation has been carried out using a strain guage weight transducer [9].

Figure 6.1 is a schematic diagram of a typical system for studying gas–solid reactions using a recording balance. The apparatus depicted is used for studying the reduction of oxide pellets by hydrogen; the hydrogen is contained in the usual cylinder depicted to the lower left of the diagram. It is customary to bring the apparatus up to the desired temperature with an inert gas (in this case, helium) flowing through it. In the apparatus depicted, helium was

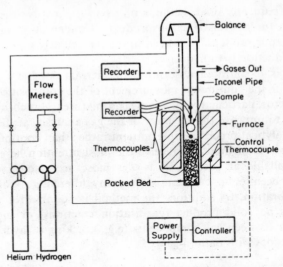

FIG. 6.1. Schematic diagram of the gravimetric apparatus for measuring gas–solid reaction rates.

also passed through the bell jar of the recording balance during reaction, to prevent ingress of hot hydrogen. Usually the reaction is carried out with a gas flow rate sufficiently high so that the progress of reaction is insensitive to the flow rate. For this reason, calibrated rotameters are of sufficient accuracy as gas flow measuring devices.

Most gas–solid reactions proceed at measurable rates only at elevated temperatures, commonly in the range 200–1400°C. The most convenient way of providing a constant high-temperature environment for carrying out gas–solid reactions is through the use of electrically heated tubular furnaces. A wide range of furnaces is available commercially for operation at temperatures from say 200°C up to 2500–3000°C. Resistance-heated tubular furnaces wound with, e.g., chromel–alumel wire may be used for temperatures up to about 1000–1100°C; a commonly used example of this is the Marshall furnace, a photograph of which is shown in Fig. 6.2. Higher temperatures, say up to 1500–1600°C, may be attained by platinum-wound furnaces or by molybdenum furnaces, which tend to be more expensive. Still higher temperatures will require resistance heating by silicon carbide or graphite, which can be used in reducing or neutral atmospheres only.

In ensuring the reproducibility of the results, it is essential that the furnace should have an adequate constant temperature zone where the reactant pellet is placed. Furthermore, the temperature of the furnace has to be maintained at the desired level within a specified accuracy, through the use of a

FIG. 6.2. Photograph of a typical tubular resistance furnace (courtesy of Varian Vacuum Division).

temperature controller. This temperature controller will then regulate the power supply from a "power pack," which then in turn is connected to the electric leads. In selecting the power pack, the controller, and the furnace from commercially available supplies, care must be taken that these meet the desired specifications with regard to (steadily) attainable temperature, accuracy of control, and compatibility of the system with the power supply available in the laboratory. The actual size of the hot zone (more precisely, the constant-temperature zone) will depend on the maximum size of the specimen to be reacted. This point will be discussed subsequently.

The sensing element of the controller (almost certainly a thermocouple, at the temperatures involved) should be placed close to the hottest part of the furnace heating element. This is preferable to placing the sensing element near the pellet, since the delay between a change in the power level to the furnace and the resulting temperature change at the pellet is so great as to render the system uncontrollable by the second method. However, this does imply that the flow rate of the gas must be held steady during a run and the apparatus must be given sufficient time to reach steady state before the run is commenced.

In the apparatus shown in Fig. 6.1 the reaction between the gas and the solid pellet is carried out in an Inconel pipe. Inconel has good oxidation resistance in air up to high temperatures and is therefore ideal for this purpose. The gases pass through a packed bed of conductive material before reaching the pellet, this bed serving to bring the gas up to the reaction temperature. Thermocouples are placed around the pellet and within the packed bed, to assist the experimenter in arriving at an isothermal region surrounding the pellet and for recording the pellet temperature during reaction. In case of highly exothermic or endothermic reactions it may be necessary to place thermocouples inside the pellet itself. Nowadays, thermocouples are available that are sufficiently fine to do this without interfering with the weighing or the reaction of the pellet. The leads for the intrapellet thermocouples may be twined around the pellet suspension wire, passing out of the reaction tube at some point above the furnace, or they may actually be the pellet suspension wires. It is desirable to connect the thermocouples to a recorder having an adjustable zero–adjustable range feature in order to detect the temperature differences of a few degrees at several hundred degrees centigrade.

With regard to the continuous weight measurement, several suitable recording balances are available (Cahn, Ainsworth, etc.) that are capable of weighing specimens ranging from about 10^{-2} g to about 100 g; some of these balances are sold complete with recorders.

The balances must be protected from the heat of the furnace and from some of the gaseous products and reactants (e.g., water vapor, sulfur dioxide). The authors are aware of at least one model that may be operated above atmospheric pressure and several that may be run below. These balances are typically sensitive to vibration and this must be borne in mind when mounting them.

In the apparatus shown in Fig. 6.1, there are baffles to prevent intrusion of hydrogen and water vapor into the balance or helium into the reaction zone. The pellet suspension wire passes through holes in these baffles. It must be fine, yet strong enough to withstand the load of the pellet at the reaction temperature. The authors found tungsten–rhenium wire, of the type commonly used as thermocouple extension wire, suitable for this purpose, under reducing conditions. Platinum would be suitable for oxidizing conditions. A tightly fitting "cage" is perhaps the best method of holding the pellet, allowing for the shrinkage that can occur with some reactions.

Those parts of the apparatus exposed to high temperatures or corrosive gases must be selected with the materials of construction in mind. This applies to frequently overlooked components such as gaskets and thermocouple gland packing.

Safety hazards arise from the frequently encountered condition of flammable gases and high temperatures. If possible, it is advisable to operate the inside of the apparatus at a few inches of water above atmospheric pressure, rather than at atmospheric pressure. In this way the likelihood of developing an explosive mixture within the apparatus is reduced and any gross leaks will be immediately obvious.

In dealing with flammable gases (hydrogen, carbon monoxide, etc.) care must be taken that these are discharged well outside the laboratory in a manner that eliminates the hazards of explosion. Reactions involving carbon monoxide, of course, present an additional hazard due to the toxicity of this gas.

Not shown in Fig. 6.1 are certain pieces of auxiliary equipment including the thermocouple cold junctions (where the temperature recorder or controller do not have built-in reference junction compensation), manometers for measuring pressure at various points in the apparatus, purification equipment for the gaseous reactant (e.g., drying tubes), the gas vent, cooling jackets or the like, downstream from the pellet (which can be used to lower the gas temperature to more manageable levels), and doors or flanges in the reaction tube (giving access to the pellet).

EXPERIMENTS FOR SOLID–SOLID REACTIONS

The commonly used experimental arrangements for studying solid–solid reactions (proceeding through gaseous intermediates) are very similar to those described on the preceding pages. However in this latter case, two additional factors have to be taken into consideration:

(i)   Since solid–solid reactions do not require the supply of reactant gas from external sources, we cannot preheat the reactant specimen in an inert gas stream to the reaction temperature as done previously; instead provision must be made for the rapid introduction of the solid reactant into the high-temperature zone. This may be done by lowering the solid pellet into the reaction zone by means of a winch and a chain, as shown in Fig. 6.3. While this arrangement may appear to be quite straightforward in principle, some practical difficulties may have to be overcome in making the system operational [10].

(ii)   In the majority of solid–solid reactions proceeding through gaseous intermediates, we are concerned with two coupled reaction systems (see Chapter 5). It follows that a simple weight change measurement alone cannot identify the progress of the individual steps of the reaction, so that in addition we also have to measure the composition of the gases exiting the system.

EXPERIMENTAL PROCEDURE

When the equipment is first built various parts (notably the recorders, the flow meters, the balance, and the thermocouples) must be calibrated and checked for proper operation. This should be repeated at periodic intervals, determined by the reliability of the equipment. Such calibrations are well known, self-evident, or covered in the manufacturers' literature and will not be discussed further.

A typical run, using a previously prepared pellet starts with weighing the pellet and measuring (using caliper gauges or a micrometer) its principal dimensions. For highly irregular pellets it might be necessary to determine the volume by a procedure such as the mercury displacement technique.

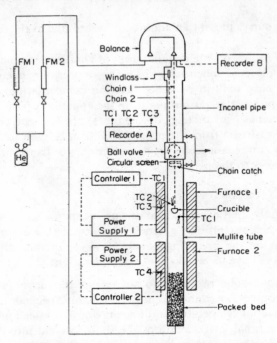

FIG. 6.3. Modified gravimetric apparatus for measuring the rate of solid–solid reactions.

The pellet is then placed in a wire cage, suspended in the reaction tube, and the thermocouples positioned around it so as to avoid contact between the thermocouples and the pellet or suspension wire. Any internal thermocouples are placed within the pellet and sealed in place at this stage. The reaction tube and balance bell jar are then sealed and the inert gas purge to the tube and bell jar is turned on. After several minutes of flushing, the furnace can be turned on and the heating up of the apparatus commenced. It is advisable to follow the weight of the pellet during the warm-up period in order to detect any weight changes as absorbed water, volatile binders, etc., are driven off.

When the desired temperature is approached, the power inputs to the various coils in the furnace are adjusted to obtain the desired axial temperature profile (an isotherm) in the vicinity of the pellet. The warm up and the attainment of the desired temperature profile will usually take at least one and perhaps as long as several hours, depending on the investigator's experience and the sophistication of his equipment. Attention must be paid to this last point if sintering, thermal decomposition, vaporization, etc., of the solid reactant can take place at the reaction temperature. In such cases it might be necessary to develop some arrangement for holding the pellet in the cooler upper part of the reaction tube and lowering it into place only when the desired isothermal zone has been achieved.

The inert gas flowing through the reaction tube is then replaced by gaseous reactant and the weight of the pellet, temperature, reaction tube pressure, and gas flow rate are monitored from that moment until the end of the reaction (indicated by no further weight change). After the reaction is completed, the furnace is shut off and the apparatus allowed to cool with a stream of inert gas passing through it. The reacted pellet can then be removed from the reaction tube, its principal dimensions measured, and further tests (such as those described in the second half of this chapter) may then be carried out.

OTHER TECHNIQUES FOR FOLLOWING GAS–SOLID REACTIONS

The reduction of oxides by hydrogen can be followed by measuring the hydrogen flow rate accurately (for example, using a wet-test meter) and determining the dew point of the gas downstream from the solid. It is of course necessary to dry the hydrogen entering the reactor in this case. Bickling [11] used an ingenious electronic dew point hygrometer to give a continuous record of the dew point of the gas.

An alternative procedure that can be used in the case of oxide reduction by hydrogen is that described by Parravano [12]. The reaction is followed by observing the pressure drop in a closed system. The hydrogen is recirculated to the reactor containing the solid pellet through a cold trap. This procedure suffers from the disadvantage that careful temperature control is necessary (throughout the whole closed system, not just the reaction zone). A second disadvantage is that the pressure change accompanying reaction introduces an unnecessary complication (the rate of reaction may be expected to depend on pressure). Modification to run such a system at a constant pressure (and measure the volume change) would seem possible.

The method can be extended to the reduction of metal oxides by carbon monoxide. or the roasting of sulfides in air or oxygen by, for instance, absorbing the gaseous product in sodium hydroxide solution.

There are many convenient devices for measuring the composition of two-component gas mixtures currently in use in gas chromatography. Foremost among these devices are thermal conductivity cells, flame ionization detectors, gas density balances, and the microcross-section detector. These sensors provide a continuous signal suitable for feeding into a chart recorder. When coupled with an accurate flow meter, such devices enable a reaction to be followed from the gas composition downstream from the pellet.

The thermal conductivity cell consists of two columns; the gas to be analyzed is passed through one of these while a reference gas (the pure gaseous reactant is the obvious choice) is passed through the other. Each column contains one or two electrically heated filaments that have a termperature-dependent resistance. The filaments are wired up as a bridge circuit such that the signal from the bridge depends on the difference between the thermal conductivity of the gas mixture being analyzed and the reference gas.

The cell is machined from a steel block and is held in a controlled ($\pm 0.2°C$) temperature bath. These detectors have the advantage of being useful for any binary gas mixtures showing more than a few percent difference in thermal conductivity. The disadvantages are temperature sensitivity (the reactor gas and reference gas must be brought to the cell temperature) and flow sensitivity. The detector requires calibration for every gas mixture to be analyzed.

The flame ionization detector works by measuring the current in a flame (into which the gas to be analyzed is passed) across which a potential difference of 200–400 V is applied by means of platinum electrodes. This sensor is not temperature sensitive and is capable of measuring very low concentrations. However, its primary usefulness is in detecting organic carbon atoms, and it is not suitable for hydrogen/water vapor, carbon monoxide/carbon dioxide, or oxygen/sulfur dioxide mixtures. Calibration is required for each gas mixture.

The microcross-section detector is a somewhat similar device, differing mainly in that ionization of the gas stream is produced by tritium adsorbed on a suitable substrate and only 75–100 V are maintained across the electrodes. To avoid stripping the trituim from the substrate, the gas temperature must be below 200°C and an AEC license is required to use such a detector. The device is most useful in analyzing mixtures of hydrogen and another gas, although it might be suitable for other gas mixtures. The current flow in these sensors is dependent on the molecular cross section (which is simply the sum of the well-known cross sections of the constituent atoms) and calibration is not required for all gas mixtures.

The gas density balance contains four electrically heated filaments connected in a bridge network, as in the thermal conductivity cell. However, at no time does the gas mixture being analyzed come in contact with the filaments, which has the advantage of preventing attack or fouling of the filaments. The reference gas stream is split on entering the cell, with roughly half passing over the filaments in the top part of the cell, and half through an identical channel in the lower part of the cell (mixing with half of the sample stream at a point downstream from where the reference stream passes over the filaments). With the reference gas and reactor gas of the same composition, the system is symmetrical and the bridge is balanced. With the reactor stream containing a component of different density than the reference stream, however, flow becomes asymetric and the bridge circuit becomes unbalanced. The detector is temperature sensitive and the same careful temperature control must be used as in the thermal conductivity cell. While attractive for handling corrosive gas mixtures, gas density balances are somewhat less sensitive than thermal conductivity cells. Care must be taken (particularly with mixtures containing hydrogen) to maintain the relative flow of reference and reactor gases at a level such that no component of the reactor gas diffuses back up the reference gas stream to the filaments.

The reader wishing to know more about these and other detectors for binary gas mixtures should consult texts on gas chromatography [13]. Obviously, a gas chromatograph can be used to analyze samples withdrawn from the downstream gas. Unfortunately, however, the time required to complete an analysis is of the order of one minute. For some fast reactions this may require numerous gas samples to be taken and stored for later analysis.

The ingenious reader will be able to think of other techniques for following reactions. However, the above, particulary the gravimetric method, are the most frequently used and therefore provide some guarantee of reliability. In most cases the design of the equipment will follow the route laid out for the gravimetric method at the end of this chapter. Several of the procedures described in this section are suitable for use in studying gas–solid reactions in packed and fluidized beds. In some cases, for instance reactions in which the solid product is vaporized at the reaction temperature, the gravimetric technique may be unworkable and downstream gas analysis must be employed.

6.4 Characterization of the Solid Structure

We saw in the earlier chapters that the rate of reaction between a solid and a gas may depend quite strongly on the size of the solid particles (grains) that make up the solid agglomerate. It is therefore appropriate to make a brief mention of the various techniques for particle size measurement. Further information may be found in the monographs by Irani and Callis [14] and by Allen [59] or in a recent review article by Davies [15].

Probably the simplest technique for particle size measurement is sieving. Conventional woven-wire screens can be used to analyze powders, most of whose particles fall into the range of 50 μm to 1 cm. The range can be extended downward to 10 μm by using electroformed screens, which are more accurate than woven wire screens up to about 200 μm. It is advisable to

(1) use a sufficient number of screens so that a smooth distribution curve can be drawn after the analysis,

(2) use a small enough sample and sufficient length of agitation time so that fines are not trapped on the upper screens by bridging, and

(3) ensure that the solid is free flowing and well dispersed (i.e., not clustered into multiparticle aggregates).

This last point is important in all particle size measuring techniques and it may be necessary to use a liquid sieving technique, together with dispersing agents, in difficult cases.

Microscopy is perhaps the most reliable (and tedious) technique, yielding information on particle shape as well as particle size distribution. Optical microscopy is suitable for the range 0.5–100 μm, whereas electron microscopes are suitable from 5 μm down to a limit that depends on the instrument, but is typically 0.001 μm. In optical microscopy or scanning electron microscopy the powder to be examined is usually made into a dilute dispersion in a suitable liquid (frequently water) or gel by working the powder into a few drops of the liquid on a microscope slide with a spatula, and then mixing the slurry with a large volume of liquid. The viscous forces involved in this "spatulation" are usually sufficient to break up aggregates of particles, but in certain cases it may be necessary to use a dispersing agent (e.g., common household detergent). A drop of the liquid is then placed on a slide, covered with a glass cover slip, and examined under the microscope. With a well-prepared dispersion, the field of view should contain a large number of individual particles. In the case of transparent particles, it is necessary to use a liquid of widely different refractive index for dispersing the particles, or to employ staining. In the case of electron microscopy, a drop of the liquid containing the dispersed particles is placed on a thin membrane (of collodion or similar organic material, coated with carbon) supported on a standard $\frac{1}{8}$-in.-diam. wire grid or on a standard scanning microscope sample holder. The liquid evaporates, leaving dispersed particles. Techniques for preparing the membranes are well known to electron microscopists and the reader is referred to a text such as Hall [16] for additional details. In some cases the sample prepared as above can be examined directly under the electron microscope. Usually, however, it is better to vacuum deposit a layer of gold, chromium, platinum, or other conducting material onto the membrane and particles. This prevents build up of charge (which could distort the image) on the sample in the electron beam and reduces any effects the beam might have on the solid. Such effects are the reduction of oxides and the melting or sintering of materials with low melting points. In the case of solids that do not scatter electrons to the required degree (solids containing only atoms of low atomic weight) a technique known as "shadow casting" is used to heighten contrast. This is vacuum deposition of metal from an angle of about 30° to (rather than normal to) the membrane surface. "Shadows" of the particles appear in the subsequent electron micrograph, not only increasing the particle contrast but allowing calculation of the particle height (from the length of the shadow and the angle of shadow casting) and estimation of its shape in three dimensions (from the shadow shape). Particles are sometimes shadow cast in two directions to enable more accurate estimation of shape.

The optical and electron microscopes may be used to determine a particle size distribution for particles that have been agglomerated into pellets. The optical technique is straightforward, involving little more than placing a fragment of the pellet under the objective lens and viewing the solid with reflected light. In some cases staining, oil immersion, or other standard microscopic techniques may be required. The procedure for the transmission electron microscope is more complex. The surface of the solid is "replicated" by one of several techniques. The simplest such technique is to place a drop of plastic solution (e.g., collodion in amyl acetate) on the surface and allow it to dry. The replica is then stripped from the surface (e.g., by floating on water), picked up on a copper grid, and shadow cast. In this way a "negative" replica (hills corresponding with valleys in the sample) is produced. Preparation of pellet fragments for examination by the *scanning* electron microscope is much easier. The fragment is mounted on a scanning microscope sample holder and a layer of metal vacuum deposited on it. The sample may then be placed in the instrument and viewed. Some scanning electron micrographs of reacted and unreacted iron oxide and nickel oxide specimens have been shown in Chapter 4.

In addition to the sometimes lengthy procedures involved in sample preparation, microscopic techniques are also tedious in that many particles need to be measured to obtain the particle size distribution. Just to obtain a mean particle size usually entails measurement of over a hundred particles, whereas over a thousand particle measurements might be desirable to obtain a particle size distribution. With the optical microscope measurements can be made at the instrument using a previously calibrated (by stage micrometer) eyepiece reticule scale or filar micrometer. An alternative is to make measurements from photomicrographs (slides, negatives, or prints), which is the standard procedure for both electron microscopes. Needless to say, care must be taken to obtain representative samples of the solid for specimen preparation and to avoid measuring the same group of particles or regions of the pellet twice.

Many automatic and semiautomatic devices are available for measuring particle size distributions at the microscope or from photomicrographs. Although these make the measurements much more rapid, their cost is not usually justified merely for investigations connected with gas–solid reactions.

Sedimentation techniques provide reliable particle size measurements in the size range 2–50 μm by gravitational settling, and 0.1–10 μm by using centrifuges. These methods are all based on the equality between the gravitational (or centrifugal) force and the drag force on a particle falling through a fluid at its terminal velocity. If flow is laminar, the fluid can be regarded as a continuum up to the particle surface, and where Brownian motion has a relatively small effect we can write

$$\varphi \, (\pi/6) \, d_p{}^3 (\rho_s - \rho_f) g = 3\pi\mu v d_p \qquad (6.4.1)$$

where ρ_s is the solid density, ρ_f the fluid density, d_p the particle diameter, μ the fluid viscosity, v the terminal velocity, and φ a factor that accounts for departure from sphericity and is frequently assumed to equal unity. In Eq. (6.4.1) g is the acceleration due to gravity or the centrifugal field in the centrifuge ($\omega^2 R$, where R is the centrifuge radius and ω the angular velocity). These conditions are satisfied for most fluid–particle systems within the size ranges given of 0.1–10 μm.

Using this equation as a starting point it is possible to calculate particle size distributions from rates of settling. Particle concentrations should be below 1%, and the method is only suitable for particles whose longest dimension exceeds its shortest by a factor of less than four. Terminal velocities are reached virtually instantaneously with particles below 50 μm in liquids. As with other particle measurement techniques, it is necessary that the particles be dispersed in the fluid, and one may have to use dispersing agents (at concentrations of roughly 0.1%) for this purpose.

The X-ray line-broadening technique is capable of high accuracy in the range of about 0.005–0.10 μm, provided the particles whose size is to be measured are individual crystals. This method is much less satisfactory if the particles are amorphous or made up of a number of crystals. Some loss of accuracy results if the particles contain many crystal defects (caused, for example by nonstoichiometry or certain impurities).

The technique is based on the broadening of X-ray diffraction lines from a crystal, which results as the crystal lattice departs from an ideal lattice. Such an ideal lattice would contain no defects and be infinite in extent. All real crystals show line broadening, either because of defect structure or simply because the lattice is finite. The smaller the crystal, the greater is the line broadening, until below the limit of approximately 0.10 μm it becomes measurable in most diffraction equipment and becomes much larger than defect broadening for most crystals. Below the limit of roughly 0.005 μm the adjacent lines in the diffraction pattern begin to merge and measurement becomes difficult. The technique yields an average particle size but not a particle size distribution.

The method consists of placing a sample of the powder in the beam of a conventional X-ray diffraction machine and measuring the width of a characteristic line for the solid. The powder is usually prepared by mixing with an organic adhesive, a few drops of the slurry being placed on a glass microscope slide and allowed to dry. The slide then serves to support the powder in the instrument. It is not necessary to disperse the powder (as is done in sample preparation for the transmission microscope or sedimentation procedures) and a fragment of an agglomerate may be glued to the slide. The sample should be sufficiently large that the detector can readily pick up its major diffraction lines. The source and detector slits should also be kept as narrow as possible, consistent with this requirement. The choice of a particular diffraction line is largely a matter of judgment based on two major considerations: the lines become broader with increasing diffraction angle (measured from the direction of the source beam), and the lines become weaker with increasing diffraction angle.

All X-ray diffraction equipment shows some line broadening simply due to instrumental characteristics (e.g., finite widths of source and detector slits) and it is necessary to determine this "instrumental broadening" and apply an appropriate correction in the calculation of particle size. Instrumental broadening is measured by means of a sample known to have a large crystal size, such as is readily available for calibration of diffraction equipment. A line close to that of the line from the particles to be measured is the best choice; in this way the instrument is calibrated simultaneously with the determination of the instrumental broadening at the appropriate diffraction angle. Instrumental broadening is usually weakly dependent on diffraction angle and is of the order of 0.1°.

X-ray line broadening measurement is a quick and accurate way of determining average particle size, within the range 0.005–0.10 μm, where it is known independently that the particles are individual crystals relatively free from defects. For a more detailed discussion of this technique, including methods of calculating particle size from the linewidth, the reader is referred to Azaroff [17] and Bartram [18].

The final particle size measurement device that we consider is the Coulter counter. This is one of a group of instruments called sensing-zone devices and its principle of operation is described as follows. A dispersion of the powder in a strong electrolyte solution is prepared, if necessary using a dispersion agent. The dispersion (which should be dilute) is allowed to flow from one chamber of the apparatus to another through an orifice somewhat larger than the largest particles in the slurry. Each chamber contains an electrode and a direct current is passed through the orifice. Each particle changes the electrical resistance of the orifice as it passes through it; the change is dependent on the size of the particle but (for nonmetallic particles) independent of the resistivity of the particle. The changes in resistance are detected as current or voltage fluctuations that pass to an electronic pulse height analyzer for transformation to particle size distribution readings.

The Coulter counter yields satisfactory results rapidly in the range 0.5–500 μm. Inaccuracies do occur, however, sometimes from the solvation of the particles by the electrolyte.

Further particle size measurement techniques are light scattering, cascade impaction, X-ray absorption and fluorescence, permeability, and adsorption. The last two are discussed further in the next section.

CHARACTERIZATION OF THE STRUCTURE OF AGGLOMERATES

Whenever possible the study of the structure of agglomerates should commence with determination of the size distribution (or at least, average size) of the particles of which the agglomerate is made up. In certain cases (e.g., if the solid is supplied to the investigator as an agglomerate that has been well sintered) particle size measurement may be out of the question. The investigator must then pass on directly to the characterization procedures discussed below

Optical and electron microscopy are probably the best techniques for studying the structure of a porous solid and determining mean pore size, pore size distribution, and the like; the principle drawback is the lengthy procedures involved in sample preparation and measuring enough pores to arrive at statistically meaningful numbers. Since microscopy has been dealt with at some length in the previous section, there is no further discussion of this topic here.

Adsorption, which was discussed in Chapter 2, provides a basis for measuring the surface area per unit mass (the specific surface) of a porous solid and also determining the pore size distribution in the pore size range 10^{-1}–10^{-3} μm, through the BET equation.

Once the specific surface of the solid is known from adsorption studies and the porosity of the model is measured by one of the techniques described below, then a mean particle size (or alternatively, a mean pore size) can be calculated by making assumptions concerning the geometry of the solid. Such an assumption might be that the solid consists of uniform spherical grains. Calculation of a mean pore size in this way is questionable, however, in cases where the grains making up the solid have some internal small pores. These "micropores" make an inordinate contribution to the specific surface and thereby cause a gross underestimation of grain size. The same problem is posed by surface roughness on the grains, although the effect is usually less severe. Calculation of mean pore diameter by assuming uniform cylindrical pores similarly leads to underestimation if the pores have surface roughness. The topic of structure determination by adsorption (and other methods) has received an extensive review by Dullien and Batra [19].

Mercury intrusion porosimetry is a convenient method of determining pore size distributions within the range 15×10^{-4}–2×10^3 μm. Let us consider a uniform cylindrical pore of radius r, which intersects the surface of a porous pellet. If the pellet is placed in a chamber that is first evacuated and then filled with mercury, the pressure required to force the mercury to enter the pore, thus overcoming surface tension forces, is given by

$$P = (2\sigma \cos \theta)/r \qquad (6.4.2)$$

where σ is the surface tension of mercury and θ the solid–mercury contact angle. Using this equation, a pore size distribution can be calculated from a plot of the volume of mercury intruded into the pellet against pressure. Notice that on increasing the pressure, the larger pores fill first, in contrast to the behavior in adsorption studies. Figure 6.4 shows a set of pore size distribution measurements using mercury porosimetry on hematite pellets reduced at various temperatures.

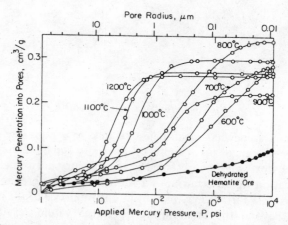

FIG. 6.4. Mercury penetration porosimeter curves for iron produced by the reduction of hematite ore with hydrogen at various temperatures [57].

The experimental technique is straightforward but there are many difficulties in interpreting the results of mercury intrusion. The contact angle θ is usually not known *a priori* and may depend on impurities on the solid surface or in the mercury. Most workers assume a contact angle of 140.°. The error introduced by this assumption is nearly always small compared with other errors involved in interpretation. "Ink bottle" pores shown in Fig. 6.5, fill only at a pressure corresponding to the diameter of the neck, and the volume of the pores is thus erroneously assigned too small a pore diameter. Such ink bottle pores will not completely empty as the mercury pressure is lowered and therefore the retraction (mercury volume versus pressure) curve will not coincide with the penetration curve.

A special problem is posed when the porous solid is an assemblage of individual grains rather than a continuum containing pores, because in these

Neck diameter

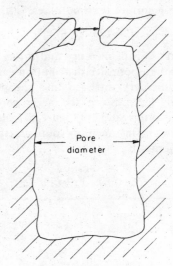

Pore
diameter

FIG. 6.5. An "ink bottle" pore.

cases it becomes somewhat meaningless to talk of a pore size distribution.
For an agglomerate made up of uniform spheres in some regular packing the
following behavior would be observed in the mercury porosimeter.

No penetration would occur until a "breakthrough" pressure is reached,
whereupon a stable surface may pass from one interstice to the next. Raising
the pressure beyond the breakthrough pressure would result in gradual
further penetration as the regions surrounding the points of contact of the
spheres are gradually filled with mercury.

This contrasts with the behavior of a continuous solid with uniform
cylindrical pores, all of the same diameter. Such a solid would fill completely
at the pressure determined by Eq. (6.4.2), and no further penetration would
occur on further increasing the pressure.

Mercury penetration into beds of uniform spheres has been studied by
Frevel and Kressley [20] and by Mayer and Stowe [21, 22]. Where it is known
a priori (for example, by microscopy) that the agglomerate is made up of
nearly spherical particles of roughly equal size, the procedures of these
workers can be used to determine a mean particle size.

Mercury intrusion porosimeter results should be regarded as semi-
quantitative, in other words, useful mainly in support of other measurements
or for comparison between closely similar structures. The previously mentioned review of Dullien and Batra [19] provides a good description of the
latest developments in this field.

Permeametry, the measurement of the rate of flow of a fluid through a porous medium under a known pressure gradient, is a technique by means of which a mean particle size (but not a particle size distribution) can be determined. The equation for the rate of fluid flow through a packed bed of uniform spheres is the semiempirical Ergun equation

$$\frac{\Delta P}{L} = \frac{150\mu V}{d_p^{\,2}} \frac{(1-\varepsilon_v)^2}{\varepsilon_v^{\,3}} + \frac{175\rho V^2}{d_p} \frac{(1-\varepsilon_v)}{\varepsilon_v^{\,3}} \tag{6.4.3}$$

where $\Delta P/L$ is the pressure gradient, μ the fluid viscosity, V the fluid velocity, ε_v the porosity, d_p the diameter of the sphere, and ρ the fluid density. The equation is applicable only when the interstices between spheres are much larger than the mean free path of the molecules of the fluid. If the fluid is a gas then the pressure drop across the bed must be sufficiently small so that V and ρ are essentially constant throughout the bed.

A mean particle size can be calculated from permeametry measurements using the Ergun equation provided that

(1) the particles making up the agglomerate are roughly spherical and are themselves nonporous,
(2) the particles fall within a narrow size range,
(3) the interstices are much larger than the mean free path of the fluid.

For most conditions encountered in permeametry measurements, the second term on the r.h.s. of the Ergun equation will be negligibly small compared with the first. The simplified equation obtained by striking out this negligible term is the Blake–Kozeny equation.

If conditions (1) and (2) are met, but the interstices are much *smaller* than the mean free path of the molecules, then the dusty gas model of Mason and coworkers [23] can be used to calculate a mean particle size from permeametry measurements. There is a major experimental difficulty involved in using permeametry at very small particle sizes, however. The fluid may "channel" through fissures in the agglomerate, which, although large compared with the interstices, may well pass unnoticed to the eye.

It should be clear from the above remarks that permeametry is an auxiliary, rather than primary, method for determining particle size. It is useful when other evidence (e.g., microscopy) confirms the applicability of the Ergun equation.

Low-angle X-ray scattering (as distinct from X-ray diffraction line broadening) has been used by Ritter and Erich [24] and Clark and Liv [25] for measurement of pore sizes in agglomerates.

Pore size distributions obtained by measurement of adsorption isotherms or mercury penetration porosimetry enable the calculation of the porosity of an agglomerate by simple integration. Indeed, the porosity can be read directly from the distribution curve if it is represented as the volume of pores greater than a certain size, per unit volume of agglomerate. We now briefly mention other techniques for determining the porosity of an agglomerate.

The simple technique of weighing an agglomerate of regular shape and then measuring its dimensions by means of caliper gauges or micrometers has already been mentioned. These measurements enable the calculation of an "observed" density for the agglomerate. The "true" density of the solid making up the agglomerate can then be found in a handbook (or by sintering the solid to zero porosity and again measuring its dimensions) and the porosity calculated from the relationship

$$\varepsilon_v = 1 - \rho_0/\rho_t \qquad (6.4.4)$$

where ε_v is the porosity, ρ_0 the observed density, and ρ_t the true density of solid making up agglomerate. The procedure can be extended to agglomerates of irregular shape if a liquid of known density can be found in which the agglomerate will sink and which will not wet the solid surface (i.e., intrude into the pores). The observed density of the agglomerate can then be calculated using Archimedes' principle.

Pycnometer techniques for determining densities of solids are well known. In essence they consist of weighing a vessel completely filled with a liquid of known density, displacing some of the liquid from the vessel with a fragment of solid of known weight, and then reweighing the vessel. Again it is essential that the liquid should not wet a porous solid. Mercury is a most suitable medium for the pycnometer liquid.

A pycnometer is a useful device for determining the true density of the solid making up the agglomerate in cases where this is not known. In this cases the pycnometer is filled with a liquid that *will* penetrate the pores under surface tension forces. It is usually necessary to dispel air trapped in the pores (e.g., by boiling the liquid or holding the solid under vacuum prior to immersion). Such liquids would also enable a true density determination by the method based on Archimedes' principle.

Pores within agglomerates may be roughly classified into three categories:

(a)   completely closed pores, which have no access from outside the agglomerate;

(b)   "dead end" pores (sometimes called "blind pores"), which have only one access (or multiple accesses in closely adjacent positions); and

(c)   open pores, which have two or more widely separated accesses.

Care should be taken in interpreting measurements on agglomerates since certain techniques will detect (or fail to detect) each category in different ways. For example, open pores will be detected by permeametry, mercury penetration porosimetry, adsorption, microscopy, and X-ray scattering. Dead end pores, however, would not be detected by permeametry and only microscopy (of sectioned agglomerate) and X-ray scattering would detect closed pores. Again, all the procedures described above for measuring porosity detect open, dead end, and closed pores if a " handbook " value is used for the true density. If, however, pycnometry is used with a wetting liquid for determining the "true" density, then the closed (and perhaps dead end) pores will not be detected. The X-ray scattering technique detects open, dead end, and closed porosity under all circumstances.

6.5 Measurement of Diffusion Coefficients

ORDINARY MOLECULAR DIFFUSION COEFFICIENTS

The experimental determination of ordinary molecular diffusion coefficients has been recently reviewed by Mason and Marrero [26] and by Westenberg [27] and is given only brief coverage here.

Figure 6.6, due to Marrero and Mason [28], summarizes the techniques most frequently used for measuring ordinary molecular diffusivities. The figure is largely self-explanatory. The three unsteady state methods obtain results by measuring the change of concentration with time, as do the two-bulb and capillary leak quasi-steady state methods. The steady state methods rely

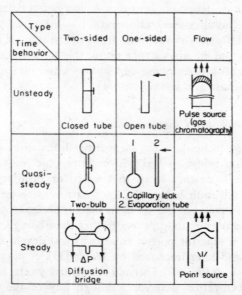

FIG. 6.6. Major experimental methods for measuring diffusion coefficients [26].

on measuring concentrations at two (or more) points within the apparatus, together with gas flow rate(s). In the evaporation tube technique the diffusivity is determined from the rate of volume (or mass) change of the liquid. In order to use the technique it is necessary to know the vapor pressure of the liquid at its temperature within the tube. Mason and Marrero judged the reliability of the closed-tube and two-bulb methods to be good (with an accuracy of better than 5%); that of the point source, gas chromatograph, open tube, and diffusion bridge procedures to be average (with an accuracy of about 5%); and the capillary leak and evaporation tube equipment to be poor.

The point source, gas chromatograph, and diffusion bridge procedures may have the advantage at extreme temperatures and pressures over the methods rated good. This is because it is easier to design the equipment so that no mechanical moving parts are subject to extreme conditions. Pakurar and Ferron [29] have used the point source method at temperatures up to 1800°K. Giddings and coworkers [30] have measured diffusivities up to 1360 atm by the chromatographic technique.

It should be noted that the tube in the gas chromatography technique contains no packing. Ideally, a small quantity of gas A is introduced instantaneously into the laminar stream of gas B that flows through the tube. The spreading of the peak (measured by a suitable gas chromatograph detector) is due to both axial and radial diffusion of A in B. The point source method employs a *continuous* injection of A into B at a *point* within the tube. The steady state radial concentration profile is measured at a point downstream and the diffusivity determined from the profile. A technique that is somewhat similar but employs a tube permeable to A and entails measuring the much more easily determined cup mixing concentration has been described by Collingham and coworkers [31].

DIFFUSIVITIES IN POROUS MEDIA

The two major techniques for measuring diffusivities in porous media are by the diffusion bridge and gas chromatographic peak broadening. The equipment differs somewhat from that used for ordinary molecular diffusivities. In the diffusion bridge the long narrow tube (see Fig. 6.6) is replaced by a porous pellet of the solid. In the gas chromatograph the column is packed with the porous solid pellets (rather than being left empty).

The diffusion bridge for porous solids was first developed by Wicke and Kallenbach [32] and later improved by Weisz [33]. A diagram of a typical apparatus is shown in Fig. 6.7. Thermal conductivity cells have usually been used as the concentration detector although any of the detectors (flame ionization, etc.) described earlier in this chapter could also be employed.

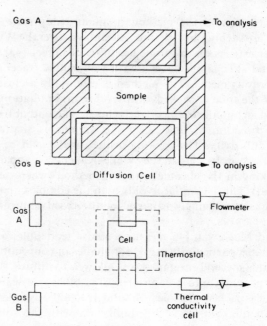

FIG. 6.7. Schematic diagram of the Wicke–Kallenbach diffusion apparatus or diffusion bridge.

Care has to be taken in designing the equipment to ensure that boundary layer effects are not present at the ends of the porous plug and that the pressure gradient across the plug is zero. Typical experimental investigations in which the Wicke–Kallenbach apparatus was used are described in the literature [34–39].

The disadvantages of the Wicke–Kallenbach apparatus are

(a)   single samples of cylindrical shape must be used,

(b)   cylindrical pellets pressed in a conventional die may display inhomogeneity, which could effect the diffusivity measurements [40];

(c)   difficulties may be encountered in sealing the various components of the cell at extreme temperatures and pressures;

(d)   diffusion into dead end pores goes undetected (the method is a steady state one) although such pores play a part in gas–solid reactions.

The chromatographic peak-broadening technique, as applied to porous solids, was put forward by Davis and Scott [41] and further developed by Schneider and Smith [42]. Analysis of the experimental results is somewhat easier if the gas component injected into the column is not adsorbed on the solid. Because several pellets are used in the column, no seals are needed

around the pellets, and the method is an unsteady state one, chromatographic peak broadening has obvious advantages over the Wicke–Kallenbach
method. Other articles on this topic are those of Leffler [43] and Eberly [44].

Bennett and Bolch [45] have used a radioactive tracer to measure diffusivities in porous media. The method is essentially a closed-tube one, because most of the tube is packed with solid with a radiation detector placed
at one end and an ampule of tracer at the other. The output from the radiation
detector as a function of time, after the ampule is shattered, is used to
calculate the diffusivity. The packing in the tube should be uniform (in the
same way as the porous plug in the Wicke–Kallenbach apparatus) as distinct
from the packing in the chromatographic method (where several individual
pellets are used). This technique avoids both the complex analysis of chromatographic peak broadening and some of the disadvantages of the Wicke–
Kallenbach apparatus.

Olsson and McKewan [46, 47] have used a technique for measuring gas
diffusivities that is particularly suited for studies in connection with gas–solid
reactions. In this method reactant gas is allowed to diffuse, through a porous
plug of solid product, into a canister that is otherwise sealed. The canister
contains loosely packed granules of solid reactant. By operating such that
the gas within the canister is in equilibrium with the solid granules, the
diffusivity of the gases within the porous plug can be deduced from the rate
of change of the weight of the canister. The work of these authors is one of
the rare instances where diffusivities were measured through a reacting
solid matrix.

In connection with their work on gas–solid reactions Hills and coworkers
[48] and Evans [3] have described novel techniques based on the Wicke–
Kallenbach apparatus. The former used hollow spheres instead of the conventional diffusion cell; the latter followed the quasi-steady state diffusion
in a Wicke–Kallenbach type diffusion cell (but with gas entrance and exit
lines closed) by observation of the movement of a mercury plug in a horizontal glass tube connected across the cell.

There are two important points in connection with determination of
diffusivities in porous media for gas–solid reaction studies:

(1) For a binary gas mixture it is frequently necessary to measure two
or even three diffusivities. For example, if diffusion in the porous medium
during the reaction is Knudsen type then it will be necessary to measure the
Knudsen diffusivity of each component. If the diffusion is in the transition
regime then the investigator will need to measure the effective molecular
diffusivity within the solid, in addition to the Knudsen diffusivities. Only
in the purely molecular diffusion regime will one diffusivity suffice.

(2) It is difficult to determine diffusivities within a reacting solid matrix.

Apart from the more mundane experimental problems of maintaining seals, ensuring isothermality, etc., at high temperatures it is clearly next to impossible to measure the effective diffusivity of gaseous reactant through a porous solid reactant or (except when reaction is irreversible) that of a gaseous product through solid product.

Both these points should be born in mind when planning diffusion studies. Here we are at the limits of present-day knowledge. Few investigators of gas–solid reactions have carried out independent measurement of even a single gas diffusivity within their solid, even fewer have done so at reaction conditions, and to the best of our knowledge, none has measured more than one diffusivity.

6.6   The Design of Experimental Studies in Gas–Solid Reaction Systems

In this section we shall attempt a synthesis of the ideas outlined in Chapters 2–4 and in the preceding sections of this chapter to outline the design of an experimental study of gas–solid reaction systems. This integration of experimental studies with the previously described mathematical model development is thought to be highly desirable, because experimental verification is an essential ingredient of any modeling work. Moreover, experimental work must form the basis of the improved mathematical representation of reaction systems.

As discussed earlier, provided the reaction is thermodynamically feasible, the overall rate at which a solid pellet reacts with a gas may be limited by

(1)   gas phase mass transfer of the reactants or the products,
(2)   pore diffusion,
(3)   chemical kinetics, including the kinetics of adsorption.

The objective of the experimental study is then to establish which of these steps or what particular combination thereof is rate controlling over the conditions of interest and then to develop a detailed inderstanding of this rate-controlling step(s). The term *conditions of interest* needs a more precise definition. If the experimental study is carried out as part of an application-oriented (industrial) program where some of the variables are fixed by certain extraneous constraints (e.g., fixed pellet size, reaction temperature, gas composition) then it might be sufficient to identify the rate-controlling step for this limited range of variables.

The conditions of interest would be rather more broadly defined for more fundamental studies where one wishes to explore a much broader range of temperatures, particle sizes, etc. Let us now proceed by considering the various possible rate-contolling mechanisms.

MASS TRANSFER

Perhaps the first step in experimental studies of gas–solid reactions is to establish whether gas phase mass transfer is rate controlling, or at least plays an important role. This may be done by reacting specimens of regular geometry (preferably spheres) with a moving gas stream at various gas velocities.

Recall Eq. (2.2.29), which provides a very simple relationship between the extent of reaction (conversion) and time:

$$X = \frac{bh_D(C_{A0} - C_{C0}/K_E)}{(1 - \varepsilon)\rho_s} \left(\frac{A_p}{V_p}\right) \frac{K_E}{1 + K_E} t \qquad (2.2.29)$$

It is seen from the form of Eq. (2.2.29) *that mass transfer control implies a linear relationship between the extent of reaction and time;* as we discussed in Chapter 2, the mass transfer coefficient appearing in Eq. (2.2.29) depends on the linear gas velocity past the particle. It follows that if measurements are carried out with identically sized particles at the same temperature but show a dependence on the linear gas velocity, this is an indication that mass transfer plays an important role.

A word of caution must be interjected here; at low gas velocities there may be an apparent dependence on the gas velocity because of *reactant starvation.* Starvation occurs when a sufficient portion of the reactant is consumed by the solid, so that in actual fact, the particle is not contacted with a gas of bulk composition C_{A0}, but the gaseous reactant concentration is less than this value. When operating in this region the effective driving force will depend on the gas velocity irrespective of whether the process is mass transfer controlled. In planning experiments care must be taken to avoid this starvation phenomenon; whether this criterion is met can be checked by a simple mass balance. All the above considerations were essentially qualitative. The real quantitative verification that a process is mass transfer controlled is by the fact the the actual extent of reaction as a function of time is then predictable from Eq. (2.2.29), where the mass transfer coefficient h_D is calculated from the appropriate correlation of the pellet geometry and Reynolds number. (A selection of such correlations was presented in Chapter 2.) Alternatively, mass transfer control may be proved conclusively, if the experimentally measured mass transfer coefficients are found to agree with those predicted on the basis of the appropriate mass transfer correlations.

Quite a number of gas–solid reactions have been shown to be mass transfer controlled over limited ranges of temperature and for moderate pellet sizes. These include the reduction of cupric oxide with hydrogen at 400–900°C studied by Themelis and Yannopoulos [49], and the chlorination of nickel oxide at 1200°C measured by Fruehan and Martonik [50].

As shown in Figs. 6.8 and 6.9 the measurements of these investigators are indeed consistent with the mass transfer correlations quoted in Chapter 2.

We stress to the reader that a blanket statement that a given reaction "is mass transfer controlled" could be very misleading unless one specifies the actual experimental conditions (temperature, particle size, gas velocity, etc.). Once mass transfer control is established for a given range of parameters, further experimental work within the same range would be of only limited usefulness because the correlations for customary geometries are well established and further confirmation of these by additional data, perhaps on "more exotic" systems, is hardly necessary.

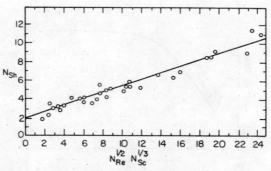

FIG. 6.8. Mass transfer by forced convection to a cupric oxide particle during reduction by hydrogen [49].

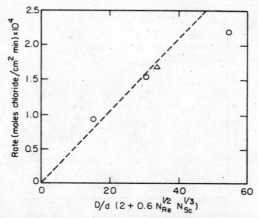

FIG. 6.9. Rate of chlorination of nickel oxide spheres at 1473°K compared with a rate calculated from a mass transfer correlation (– – –) [50], $\bigcirc$, NiO dense; $\triangle$, NiO porous.

In many investigations, e.g., in development work, it is sufficient to establish that the process is mass transfer controlled within the desired range of variables and then a design can be developed through the use of the appropriate correlations.

In other fundamental studies *one may wish to exclude mass transfer effects deliberately* so that some of the rates of other steps may be determined with some accuracy.

Mass transfer effects may be minimized by operating the gas–solid reaction system at sufficiently high gas velocities, so that any further increase in the gas velocity does not produce an increase in the overall reaction rate, as

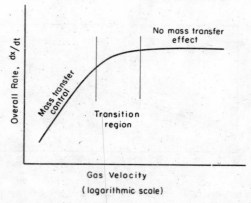

FIG. 6.10. The minimization of external mass transfer effects by operation at high gas velocities.

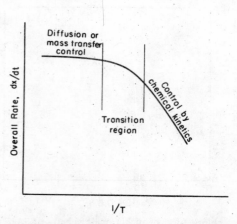

FIG. 6.11. The mimimization of external mass transfer or pore diffusion effects by operation at lower temperatures.

shown in Fig. 6.10. Under these conditions some factor other than mass transfer would become rate controlling. While this procedure appears attractive in principle, it may not be feasible in practice, particularly as the mass transfer coefficient is proportional to the square root of the gas velocity and high gas velocities would interfere with the operation of conventional recording balances.

Mass transfer effects may also be minimized by operating the system at lower temperatures; since mass transfer involves gaseous diffusion through a boundary layer, which is not an activated process, while the chemical reaction between the gas and the solid generally has a substantial activation energy, operation at progressively lower temperatures will eventually lead to control by chemical kinetics, as shown in Fig. 6.11.

KINETIC PARAMETERS

In the study of gas–solid reactions it is of particular importance to devise a satisfactory means for determining the kinetic parameters, because unlike gas phase mass transfer and pore diffusion, the kinetic parameters cannot be predicted or estimated, but must be measured esperimentally.

In the experimental determination of kinetic parameters it is very important to ensure that the measurements are carried out *under conditions such that the overall rate is indeed controlled by chemical kinetics*, i.e., pore diffusion and gas phase mass transfer do not play an appreciable role. The apparent falsification of kinetics measurements, when these criteria are not met, has been discussed at some length in Chapters 3 and 4. It may be worthwhile to reemphasize here that one of the principal attractions of the structural models for gas–solid reactions described in the earlier chapters is that they allow us to separate the effects of chemical kinetics, pore diffusion, and gas phase mass transfer.

If we consider the formulation developed in Chapter 4 for the reaction of porous solids, we may recall that the following approximate expression was suggested for relating the extent of reaction X to the dimensionless time t^*:

$$t^* = g_{F_g}(X) + \hat{\sigma}[p_{F_p}(X) + (2X/N_{Sh}^*)] \tag{6.6.1}$$

where

$$\hat{\sigma} = \frac{V_p}{A_p}\sqrt{\frac{(1-\varepsilon)kC_{A0}^{n-1}F_p}{2D_e}\left(\frac{A_g}{F_g V_g}\right)\left(1+\frac{1}{K_E}\right)} \tag{6.6.2}$$

is the dimensionless reaction modulus.

Inspection of Eq. (6.6.1) shows that in a mathematical sense $\hat{\sigma} \to 0$, $N_{Sh}^* \to \infty$, represent the criteria for control by chemical kinetics. In a practical, physical sense this means that experiments with small pellets ($V_p/A_p \to 0$)

at low temperatures ($k \to 0$) and high gas velocities would meet these criteria. In practice N_{Sh} need not be very large to avoid limitation by external mass transfer.

Many earlier investigators sought to ensure operation in the chemically controlled regime by performing experiments at low temperatures; in the majority of cases this is not a very satisfactory way to proceed because it may confine the study to too narrow a temperature range and would not give reliable information on the activation energy.

It follows that measurements with fine powders or with compacts having a short diffusion path (e.g., thin disks) at high enough gas velocities to exclude mass transfer effects would represent the preferable approach for determining the kinetic parameters.

As seen from Eq. (6.6.1), when $\hat{\sigma} \to 0$ the following asymptotic solution will be valid:

$$t^* = g_{F_g}(X) = 1 - (1 - X)^{1/F_g} \tag{6.6.3}$$

As shown in Chapter 4, for small values of $\hat{\sigma}$ the following approximate solution may be derived by series expansion:

$$1 - (1 - X)^{1/F_g} = (1 - C_1 H R_p^2)(bk C_{A0}/r_g \rho_s)t \tag{6.6.4}$$

where

$$H = \hat{\sigma}^2/R_p^2 \quad \text{and is independent of } R_p \tag{6.6.5}$$

Thus, $C_1 = \frac{2}{3}$ for $F_p = 1$, $F_g = 1$; $C_1 = \frac{2}{5}$ for $F_p = 3$, $F_g = 1$, etc.

The kinetic parameters may then be obtained from measurements conducted in the region of small σ values (say, $\hat{\sigma} < 0.3$) in the following manner:

By conducting experiments under otherwise fixed conditions, for small but varying pellet size R_p a series of linear plots may be prepared according to Eq. (6.6.4). The slopes of the $g_{F_g}(X)$ versus t plots (or at least their linear portions) may be designated S; thus we have

$$S = (bk C_{A0}/r_g \rho_s) - (C_1 H bk C_{A0}/r_g \rho_s)R_p^2 \tag{6.6.6}$$

On plotting S versus R_p^2 the intercept with the $R_p^2 = 0$ axis provides k/r_g since C_{A0} and ρ_s are known.

This procedure may be used for obtaining the intrinsic rate constants and by conducting similar experiments at different temperatures we can deduce the activation energy associated with this intrinsic rate constant.

This procedure is illustrated in Fig. 6.12, using experimental data reported on the reduction of nickel oxide with hydrogen. Figure 6.12a shows the actual experimental data on a plot of $g_{F_g}(X)$ against the dimensional (i.e., real) time; also indicated on this plot is the theoretical line, corresponding to a zero diffusion path length. Figure 6.12b shows the linear plot according to Eq. (6.6.6) from which the slope S and thus the quantity k/r_g may be deduced.

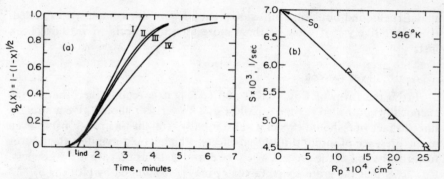

FIG. 6.12. Illustration of the extrapolation of kinetic data to zero particle size in order to minimize pore diffusion and external mass transfer effects [51]. $P = 1$ atm, $T = 546°K$; R_p (cm): (I) 0, (II) 0.0343, (III) 0.0440, (IV) 0.0503.

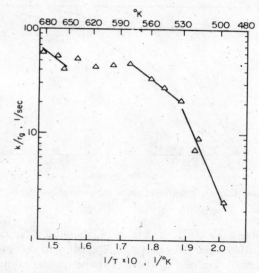

FIG. 6.13. Arrhenius plot of reaction rate data on the reduction of nickel oxide by hydrogen [51].

Figure 6.13 shows a plot of the k/r_g values obtained for the reduction of nickel oxide, through the use of the procedure outlined above. It is readily observed that the temperature dependence of the reaction rate constant obeys a rather involved relationship that indicates the existence of various solid state transitions for NiO.† Had the measurements been conducted

† These transitions have been reported by other investigators. For a detailed discussion of these pheomena the reader is referred to [3, 10].

under such conditions that diffusional effects were not completely excluded, it is quite likely that some of these more intricate effects would have been lost.

GRAIN SHAPE FACTOR

From the form of Eqs. (6.6.3) and (6.6.4) it is readily seen that in the chemically controlled regime a plot of $g_{F_g}(X)$ versus t should give a straight line; in fact one should expect a substantially straight-line relationship even in the presence of minor diffusional effects, which would manifest themselves towards the end of the process when the effective diffusion path is the greatest. It follows that the grain shape factor F_g may be evaluated by plotting $g_{F_g}(X)$ against time for various trial values of F_g and then selecting the value of F_g that provides the best straight-line relationship.

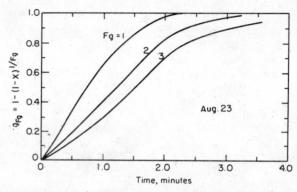

FIG. 6.14. Example of the determination of the grain shape factor. F_g by curve fitting [51].

The procedure is illustrated in Fig. 6.14, showing a typical set of results reported in [51], which indicates that $F_g = 2$ appears to show the best data fit. Note that the assumption of a cylindrical shape for the grains would be unrealistic on physical grounds; however, a more appropriate interpretation of the value of F_g found would be to state that the shape was intermediate between a slab and a cube (sphere).

PORE DIFFUSION

In the preceding we have discussed the planning of experiments with a view of delineating the effect of gas phase mass transfer and chemical kinetics. All that remains is to comment briefly on pore diffusion.

If either the solid reactant or the solid product are porous, pore diffusion could play an important role in determining the overall reaction rate. On the preceding pages some implicit comments were made on how pore diffusion effects may be eliminated or minimized by performing measurements for conditions corresponding to $\hat{\sigma} \to 0$, which in a physical sense would correspond either to low chemical reaction rates or to short diffusion paths, so that the relative importance of pore diffusion effects would be small.

In contrast to the case of the kinetic parameters, which must be determined directly through measurements on reacting systems, *it is generally preferable to perform independent measurements with a view of obtaining values of the pore diffusion coefficient.* In investigations aimed at the fundamental study of gas–solid reactions it is desirable to carry out independent measurements to characterize the nature of the reactant and product solids, such as described in Section 6.4, to be accompanied by the direct determination of the pore diffusion coefficient, using the techniques presented in 6.5.

Clearly, a sense of proportion has to be retained here as the extent to which such studies are justified will have to depend on the relative importance of diffusion (and structural changes) in the overall process sequence. While it is less desirable than direct measurement, the numerical values of the pore diffusion coefficient D_e may also be deduced from actual kinetic measurements conducted in the region where the overall rate is limited by pore diffusion.

As shown in Chapter 4, when $\hat{\sigma} \to \infty$,

$$p_{F_p}(X) = t^*/\hat{\sigma}^2 \qquad (6.6.7)$$

where

$$p_{F_p}(X) = \begin{cases} X^2 & \text{for} \quad F_p = 1 \\ X + (1 - X)\ln(1 - X) & \text{for} \quad F_p = 2 \\ 1 - 3(1 - X)^{2/3} + 2(1 - X) & \text{for} \quad F_p = 3 \end{cases} \qquad (6.6.8)$$

For large values of $\hat{\sigma}$, say $\hat{\sigma} > 3.0$, the following approximate relationship holds:

$$p_{F_p}(X) = (t^*/\hat{\sigma}^2) - (1/2\hat{\sigma}^2) \qquad (6.6.9)$$

or

$$p_{F_p}(X) = \frac{2bF_p D_e C_{A0}}{R_p^2(1 - \varepsilon)\rho_s(1 + 1/K_E)} t - \frac{1}{2\hat{\sigma}^2} \qquad (6.6.10)$$

Thus on plotting $p_{F_p}(X)$ against t, the value of D_e may be calculated from the slope. While it would be desirable to conduct independent measurements of D_e, the procedure outlined here provides a convenient means for ascertaining whether these independent measurements are, in fact, consistent with the experimental data (a further discussion of this procedure is available in the literature [51]). A similar argument was developed by Weisz and Prater [52].

and Weisz and Goodwin [53], who have shown that the effective diffusivity in a porous pellet can be determined by measuring the reaction rate of gas–solid reaction controlled by pore diffusion, using the combustion of coke deposited in silica–alumina catalyst pellets. Kawasaki *et al.* [54] found reasonable agreement between estimated values of the diffusivity and those calculated from the rates of iron oxide reduction by gases controlled by product layer diffusion.

CLOSURE

We summarize the recommended procedure for the planning of experimental studies of gas–solid reactions as follows:

(1)   Establish the thermodynamic feasibility of the reaction, or more precisely define the range of temperatures, pressures, and reactant compositions where the reaction is feasible.

(2)   Determine the range of variables (temperature, particle size, gas velocity, etc.) over which the reaction is mass transfer controlled. For problems motivated by industrial research objectives the definition of these limits may be adequate for design purposes under certain circumstances.

(3)   If the process is controlled by factors other than mass transfer over the range of variables of interest, conduct experiments corresponding to the $\hat{\sigma} \to 0$ region, with a view of determining the kinetic parameters, viz., k/r_g and F_g. As discussed earlier, this may be done by working with small particles, thin disks, fine powders, and at low temperatures.

(4)   If pore diffusion is found to play a role in the overall rate, it is desirable to conduct independent studies aimed at the characterization of the porous solid through the use of electron microscopy, pore size distribution measurements, etc. These studies should be accompanied by the direct determination of the pore diffusion coefficient.

The pore diffusion coefficient thus obtained can then be checked for consistency with the pore diffusion coefficient that may be deduced from actual kinetic measurements conducted in the region of large $\hat{\sigma}$ values.

The parameters obtained through the measurements (and calculations) described in (2)–(4) may then be used for predicting the behavior of the system in any intermediate regime, i.e., where mass transfer, pore diffusion, and chemical kinetics may all play a major role.

Let us conclude this chapter by presenting a substantial example to illustrate how these considerations apply to the actual design of an experimental study of a gas–solid reaction.

Example 6.6.1   Design an apparatus to study the reduction of spherical ferric oxide pellets in hydrogen. The maximum diameter of the pellet to be studied is 2 cm. The minimum diameter of the pellet to be studied is 0.4 cm. It is not anticipated that reactions will be carried out below 400°C or above 1000°C. From the work of previous investigators it is not expected that the rate of reduction will exceed 6.2% per min. Work will be carried out at or near atmospheric pressure. Errors exceeding 2% are not tolerable and for this reason water vapor in the reactor should be kept below 2%. The vent pipe for the system must be 50 ft long.

The density of ferric oxide is 5.12 g/cm³. It follows that the most massive sphere to be studied is

$$5.12 \text{ g/cm}^3 \times \tfrac{1}{6}\pi \times 2^3 \text{ cm}^3 = 21.5 \text{ g}$$

(In fact, 2-cm-diam. spheres would typically weigh less than this due to porosity.) 30% by weight of Fe_2O_3 is oxygen, and therefore the maximum weight loss is $21.5 \times 0.3 = 6.43$ g. It is a simple matter to calculate that the smallest pellet will suffer a weight loss of 0.0516 g on reduction. A balance can now be selected. It should have a capacity of at least 30 g (allowing a few grams over the 21.5 g heaviest pellet for suspension wire, etc.) and be accurate to 0.5 mg if the reduction of the smallest pellet is to be studied within an error of 1%.

Two-inch schedule-40 pipe (in stainless steel or Inconel) would be a suitable reaction tube having an i.d. of 5.25 cm = 2.6 times the largest pellet diameter. The o.d. of such a pipe is 2.375 in., which would fit within resonable clearance inside a 2.5-in. i.d. tubular furnace. A reasonable heated length for the furnace is 5×2.5 in. = 12.5 in. Chromel wire should be adequate for the furnace coils. Sufficient information has now been obtained to order a tubular furnace.

If the most massive sphere (oxygen content = 6.43 g) reduced at the maximum rate of 6.2% per min, the amount of hydrogen required for reduction would be

$$\frac{6.43 \text{ g}}{16 \text{ g (g-mole)}^{-1}} \times 0.062/\text{min} \times \frac{1}{60} \text{ min/sec} = 4.16 \times 10^{-4} \text{ g-mole/sec}$$

If the hydrogen for reduction is to be no more than 2% of the total hydrogen flow through the reactor (i.e., water vapor in the hydrogen is tolerable only up to the limit of 2%), then

$$\text{total hydrogen flow} = (4.16 \times 10^{-4})/0.02 \text{ g-mole/sec}$$

$$= 208 \times 10^{-4} \text{ g-mole/sec} \equiv 30.5 \text{ liters/min}$$

at standard conditions (1 atm and 25°C).

It is now necessary to calculate the mass transfer coefficient between pellet and gas at this flow rate. It is a relatively simple matter to show that the mass transfer coefficient increases with increasing temperature and decreasing sphere size. Mass transfer between pellet and gas will pose a resistance to overall reaction then, at the lowest temperature (400°C) and for the largest pellet (2 cm diameter), if at all. It follows that we need to calculate the mass transfer coefficient only for this one condition. At 400°C and 1 atm,

$$\text{molar density of hydrogen} = P/RT = 0.181 \times 10^{-4} \text{ g-mole/cm}^3$$

Thus

$$\text{actual gas flow rate} = 208 \times 10^{-4}/0.181 \times 10^{-4} \text{ cm}^3/\text{sec} = 1150 \text{ cm}^3/\text{sec}$$
$$\text{cross-sectional area of reaction tube} = \tfrac{1}{4}\pi \times 5.25^2 \text{ cm}^2 = 21.6 \text{ cm}^2$$

Hence

$$\text{gas velocity} = 1150/21.6 \text{ cm/sec} = 53.2 \text{ cm/sec}$$

At 400°C and 1 atm,

$$\text{hydrogen density} = 0.362 \times 10^{-4} \text{ g/cm}^3$$
$$\text{viscosity} = 1.53 \times 10^{-4} \text{ poise}$$
$$\text{Reynolds number} = 2 \times 53.2 \times 0.362 \times 10^{-4}/1.53 \times 10^{-4} = 25$$

The mass transfer coefficient has already been calculated in Chapter 2 using the Ranz–Marshall correlation; its value is 9.0 cm/sec. Therefore the mass transfer rate from gas to pellet under conditions of mass transfer limitation is

$$\pi h_D C_{A0} D^2 = 3.15 \times 9.0 \text{ cm/sec} \times 0.181 \times 10^{-4} \text{ g-mole/cm}^3 \times 4 \text{ cm}^2$$
$$= 20.5 \times 10^{-4} \text{ g-mole/sec}$$

But this is almost five times the maximum anticipated reaction rate of 4.16×10^{-4} g-mole/sec. We therefore conclude that a gas flow rate of 30.5 liters/min is sufficiently large (in the apparatus described) that mass transfer between pellet and gas stream plays only a small part in determining the overall progress of reaction. Under these conditions, negligible error is introduced by using one of the correlations of Chapter 2 for mass transfer coefficients in calculations.

The gas flow rate of 30.5 liters/min of hydrogen (at standard conditions) enables a suitable flow meter to be ordered. A pressure drop of 1 in. of water along the 50-ft vent pipe is reasonable; given this, the vent pipe diameter is calculated as follows. The flow is almost certainly laminar under these conditions. Let us assume that most of the gas in the vent pipe is at room temperature (25°C) and that the small amounts of water vapor can be ignored. At this temperature the viscosity of hydrogen is 98×10^{-6} poise, and the density $= 0.818 \times 10^{-4}$ g/cm^3. Poiseuille's equation gives

$$(\text{vent pipe radius})^4 = \frac{\text{flow rate} \times 8 \times \text{viscosity}}{\pi \times \text{pressure gradient}}$$

$$\text{pressure gradient} = 0.02 \text{ in. H}_2\text{O/ft} \equiv \frac{0.02}{13.6 \times 12 \times 2.54} \text{ in. Hg/cm}$$

$$\frac{0.2 \times 3.386 \times 10^4}{13.6 \times 12 \times 2.54} \text{ gm/cm}^2 \text{ sec}^2 = 1.64 \text{ gm/cm}^2 \text{ sec}^2$$

Thus

$$\text{vent pipe radius} = \left(\frac{30.5 \times 10^3 \text{ cm}^3/\text{sec} \times 8 \times 0.89 \times 10^{-4} \text{ gm/cm sec}}{60 \times \pi \times 1.64 \text{ g/cm}^2 \text{ sec}^2}\right)^{1/4}$$

$$= 0.515 \text{ cm}$$

i.e.,

$$\text{vent pipe diameter} = 0.405 \text{ in.}$$

$$\text{velocity in pipe} = \frac{30.5 \times 10^3}{60 \times \pi \times 1.43} \text{ cm/sec} = 113 \text{ cm/sec}$$

Thus

$$\text{Reynolds number} = \frac{0.515 \times 113 \times 0.818 \times 10^{-4}}{0.89 \times 10^{-4}} = 53.6$$

which is well below the Reynolds number of 2100 at which flow becomes turbulent, justifying our initial assumption of laminar flow.

A baffle with a 0.4-cm orifice is satisfactory, as can be seen by the following calculation. The pressure drop due to such a baffle is given by

$$\Delta P = \left(\frac{F\sqrt{\rho/2}}{C_d S_o}\right)^2$$

for orifices of the size of interest. Here F is the volumetric flow rate through the orifice, ρ the gas density, C_d the discharge coefficient, S_o the orifice area, and ΔP the pressure drop. Under reaction conditions

$$\text{Reynolds number} = \frac{4F\rho}{\pi_\mu D_o} \doteq \frac{4 \times 1150 \text{ cm}^3/\text{sec} \times 0.362 \times 10^{-4} \text{ g/cm}^3}{\pi \times 1.53 \times 10^{-4} \text{ g/cm sec} \times 0.1 \text{ cm}}$$

$$= 870$$

At this Reynolds number, $C_d \simeq 0.6$ [58]

$$\Delta P = \left(\frac{1150 \times \sqrt{0.362 \times 10^{-4}/2}}{0.6 \times \pi \times 0.2^2} \right)^2$$

$$= 0.42 \times 10^{-4} \text{ g/cm sec}^2$$

$$= 0.124 \text{ in. of Hg} = 1.7 \text{ in. of water}$$

Finally the heating bed must be designed. Let us carry out calculations to determine if a bed of 1-cm spheres with a void fraction of 0.3 would be suitable.

Unit volume of the bed contains $0.7/\tfrac{4}{3}\pi r^3$ spheres (of radius r), which have a total surface area of

$$0.7/\tfrac{4}{3}\pi r^3 \times 4\pi r^2 = 2.1/r = \text{surface area per unit volume of bed} = 4.2 \text{ cm}^2/\text{cm}^2$$

According to Bird et al. [55, p. 411] the heat transfer coefficient between gas and solid in such a bed is given by

$$h = \frac{k}{D} (0.91 N_{\text{Re}}^{0.49} N_{\text{Pr}}^{1/3}) \qquad \text{for } N_{\text{Re}} < 50$$

where $N_{\text{Re}} = v\rho/a\mu$ and N_{Pr} is the Prandtl number. In the above equations h is the heat transfer coefficient, k the thermal conductivity of hydrogen, D the bed particle size, v the superficial gas velocity, a the bed surface area per unit volume, and μ the viscosity.

Evaluating the Reynolds number at 400°C, we obtain

$$N_{\text{Re}} = 53.2 \times 0.362 \times 10^{-4}/4.2 \times 1.53 \times 10^{-4} = 3.0$$

thermal conductivity of hydrogen at 400°C [56]

$$= 6.3 \times 10^{-4} \text{ cal/cm sec °K}$$

$$\text{Prandtl number} \simeq \frac{\bar{C}_p/R}{\bar{C}_p/R + 1.25} = \frac{2.5}{3.75} = 0.67$$

where $\bar{C}_p$ is the molar specific heat. Therefore

$$h = \frac{6.3 \times 10^{-4} \text{ cal}}{\text{cm. sec °K 1 cm}} (0.91 \times 3.0^{0.49} \times 0.67^{1/3})$$

$$= 8.6 \times 10^{-4} \frac{\text{cal}}{\text{cm}^2 \text{ sec °K}}$$

The characteristic length for heat transfer in the bed is

$$Z_0 = F_\rho C_\rho / haS = v\rho C_\rho / ha$$

where S is the bed cross-sectional area and

$$Z_0 = \frac{53.2 \times 0.362 \times 10^{-4} \times 2.5 \times 1.987}{8.6 \times 10^{-4} \times 4.2 \times 2} \text{ cm} = 1.32 \text{ cm}$$

From this it follows that a bed of 1-cm spheres would serve provided $2\frac{1}{2}$–3 in. of bed (about 5–6 times Z_0) lie in the heated length of the furnace. The pellet center would typically be about 2 in. above the top of the bed and the heated zone should extend a few pipe diameters (say, 6 in. in this example) above the pellet center (in order to ensure that the radiation incident on the pellet is nearly uniform). It follows that the heated length should be $10\frac{1}{2}$–11 in. or more. Since we have already given a heated length for the furnace of about $12\frac{1}{2}$ in. our originally described furnace will be adequate.

This completes the major design of the apparatus.

Notation

A_g, A_p	external surface area of grain, pellet
b	stoichiometric coefficient
C_{A0}, C_{C0}	bulk concentration of gaseous reactant, product
C_1	defined following Eq. (6.6.5)
D	pellet diameter
D_e	effective diffusivity within porous solid
d_p	pellet diameter (particle diameter)
F_g, F_p	shape factor for grain, pellet
$g_{F_g}(X)$	defined by Eq. (3.2.11)
H	defined by Eq. (6.6.5)
h_D	mass transfer coefficient
K_E	equilibrium constant
k	reaction rate constant
L	length
N_{Sh}, N_{Sh}^*	Sherwood number, modified Sherwood number
n	reaction order
P	pressure
$p_{F_p}(X)$	defined by Eqs. (4.3.7)–(4.3.9)
R_p	radius (or thickness) of pellet
r	pore radius
r_g	grain radius
S	defined by Eq. (6.6.6)
t	time
V	fluid velocity
V_g, V_p	volume of grain, pellet
$\mathbf{V}$	terminal velocity

X	extent of reaction
ε	porosity
ε_v	void fraction
φ	shape factor
μ	viscosity
ρ_s, ρ_f	solid density, fluid density
ρ_0	observed solid density
ρ_t	true solid density
σ	surface tension
$\hat{\sigma}$	defined by Eq. (6.6.2)
θ	contact angle

References

1. Y. Iida and K. Shimada, *Bull. Chem. Soc. Japan* **33**, 790 (1960).
2. J. Deren and J. Stoch, *J. Catal.* **18**, 249 (1970).
3. J. W. Evans, Ph.D. Thesis, State Univ. of New York at Buffalo (1970).
4. W. A. Knepper (ed.), "Agglomeration," Wiley (Interscience), New York, 1962.
5. T. Rigg, *Can. J. Chem. Eng.* **42**, 247 (1964).
6. W. K. Lu and G. Bitsianes, *Can. Met. Quart.* 7, 3 (1968).
7. J. C. Yannopoulos and N. J. Themelis, *Can. J. Chem. Eng.* **43**, 173 (1965).
8. P. F. J. Landler and K. L. Komarek, *Trans. AIME* **236**, 138 (1966).
9. W. M. McKewan, *Trans. AIME* **224**, 387 (1962).
10. C. I. Lin, Ph.D. Thesis, State Univ. of New York at Buffalo (1974).
11. C. R. Bickling, Ph.D. Thesis, Univ. of Wisconsin (1950).
12. G. Parravano, *J. Amer. Chem. Soc.* **74**, 1194 (1952).
13. A. I. M. Keulemans, "Gas Chromatography," 2nd ed. Van Nostrand-Reinhold, Princeton, New Jersey, 1959.
14. R. R. Irani and C. F. Callis, "Particle Size: Measurement, Interpretation and Application." Wiley, New York, 1963.
15. R. Davies, *Ind. Eng. Chem.* **62** (12), 87 (1970).
16. C. E. Hall, "Introduction to Electron Microscopy." McGraw-Hill, New York, 1963).
17. L. V. Azaroff, "Elements of X-ray Crystallography," McGraw-Hill, New York, 1968.
18. B. E. Bartram, "Handbook of X Rays" (E. F. Kaeble, ed.), Chapter 17. McGraw-Hill, New York, 1967.
19. F. A. L. Dullien, and V. K. Batra, *Ind. Eng. Chem.* **62**, 25 (1970).
20. L. K. Frevel and L. J. Kressley, *Anal. Chem.* **35**, 1492 (1963).
21. R. P. Mayer and R. A. Stowe, *J. Colloid Sci.* **20**, 893 (1965).
22. R. P. Mayer and R. A. Stowe, *J. Phys. Chem.* **70**, 3867 (1966).
23. E. A. Mason, A. P. Malinauskas, and R. B. Evans, III, *J. Chem. Phys.* **46**, 3199 (1967).
24. H. L. Ritter and L. C. Erich, *Anal. Chem.* **20**, 665 (1948).
25. G. L. Clark and C. H. Liv, *Anal. Chem.* **29**, 1539 (1957).
26. E. A. Mason and T. R. Marrero, *Advan. At. Mol. Phys.* **6**, 155 (1970).
27. A. A. Westenberg, *Advan. Heat Transfer* 3, 253–302 (1966).
28. T. R. Marrero and E. A. Mason, Nat. Bur. Std. NSRDS (1971).
29. T. A. Pakurar and J. R. Ferron, *Ind. Eng. Chem. Fundam.* **5**, 553 (1966).
30. Z. Balenovic, M. N. Meyers, and J. C. Giddings, *J. Chem. Phys.* **52**, 915 (1970).

31. R. E. Collingham, P. L. Blackshear, and E. R. G. Eckert, *Chem. Eng. Prog. Symp. Ser., No.* 102 **66** 141 (1970).
32. E. Wicke and R. Kallenbach, *Kolloid Z.* **97**, 135 (1941).
33. P. B. Weisz, *Z. Phys. Chem.* **11**, 1 (1957).
34. C. N. Satterfield and S K. Saraf, *Ind. Eng. Chem. Fundam.* **4**, 451 (1965).
35. R. N. Foster, J. B. Butt, and H. Bliss, *J. Catal.* **7**, 179 (1967).
36. D. S. Scott and K. E. Cox, *Can. J. Chem. Eng.* **38**, 201 (1960).
37. N. Wakao and J. M. Smith, *Chem. Eng. Sci.* **17**, 825 (1962).
38. J. P. Henry, R. S. Cunningham, and C. J. Geankoplis, *Chem. Eng. Sci.* **22**, 11 (1967).
39. R. S. Cunningham and C. J. Geankoplis, *Ind. Eng. Chem. Fundam.* **7**, 535 (1968)
40. C. N. Satterfield and P. J. Cadle, *Ind. Eng. Chem. Process Design Develop.* **7**, 256 (1968).
41. B. R. Davis and D. S. Scott, *Symp. Fundam. Heat Mass Transfer, 58th Ann. Meeting AIChE*, Philadelphia, Pennsylvania (1965).
42. P. Schneider and J. M. Smith, *AIChE J.* **14**, 762 (1968).
43. A. J. Leffler, *J. Catal.* **5**, 22 (1966).
44. P. E. Eberly, *Ind. Eng. Chem. Fundam.* **8**, 25 (1969).
45. E. R. Bennett and W. E. Bolch, *Anal. Chem.* **43**, 55 (1971).
46. R. G. Olsson and W. M. McKewan, *Trans. Met. Soc. AIME* **236**, 1518 (1966).
47. R. G. Olsson and W. M. McKewan, *Met. Trans.* **1**, 1507 (1971).
48. F. R. Campbell, A. W. D. Hills, and A. Paulin, *Chem. Eng. Sci.* **25**, 929 (1970).
49. N. J. Themelis and C. J. Yannopoulos, *Trans. AIME* **236**, 414 (1966).
50. R. J. Fruehanm and L. J. Martonik, *Met. Trans.* **4**, 2793 (1973).
51. J. Szekely, C. I. Lin, and H. Y. Sohn, *Chem. Eng. Sci.* **28**, 1975 (1973).
52. P. B. Weisz and C. D. Prater, *Advan. Catal.* **6**, 143 (1954).
53. P. B. Weisz and R. P. Goodwin, *J. Catal.* **2**, 397 (1963); **6**, 227 (1966).
54. E. Kawasaki, J. Sanscriate, and T. J. Walsh, *AIChE J.* **8**, 48 (1962).
55. R. B. Bird, W. E. Stewart, and E. N. Lightfoot, "Transport Phenomena." Wiley, New York, 1960.
56. R. A. Svehla, Tech. Rep. R-132, Lew Res. Center, N.A.S.A., Cleveland (1962) (N63-22862).
57. E. T. Turkdogan *et al.*, *Met. Trans.* **2**, 3189 (1971).
58. W. L. Badger and J. T. Banchero, "Introduction to Chemical Engineering." McGraw-Hill, New York, 1955.
59. T. Allen, "Particle Size Measurement," 2nd ed., Chapman and Hall, London and Wiley, New York, 1975.

Chapter 7

Gas–Solid Reactions in Multiparticle Systems

7.1 Introduction

In the preceding chapters we discussed the experimental and analytical techniques available for the characterization of reaction systems involving single particles. On the basis of this material we can relate the rate of reaction of a single particle to the structural parameters, the kinetic parameters, and the factors that determine gas–solid heat and mass transfer.

Our objective in this chapter is to develop techniques for applying the information obtained on single-particle systems to the description of multiparticle assemblies. This extension of the work is of course highly desirable because most systems of practical interest consist of multiparticle assemblies. Some typical multiparticle contacting arrangemens are sketched in Fig. 7.1, and as seen, these include packed beds, moving beds, fluidized systems, rotating kilns (both horizontal and vertical), and transfer line reactors. A full discussion of the relative merits of these gas–particle contacting arrangements will be presented at the end of the chapter.

In translating kinetic data obtained for single particles to multiparticle systems, three major groups of problems have to be faced.

(a) Physical nature of the assembly

While the laws governing chemical kinetics and pore diffusion will remain unaffected by the presence of the other particles and the modification in the flow field, nonetheless the nature of the particle assembly may affect the overall kinetics quite markedly.

In describing the behavior of multiparticle assemblies such as packed beds, fluidized beds, and transport line reactors, attention must be paid to the fluid flow aspects of the problem. The design criteria may depend quite critically

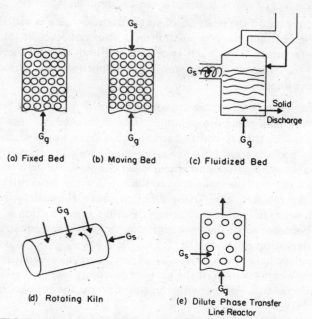

FIG. 7.1. Typical multiparticle gas–solid contacting equipment.

on the relationship between pressure drop and gas flow rate in a packed bed, on the minimum fluidization velocity, or on the terminal falling velocity of the particles in transport line reactors. In packed beds or fluidized beds the individual film coefficients may be affected by the presence of the other particles; in packed beds operating at high temperatures, thermal radiation may become significant to an extent which will in turn depend on the size and packing of the particles.

None of these effects manifest themselves in single-particle studies but have to be taken into consideration in the design of reactors involving particle assemblies.

(b) Contact time or residence time in the reaction environment

In single-particle studies the reaction time, i.e., the time period for which the particle and the reactant gas stream are in contact, is a well-defined quantity. In contrast, in many gas–solid reaction systems involving particle assemblies, only the *average* contact time is known and there may be a considerable spread or distribution of residence times. This is particularly true in the case of fluidized beds, kilns, and certain moving-bed type operations. Under these conditions proper allowance must be made for this spread of residence times, which could apply to both the solid and the gas streams.

(c)   The effect of spatially variable gas composition

In experiments on the reaction of single particles, care is usually taken to maintain the reactant gas composition at a constant level in the bulk, i.e., to avoid "starvation." For such systems the overall driving force is clearly defined and the "rate controlling step" or the appropriate asymptotic regime is readily identified.

The situation is rather more complicated in the majority of reaction systems involving multiparticle assemblies. This difference is illustrated in Fig. 7.2 where it is seen that, in contrast to the behavior of single-particle systems, in a moving bed the solid particles are exposed to a gas stream which may have a variable temperature and reactant gas composition. In a mathematical sense this means that the boundary conditions for the differential equations describing the reaction have to be different from those describing single-particle behavior; in fact, the equations describing the reaction of single particles may become the boundary conditions, or subsidiary equations, for the overall conservation relationships.

In general we have to deal with a set of simultaneous differential equations written for the conservation of the gaseous reactants and the conservation of the solid reactants, together with the appropriate thermal energy balance equations.

It follows from the foregoing points (a)–(c) that in general multiparticle systems are a great deal more complex than the previously described single-particle situations; while the description of the behavior of single particles is an essential building block in representing packed beds or fluidized beds, the generalization to these systems from data on single particles is far from trivial.

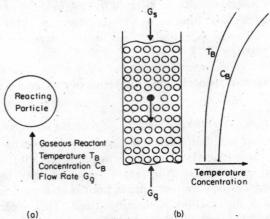

FIG. 7.2. The difference between single- and multiparticle behavior due to variation in gaseous reactant concentration and temperature in the multiparticle case.

7.2 Some Properties of Continuous Flow Systems

Let us consider a vessel or reactor through which a fluid or solid stream is being passed. If the volumetric flow rate of the stream is v and the volume of the vessel is V, then the average time or the nominal holding time of the material in the vessel is given by

$$\bar{t}_R = V/v \tag{7.2.1}$$

However, it is frequently found that some elements of the stream may spend a longer period of time than $\bar{t}_R$ in the system, whereas other elements are retained for a shorter time period.

This *distribution, or spread of residence times* is an important characteristic of the system and may have a profound effect on its performance as a reactor or contacting device. In many cases the actual flow pattern through the system is conveniently determined by the addition of a tracer at the inlet, the concentration of which is then monitored at the outlet. The analysis of the response curves then provides the desired information.

In the simplest case two basic modes of tracer addition may be considered:

(a) *Stepwise* addition of a tracer, which is then continued for a long period of time or,

(b) The addition of a tracer *pulse* over a brief time period which is small compared to the nominal holding time $\bar{t}_R$.

The response curves corresponding to case (a) have been traditionally termed F diagrams and the response to the pulse addition are called C diagrams [1].

Let us define the following quantities: c_i, tracer concentration at the inlet; c, concentration of the tracer in the exit stream; t, time; $\theta = t/\bar{t}_R$, dimensionless time; Q, mass of tracer added (in the generation of the C diagrams); $F(\theta) = c/c_i$, dimensionless tracer concentration at the exit in the case of the F diagrams; and $C = c/(Q/V) = E(\theta)$, dimensionless tracer concentration at the exit in the case of the C diagrams (sometimes called the exit age distribution).

In Figs. 7.3–7.6 we shall present the characteristic F and C curves for four types of behavior frequently encountered in practice. While the figures shown are idealizations to some degree, nonetheless they represent useful standards to which real systems may be compared.

Figure 7.3 shows the F and C diagrams that characterize plug flow behavior. It is seen that for plug flow, the tracer front (F diagram) or the tracer pulse (C diagram) passes through the system without any distortion. It follows that all the elements of the stream travel through the system at the same speed and arrive at the exit at the "expected time." Under these conditions there is no spread of the residence times; thus the residence time of the solid (or fluid) stream is a uniquely defined quantity.

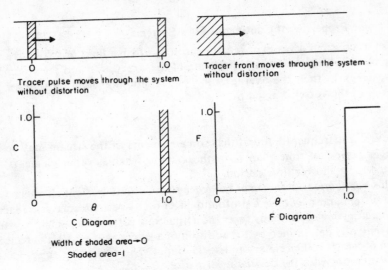

Tracer pulse moves through the system without distortion

Tracer front moves through the system without distortion

C Diagram

F Diagram

Width of shaded area→0
Shaded area=1

FIG. 7.3. F and C diagrams for plug flow.

While plug flow is an idealization, fluid flow through long pipes, gas flow through packed beds under certain conditions, and some transfer line reactors may be regarded as plug flow systems as a reasonable approximation.

Figure 7.4 depicts the F and C diagrams for systems which are *perfectly mixed*. Perfect mixing implies that upon entering the system the tracer is instantaneously dispersed uniformly throughout the reaction vessel. For a step change in tracer concentration as given by the F diagram, this will mean that tracer elements will start appearing at the exit immediately upon introduction at the inlet; however, as shown by the form of the equation describing the F curve, it will take an infinitely long time before F attains unity ($F \simeq 0.98$ for $\theta \simeq 4$).

It follows from this discussion that for completely mixed systems there is a large spread in the residence times; some elements of the stream spend very little time in the vessel, whereas other elements may be retained for much longer time periods than $\bar{t}_R$. This property of the system can be a decided disadvantage under certain circumstances because it can lead to prolonged overall holding times to ensure the complete reaction of the reactant stream. Completely mixed systems are closely approximated by stirred tank reactors such as are employed in liquid–liquid extraction, liquid–liquid reactions, and continuous leaching operations. Relatively few gas–solid systems may be regarded as completely mixed although the particulate phase of some fluidized beds does approach this state.

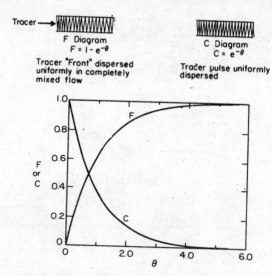

FIG. 7.4. F and C diagrams for complete mixing.

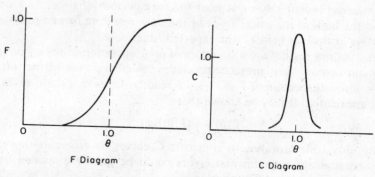

FIG. 7.5. F and C diagrams for the case of plug flow with some axial mixing.

Figure 7.5 shows the F and C diagrams for systems that are in plug flow but with some axial mixing. It is seen that for the F diagram the tracer front arrives at the exit some time before $t = \bar{t}_R$ and then there is a rapid increase in the value of F until it approaches unity at $t > \bar{t}_R$. The corresponding C diagram shows a spread about $\theta = 1$ while the area under the curve remains equal to unity.

Finally, Fig. 7.6 shows physical situations and the corresponding F and C diagrams for systems with dead volume regions and axial dispersion. It is seen that under these conditions part of the vessel is inactive, i.e., it retains

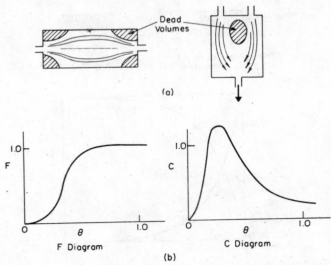

FIG. 7.6. F and C diagrams for the case of a reactor with dead volumes.

some material indefinitely or at least for long periods of time. As a consequence, the bulk of the tracer front or the peak resulting from a pulse input will arrive at the exit before "the expected time."

Dead volume regions are a symptom of poor operation or poorly designed equipment because their presence precludes the efficient use of the total reactor volume. Inspection of Figs. 7.3–7.6 readily shows that the C and the F curves are related. It may be shown that

$$C(\theta) = d[F(\theta)]/d\theta$$

It is not advisable however, to construct C curves by differentiating the F curves because any experimental errors could be greatly amplified by the graphical or numerical differentiation.

The idealized flow situations depicted in Figs. 7.3–7.6 provide a good qualitative indication of the flow field within the system. There are other instances however, when a more quantitative description is needed. Such a quantitative description may be provided by combining these idealized flow models into composite mixed models.

MIXED MODELS

There are many real systems whose tracer response cannot be represented by any one of the idealized response curves sketched in Figs. 7.3–7.6. One convenient way of representing such situations is to consider a *network* which

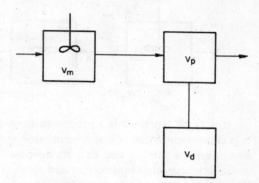

FIG. 7.7. A mixed model for correlating residence time distribution data. $V = V_m + V_p + V_d$.

contains elements that are completely mixed, contribute plug flow regions, and correspond to dead zones.

One such arrangement is sketched in Fig. 7.7; it is seen that the total volume of the reactor or vessel is made up of V_p, a plug flow region; V_m, a completely mixed region; and V_d, the dead zone.

A mixed model of this type may be used to represent C diagrams of the type sketched in Fig. 7.8, the legend of which provides an illustration of how the relative values of V_p, V_m, and V_d may be evaluated. We note that even Fig. 7.8 is an overidealized response curve and that the exact representation of many "real life" C curves would require quite an elaborate network of units.

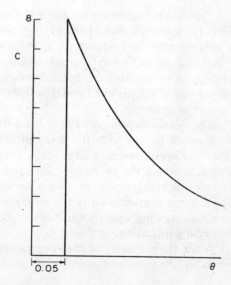

FIG. 7.8. Idealized C diagram for a system containing a region of plug flow, a region of dead volume, and a perfectly mixed region. $V_p/V = 0.05$, $V/V_m = 8$. Since $(V_p + V_m + V_d)/V = 1$, $V_0/V = 1 - 0.125 - 0.05 = 0.825$.

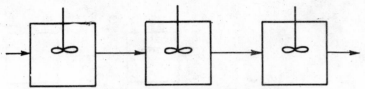

FIG. 7.9. A series of completely mixed units.

It may be worthwhile to mention here that under certain circumstances it is convenient to represent systems that are intermediate between plug flow and complete mixing and exhibit no dead volumes by assuming that they constitute several completely mixed units in series, as sketched in Fig. 7.9. We note that an infinite number of completely mixed cells would constitute a plug flow system and that it is easy to show that the C curve for n completely mixed cells of equal volume is given by

$$C_n = \frac{n^n \theta^{n-1} e^{-n\theta}}{(n-1)!}$$

It is stressed to the reader that Section 7.2 constitutes just a brief introduction to the important but complex field of continuous flow systems. Another aspect of flow patterns, namely axial and radial dispersion, will be discussed in Section 7.3; for further reading on the behavior of continuous flow systems, references [2–4] should be consulted.

7.3   Fixed Bed Systems

Typical arrangements for fixed bed contacting are sketched in Fig. 7.10. Here (a) denotes a packed bed arrangement where the solids are charged into a column and the reactant gas is then passed through them. As a result a *reaction front* is formed, which gradually moves along the column. The analogy between this moving reaction front in a macroscopic system and the movement of a distributed reaction front within porous solid particles should be readily apparent.

The system depicted in Fig. 7.10a corresponds to a batch process and as such would be unwieldy for many practical applications, particularly those involving operations on a large scale. The sketch given in Fig. 7.10b describes a system where the solids are conveyed horizontally on a moving grid or grate while being contacted with a gas stream blown through the bed. This operation is continuous and the arrangement shown is favored in many materials processing operations, including the sintering of ores, incineration, and coal gasification.

While the geometry of the systems depicted in Fig. 7.10 differs quite appreciably, the actual movement of the reaction front will be identical, provided

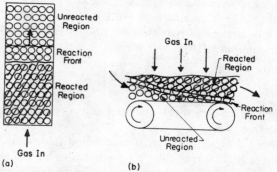

FIG. 7.10. Gas–solid contacting in a fixed-bed arrangement. (a) Packed bed, (b) packed bed with cross flow contacting.

that in (b) the observer moves along with the grate. If t_R denotes the time taken for the reaction front to travel the depth of the bed in (a), then for (b)

$$t_R \equiv W/U \tag{7.3.1}$$

where W is the length of the bed and U is the linear velocity of the grate. This correspondence between systems (a) and (b) provides the principal justification for most of the work done on transient fixed bed systems, at least from the standpoint of practical relevance.

The principal physical phenomena with which we are concerned in the representation of fixed bed reactors are fluid flow, heat transfer, and mass transfer.

7.3.1 Fluid Flow in Packed Beds

The assembly of particles in packed beds represents a rather complex geometry and, while a number of interesting studies have been made of the flow field in packed beds [5–7] for most practical problems we have to resort to empirical relationships between the pressure drop and the flow rate of the fluid. Of the numerous correlations proposed for relating pressure drop to flow rate, the Ergun equation [8] is perhaps the most widely accepted:

$$\frac{\Delta P}{L} = 150 \frac{(1 - \varepsilon_v)^2}{\varepsilon_v^3} \frac{\mu U_0}{(\varphi_s d_p)^2} + 1.75 \frac{(1 - \varepsilon_v)}{\varepsilon_v^3} \frac{\rho_f U_0^2}{\varphi_s d_p} \tag{7.3.2}$$

where ΔP is the pressure drop, L the depth of the bed, ε_v the void fraction, μ the viscosity, φ, a shape factor, d_p the particle diameter, ρ_f the fluid density, and U_0 the linear fluid velocity in the empty column. Here d_p is the diameter of a sphere of equal volume to the particle and

$$\varphi_s = \frac{\text{Surface area of sphere of equal volume to the particle}}{\text{Surface area of particle}}$$

Clearly $\varphi_s = 1$ for spheres and the typical shape factors for naturally occurring materials would range from a high of ~ 0.9 to a low of ~ 0.5 [9, 10].

When there is a range of particle sizes, the use of the *surface area mean diameter* is recommended [10]:

$$\bar{d}_p = \left[\sum \frac{X_{n'}}{d_{p,n}} \right]^{-1} \tag{7.3.3}$$

where $\bar{d}_p$ is the surface area mean diameter and X_n' is the weight fraction of the solids of diameter $d_{p,n}$.

Inspection of Eq. (7.3.2) shows that at low fluid velocities the viscous term predominates (the first term of the r.h.s.) whereas at high fluid velocities, the pressure drop is proportional to the square of the fluid velocity and the inertial term predominates. It has been suggested that this transition is accompanied by a change in the nature of the flow due to the passage of the bed from the laminar to the turbulent tregime. However, Blick [11] has suggested that the inertial term is due to the sudden expansion of the gas after its passage through constricted flow areas rather than to the onset of turbulence.

VALIDITY OF THE ERGUN EQUATION—NONIDEAL SYSTEMS

Strictly speaking, the Ergun equation is valid for isothermal systems, incompressible fluids, and for beds which are spatially uniform. When the system is nonisothermal or when the pressure drop across the bed is comparable to the absolute pressure at the inlet, the use of Eq. (7.3.2) may introduce quite serious errors. When there are large variations in the absolute temperature and pressure within the system, the Ergun equation has to be put in a differential form and then *integrated along the appropriate temperature path* [12].

By introducing the mass velocity $G_0 \equiv U_0 \rho$ and the specific volume $\bar{v} = 1/\rho$, the mechanical energy balance may be written as

$$(G_0^2/\varepsilon_v^2)\bar{v}\, d\bar{v} + \bar{v}\, dP + F_{Fr}'\, dz = 0 \tag{7.3.4}$$

where F_{Fr}' is the friction term and z is the axial coordinate in the direction of flow.

From the Ergun equation, we have

$$F_{Fr}' = C_1 \mu G_0 \bar{v}^2 + C_2 G_0^2 \bar{v}^2 \tag{7.3.5}$$

where

$$C_1 = 150(1 - \varepsilon_v)^2/\varepsilon_v^3(\varphi_s d_p)^2 \tag{7.3.6}$$

and

$$C_2 = 1.75(1 - \varepsilon_v)/\varepsilon_v^3 \varphi_s d_p \tag{7.3.7}$$

F'_{Fr} may now be substituted into Eq. (7.3.4), which upon integration from the upstream end (1) may be written as

$$\frac{G_0^2}{\varepsilon_v^2} \ln \frac{\bar{v}}{\bar{v}_1} + \int_1 (C_1 \mu G_0 + C_2 G_0^2)\, dz = -\int_1 \frac{dP}{\bar{v}} \qquad (7.3.8)$$

For isothermal systems the r.h.s. of (7.3.8) is readily evaluated by using the ideal gas law

$$P\bar{v} = P_1 \bar{v}_1 = mR'T$$

Thus we have (assuming uniform ε_v, φ_s, d_p, and μ)

$$\frac{G_0^2}{\varepsilon_v^2} \ln \frac{P_1}{P_2} + [C_1 \mu G_0 + C_2 G_0^2]\, L = \frac{P_1^2 - P_2^2}{2P_1 \bar{v}_1} \qquad (7.3.9)$$

where the subscript 2 refers to a distance L from the inlet which is designated by subscript 1. For nonisothermal systems where the temperature profile is known, a trial and error method has to be used:

(1)   The axial pressure profile $P = P(z)$ is assumed and the corresponding values of $\bar{v}$ are calculated.

(2)   The integral on the r.h.s. of Eq. (7.3.8) is then evaluated numerically, providing a second estimate of the pressure profile.

(3)   The procedure is repeated until two successive iterations agree to within a desired degree of accuracy.

Two practical points may be raised at this juncture:

(a)   In the majority of practical cases the temperature profile will not be known a priori: thus the procedure outlined here has to be performed in conjunction with the solution of appropriate differential heat and mass balance relationships.

(b)   When the local variations in temperature are not too great, i.e., in the absence of sharp peaks in temperature, a good approximation may be obtained for the *overall* pressure drop–flow rate relationship by evaluating all physical properties at some mean temperature of the bed.

NONIDEAL SYSTEMS PREFERENTIAL FLOWS AND DISPERSION

Preferential flows

In the preceding discussion we presented relationships between the pressure drop and the flow rate for both isothermal and nonisothermal systems. Implicit in all these relationships was the assumption that the resistance to flow was uniform in any given cross section of the bed; thus the overall mass flow rate could be considered uniformly distributed at any given axial position. These assumptions are not necessarily valid in many practical situations, as it is known that local variations in porosity or particle size in packed beds may give rise to flow maldistribution. Such flow maldistribution may in turn cause hot spots or general malfunctioning of the equipment.

It has been shown (see e.g., [13]) that lateral variations in porosity will necessarily occur in the vicinity of the walls, even in the case of very carefully packed beds. When there are regions in the bed which have different porosities, particle sizes, or even laterally distributed temperatures, then the one-dimensional Ergun equation (7.3.2) will no longer apply.

In recent work it has been shown [14] that the flow of incompressible fluids in a two-dimensional packed bed containing lateral inhomogeneities in the resistance to flow may be represented by the following equations:

$$\frac{\partial G_x}{\partial x} + \frac{\partial G_z}{\partial z} = 0 \quad \text{(equation of continuity)} \tag{7.3.10}$$

and

$$-\frac{\partial P}{\partial x} = \frac{C_1 \mu}{\rho} G_x + \frac{C_2}{\rho} G_x |G_x| \tag{7.3.11}$$

(mechanical energy balance)

$$-\frac{\partial P}{\partial z} = \frac{C_1 \mu}{\rho} G_z + \frac{C_2}{\rho} G_z |G_z| \tag{7.3.12}$$

where x and z denote the coordinates and G_x and G_z are the x and z components of the mass velocity.

The equivalent expressions in cylindrical coordinates may be written as

$$-\frac{\partial P}{\partial z} = \frac{C_1 \mu}{\rho} G_z + \frac{C_2}{\rho} G_z |G_z| \tag{7.3.13}$$

and

$$-\frac{\partial P}{\partial r} = \frac{C_1 \mu}{\rho} G_r + \frac{C_2}{\rho} G_r |G_r| \tag{7.3.14}$$

Equations (7.3.11) and (7.3.12) or (7.3.13) and (7.3.14) may be solved by introducing the concept of the vorticity and the stream function [15]; the appropriate finite difference form of vorticity transport and stream equations may then be solved by successive overrelaxation.

Some computed results are given in Stanek and Szekely [14, 16]. Figure 7.11 shows the streamline pattern for flow through a two-dimensional bed where the void fraction in the shaded area is 0.3, while the void fraction in the remainder of the bed is 0.425. The distortion of the streamlines is readily apparent and it is noted that "cross flow effects" are quite noticeable even some

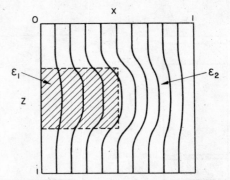

FIG. 7.11. Flow maldistribution in a packed bed in which there is a region of reduced porosity ($\varepsilon_1 < \varepsilon_2$).

distance from the boundary separating the regions of different porosity. As shown in [16], not only lateral variation in the void fraction or in the effective particle size will cause flow maldistribution, but lateral variations in temperature may also result in nonuniformities of flow. A particular case in this latter category is a tubular reactor which is being cooled from the outside.

A further, more detailed discussion of flow maldistribution phenomena is available in recent papers by Stanek and Szekely [93] and Szekely and Poveromo [94].

Dispersion

Let us consider the flow of a fluid through a packed bed in which there are no macroscopic lateral inhomogeneities of the type just discussed. As mentioned earlier, in the absence of marked temperature gradients in such systems, the pressure drop and the gas flow rate are readily related by the one-dimensional Ergun equation (7.3.2). We note however, that the assumption of plug flow or piston flow is not necessarily correct even under these circumstances.

For readable discussions of plug flow and the concept of dispersion the reader is referred to the literature [1, 2, 17]. Here we shall confine ourselves to a brief illustration of axial and radial dispersion. Let us consider the behavior of a tracer introduced into a stream flowing through a packed bed reactor, as sketched in Fig. 7.12. Here (a) shows the *axial dispersion* or spread by the time it reaches the exit of a tracer pulse introduced uniformly over the cross-sectional area of the column at a given axial position. Figure 7.12b depicts the *radial dispersion* of a tracer introduced continuously at a point source.

Both the axial spread or dispersion shown in (a) and the radial dispersion sketched in (b) constitute nonideal flow behavior, some aspects of which were

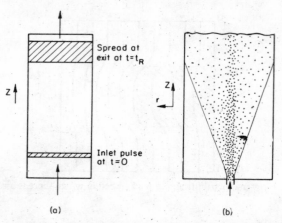

FIG. 7.12. (a) Axial dispersion of pulse and (b) radial dispersion of a continuously introduced stream.

discussed in Section 7.2. In a physical sense the dispersion shown constitutes an additional mechanism for the transport of matter, in addition to bulk flow and molecular transport.

Such dispersion has been represented in terms of *eddy diffusion* by postulating that the laws of diffusion would apply, but the constant of proportionality between the flux and the concentration gradient is no longer a molecular property but a property of the system. This quantity is termed the *eddy diffusivity* or the dispersion coefficient.

Thus the transient axial dispersion of the tracer sketched in Fig. 7.12a may be represented by an equation of the type

$$\frac{\partial C_t}{\partial t} = E_z \frac{\partial^2 C_t}{\partial z^2} - U_z \frac{\partial C_t}{\partial z} \qquad (7.3.15)$$

whereas the steady state radial dispersion sketched in Fig. 7.12b may be described by

$$E_r \left| \frac{\partial^2 C_t}{\partial r^2} + \frac{1}{r} \frac{\partial C_t}{\partial r} \right| - U_z \frac{\partial C_t}{\partial z} = 0 \qquad (7.3.16)$$

where C_t is the tracer concentration and E_z and E_r are the axial and radial eddy diffusivities. We note that in writing Eq. (7.3.16) we assumed that axial dispersion may be neglected.

We note that in general in the establishment of differential material balances for packed bed reactors, account should be taken of the eddy diffusion mechanism described. Over the years numerous measurements have been made of both radial and axial dispersion coefficients so that the effects of dispersion are readily assessed.

The experimentally measured dispersion coefficients are usually represented on plots of the particle Peclet number against the particle Reynolds number, defined as follows:

$$N_{Pe_{E,z}} = d_p U_z / E_z \quad \text{and} \quad N_{Pe_{E,r}} = d_p U_z / E_r; \quad N_{Re_p} = d_p \rho U_z / \mu$$

where $N_{Pe_{E,z}}$ and $N_{Pe_{E,p}}$ are the axial and radial Peclet numbers for eddy diffusion and d_p is the particle diameter.

Figure 7.13 shows experimental measurements reported by Balla and Weber [18] on a plot of $N_{Pe_{E,z}}$ against N_{Re_p}, the particle Reynolds number; also shown are the data of some other investigators. All these results show a consistent trend in that at low values of the Reynolds number, the Peclet number is also small and within this region molecular diffusion is the predominant dispersion mechanism and hence $N_{Pe_{E,z}} \propto N_{Re_p}$. For Reynolds numbers larger than about 1–10 the axial Peclet number tends to an asymptotic value of about 2. In an elegant paper, Deans and Lapidus [19] have shown that this asymptote corresponds to perfect mixing within the voids.

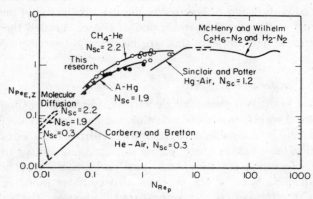

FIG. 7.13. Experimental data on axial dispersion reported by Balla and Weber [18].

The data on radial dispersion are less extensive but as shown by Wilhelm [17] the experimental results represented on a plot of $N_{Pe_{E,r}}$ against N_{Re_p} show a similar trend to that seen in Fig. 7.13 with the exception that the numerical value of the asymptote is about 10–12.

Thus for radial dispersion

$$N_{Pe_{E,r}} \to 11 \quad \text{for} \quad N_{Re_p} > 10, \quad N_{Re_p} = \rho U d_p / \mu \qquad (7.3.17)$$

PRACTICAL IMPORTANCE OF AXIAL DISPERSION

In assessing the practical importance of axial dispersion, let us examine Eq. (7.3.15). It is readily seen that axial dispersion will be important only when the term $E_z \, \partial^2 C_t / \partial z^2$ is not negligible compared to $U_z \, \partial C_t / \partial z$. By making these quantities dimensionless we can state that the effect of axial dispersion will be negligible when

$$\frac{E_z}{U_z L} \frac{\partial^2 C}{\partial z'^2} \ll \frac{\partial C}{\partial z'} \quad \text{or} \quad \frac{d_p}{L} \frac{1}{N_{Pe_{E,z}}} \frac{\partial^2 C}{\partial z'^2} \ll \frac{\partial C}{\partial z'}$$

where $z' = z/L$ and L is the length of the reactor.

Since $L \gg d_p$ and $N_{Pe_{E,z}} > 1$ for reasonable fluid velocities, i.e., for $N_{Re_p} \approx 10$, axial dispersion is unlikely to be significant. This is the case in the majority of noncatalytic gas–solid reactions of importance. Axial dispersion could, however, be significant for cases where the bed is only a few particle diameters deep or when the linear fluid velocity is small.

PRACTICAL IMPORTANCE OF RADIAL DISPERSION

Noncatalytic gas–solid reactions in mixed bed systems usually involve the movement of a reaction front in the direction of the flow and radial gradients of concentration are usually not very signfiicant. It follows that radial dispersion usually plays an insignificant role in mass transfer problems. However, radial eddy diffusion of heat (eddy thermal conductivity) may play an important role in reactors that are heated or cooled through the bounding walls. An interesting example of this type has been presented by Amundson [20].

7.3.2   Heat Transfer in Fixed Beds

In contrast to the behavior of single particles (discussed in Chapter 2) for which analytical solutions may be available, the heat transfer coefficients in fixed beds have to be determined from empirical corrrelations. For convective fluid-to-particle heat transfer these relationships are very similar to those described earlier for single particles. Additional problems do arise, however, in the case of packed beds, where we must also consider radiative particle–particle heat transfer and convective heat transfer between the wall and the bed.

FLUID-TO-PARTICLE HEAT TRANSFER

The transfer coefficients for fluid-to-particle heat transfer may be estimated with the aid of the correlation proposed by Rowe *et al.* [21]

$$N_{Nu} = A + B N_{Re_p}^n N_{Pr}^{2/3} \tag{7.3.18}$$

where

$$A = \frac{2}{1 - (1 - \varepsilon_v)^{1/3}}, \qquad B = \frac{2}{3\varepsilon_v}, \qquad n \text{ is given by: } \frac{2 - 3n}{3n - 1} = 4.65 N_{Re}^{-0.28}$$

$$N_{Nu} = \frac{h \, d_p}{k_f}, \qquad N_{Pr} = \frac{c_p \mu}{k_f}$$

Equation (7.3.18) is based on mass transfer measurements; since there exists a much wider range of data on mass transfer correlation, it is recommended that the adaptation of these relationships to heat transfer be used in calculations of fluid-to-particle heat transfer calculations. A discussion of mass transfer correlations will be given subsequently.

BED-TO-WALL HEAT TRANSFER

The transfer coefficient describing heat transfer between the wall of an empty (i.e., nonpacked) tube and a turbulent fluid is readily described with the aid of the Dittus–Boelter equation [22]

$$N_{\text{Nu}} = h_c d_c / k_f = 0.023 N_{\text{Re}_c}^{0.8} N_{\text{Pr}}^{0.3} \qquad (7.3.19)$$

where h_c is the convective heat transfer coefficient between the fluid and the wall of the column, d_c is the diameter of the column, $N_{\text{Re}_c} = U_0 \rho d_c / \mu$, and $N_{\text{Pr}} = c_p \mu / k_f$, where c_p is the specific heat of the fluid. Equation (7.3.19) is valid for $\text{Re}_c > 10{,}000$ and when the ratio column length/column diameter is larger than 60.

For packed beds it has been found that the heat transfer coefficient from the wall is enhanced by the presence of the packing. To account for this effect, Beek [23] proposed the following correlation:

$$\frac{h_c}{\rho U c_p} = A N_{\text{Re}_p}^{-2/3} N_r^{-2/3} + B N_{\text{Re}_p}^{-1/5} N_{\text{Pr}}^{-3/5} \qquad (7.3.20)$$

where the dimensionless group on the l.h.s. is the Stanton number. A and B are coefficients defined as follows:

for cylindrical particles: $A = 2.58; \quad B = 0.094$
for spherical particles: $A = 0.203; \quad B = 0.220$

RADIATIVE HEAT TRANSFER

In packed bed systems at moderate temperatures convection is the predominant mode of heat transfer; however, thermal radiation may become significant at elevated temperatures, in particular when the gas flow rate through the system is small. Following the work of Baddour and Yoon [24], Downing [25] proposed the following expression for the effective thermal conductivity of a packed bed at high temperatures in the absence of fluid motion:

$$k_{\text{eff}} = \frac{(1 - \varepsilon_v) k_s k T^3}{k_s + k T^3} + k T^3 \qquad (7.3.21)$$

where $k = 0.69\bar{\varepsilon}d_p/10^8$, k_s is the thermal conductivity of the solids in Btu/hr ft °R, $\bar{\varepsilon}$ is the emissivity, and k_{eff} is the effective thermal conductivity in Btu/hr ft °R. Equation (7.3.21) is dimensional, thus the units of the parameters have to be specified as given here.

7.3.3 Mass Transfer

Mass transfer in packed beds has been the subject of investigation by many workers but only the more recent studies receive mention here. Previous work is discussed in these studies, particularly that of Rowe et al. [21] which has been mentioned earlier.

The reader will recall from Eqs. (2.2.1) and (2.2.2) that the mass flux from a fluid to a solid surface (or vice versa) is given by the product of the mass transfer coefficient and the concentration driving force. For systems involving single particles this driving force is well defined as the difference between the concentration in the bulk of the gas and at the surface of the solid (or some suitable modification thereof in which account is taken of bulk flow effects). At any point within a packed bed system the correct value for the concentration driving force is the difference between the reactant concentration in the void space at that particular point in the bed and the concentration at the surface of the solid particle adjacent to this fluid element. As noted earlier, this reactant concentration in the fluid is usually not known a priori but has to be calculated with the aid of subsidiary equations.

Six commonly used correlations are now given and are also presented in graphical form in Fig. 7.14.

Wilson and Geankoplis[26]†

$$j_D \varepsilon_v = 1.09(N_{Re_p})^{-2/3}, \qquad 0.0016 < N_{Re_p} < 55 \tag{7.3.22}$$

$$j_D \varepsilon_v = 0.250(N_{Re_p})^{-0.31}, \qquad 55 < N_{Re_p} < 1500 \tag{7.3.23}$$

Gupta and Thodos [27]

$$j_D \varepsilon_v = 0.010 + \frac{0.863}{N_{Re}^{0.58} - 0.483}, \qquad 1 < N_{Re_p} \tag{7.3.24}$$

Malling and Thodos [28]

$$\varepsilon_v^{1.19} j_D = \frac{0.763}{N_{Re_p}^{0.472}}, \qquad 185 < N_{Re_p} < 8500 \tag{7.3.25}$$

Pfeffer [29]

$$j_D = 1.26 \left[\frac{1 - (1 - \varepsilon_v)^{5/3}}{w}\right]^{1/3} N_{Re_p}^{-2/3}, \qquad N_{Re_p} < 100 \tag{7.3.26}$$

where $w = 2 - 3\tilde{\gamma} + 3\tilde{\gamma}^5 - 2\tilde{\gamma}^6$ and $\tilde{\gamma} = (1 - \varepsilon_v)^{1/3}$.

† The j_D factor appearing in Eqs. (7.3.22)–(7.3.26) has been defined in Section 2.2.

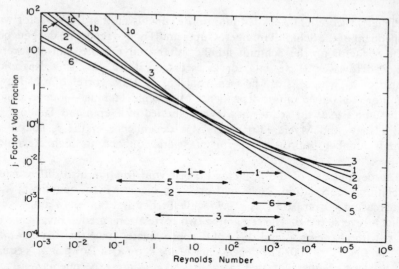

FIG. 7.14. Mass transfer correlations for fixed beds reported by various investigators. The realm of validity claimed by the investigators is indicated, 1a, $N_{Sc} = 1$; 1b, $N_{Sc} = 1000$; 1c, $N_{Sc} = 10,000$. Curve 1, Rowe et al. [21]; 2, Wilson and Geankopolis [26]; 3, Gupta and Thodos [27]; 4, Malling and Thodos [28]; 5, Pfeffer [29]; 6, Bradshaw and Myers [30].

Bradshaw and Myers [30]

$$j = 2.25(N'_{Re_p})^{-0.501} \qquad (7.3.27)$$

where $N'_{Re_p} = N_{Re_p}/(1 - \varepsilon_v)$.

Rowe et al. [21]

$$N_{Sh} = A + BN_{Sc}^{2/3} N_{Re_p}^n \qquad (7.3.28)$$

where A, B, and n have been defined following Eq. (7.3.18).

The correlation of Rowe et al. is different from the others in that it involves a dependence of the j_D factor on the Schmidt number. In fact, Rowe and Claxton assumed an equation of the form of Eq. (7.3.28) (i.e., assumed a dependence of the j factor on N_{Sc}) and then proceeded to determine the values of A, B, and n which would best fit their data. To test the assumed dependence of j_D on N_{Sc} requires measurement at low Reynolds numbers and at low Schmidt numbers. Unfortunately the low Reynolds number experiments carried out by these workers were confined to water ($N_{Sc} \simeq 1400$).

Wilson and Geankoplis carried out measurements at very low Reynolds numbers but again, the measurements were confined to liquids (N_{Sc} in the range 950 to 70,600) and their observation that the j_D factor is independent of Schmidt number may not apply at the low Schmidt numbers typical for gases (where $N_{Sc} \simeq 1$).

Pfeffer's theoretical work led him to the conclusion that the j_D factor was independent of the Schmidt number. Gupta and Thodos report no dependence of the j_D factor on the Schmidt number after studying the data of several workers. However, these data for gases extend down only to a Reynolds number of 14, where, as can be seen from Fig. 7.14, the correlation of Rowe *et al.* indicates only a minor dependence on Schmidt number—a dependence of the same size as the scatter in the data studied by Gupta and Thodos.

Bradshaw and Myers [30] observed a dependence of the j_D factor on Schmidt number but their experiments were confined to high Reynolds numbers.

The dependence of the j_D factor on bed void fraction also differs from correlation to correlation. Rowe *et al.* predict a decrease in the quantity $j_D \times \varepsilon_v$ for increasing ε_v over $0.35 \leq \varepsilon_v \leq 0.75$ while Pfeffer and Bradshaw and Myers predict a reverse trend. The remaining two correlations predict invariance of $j_D \times \varepsilon_v$ with changing ε_v and this holds approximately true for Rowe *et al.*'s equation for $N_{Re} > 100$. The dependence of $j \times \varepsilon_v$ on ε_v in Rowe *et al.*'s equations was not adequately tested by these workers since experiments were not carried out at low Reynolds and low Schmidt numbers.

An interesting electrochemical technique for studying transfer coefficients in packed beds is described by Jolls and Hanratty [31] whose data could be fitted by the following equations:

$$N_{Sh} = 1.59 N_{Re_p}^{0.56} N_{Sc}^{1/3}, \qquad 140 < N_{Re} \qquad (7.3.29)$$

$$N_{Sh} = 1.44 N_{Re_p}^{0.58} N_{Sc}^{1/3}, \qquad 35 < N_{Re} < 140 \qquad (7.3.30)$$

for a packed bed of void fraction 0.41.

Some experimental investigations (e.g. that of Rowe *et al.* and of Jolls and Hanratty) have been carried out using an inert bed containing one active sphere. While such experiments are generally more precise than those where the whole bed is active, there do appear to be small differences in the behavior of the two types of bed which diminish the usefulness of correlations obtained from inert beds. In addition, inert bed measurements suffer from the disadvantage that a local (within a small region of the bed) rather than average coefficient is obtained and such variables as ·position of the active sphere within the bed and the precise arrangement of its nearest neighbors would have to enter the picture.

7.4   Formulation for Gas–Solid Reactions in Multiparticle Systems

ISOTHERMAL SYSTEMS

Let us consider an infinitesimally thin section of a fixed bed reactor through which a reactant gas is flowing, as sketched in Fig. 7.15. In an isothermal regime and for equimolar counterdiffusion, the system may be de-

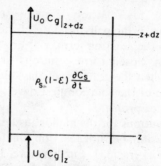

FIG. 7.15. Schematic diagram of the mass balance over an infinitesimal reactor element.

scribed by establishing a mass balance on the reactant gas and on the reactant solid. If we neglect axial dispersion

$$U_0 \frac{\partial C_g}{\partial z} + (\varepsilon_v) \frac{\partial C_g}{\partial t} = - h_D a [C_g - C_{g,s}] \qquad (7.4.1)$$

where U_0 is the linear gas velocity, C_g the gaseous reactant concentration (moles/unit interstitial volume), h_D the mass transfer coefficient, a the surface area/unit volume of the bed, and $C_{g,s}$ the reactant concentration at the surface of the particle. A similar equation must be written for the gaseous product if the reaction is reversible. Note that except at elevated pressures the second term on the l.h.s. may be neglected.

Mass balance on the reactant solid may be written as

$$h_D a [C_g - C_{g,s}] = - \frac{(1 - \varepsilon_v)}{b} \frac{\partial C_s}{\partial t} \qquad (7.4.2)$$

where b is the stoichiometric coefficient and C_g the concentration of the solid reactant (moles of solid reactant/unit volume of porous solid).

In general the rate at which the solid reacts, i.e., the time derivative appearing on the r.h.s. of Eq. (7.4.2), will depend on a number of factors, such as the solid structure, the order of the surface reaction, and pore diffusion. Thus we may write

$$- \frac{1}{C_{s,0}} \frac{\partial C_s}{\partial t} \equiv \frac{dX}{dt} = f(C_s, C_{g,s}, D_e, \text{etc.}) \qquad (7.4.3)$$

The initial and boundary conditions for these equations would typically specify the reactant gas concentration at the inlet, the initial gaseous reactant profile within the bed, and the initial conversion of the solid reactant. Thus we have

$$C_g = C_{ig}, \quad z = 0; \qquad C_g = C_{ig}'(z), \quad t = 0; \qquad C_s = C_s'(z), \quad t = 0 \qquad (7.4.4)$$

It is noted that not even this relatively simple system of equations has a general analytical solution, principally because of the possible nonlinearities that may be contained in the reaction term, viz. Eq. (7.4.3).

Analytical or at least closed form solutions are available for certain asymptotic cases; thus when the reaction of the solid is mass transfer controlled, use may be made of the solution first developed by Furnas [32] for the equivalent heat transfer problem.

Other closed form solutions for the analogous heat transfer problem (including intraparticle diffusion, but not reaction kinetics) have been proposed by Amundson and co-workers [33]. Furthermore, in a more recent paper Moriyama [34] discussed solutions to Eqs. (7.4.1)–(7.4.4) for shrinking core type kinetics.

It seems desirable to devise a method that would allow solutions to be generated readily for the system of Eqs. (7.4.1)–(7.4.4) while making use of the general kinetic expressions developed in Chapter 5. The discussion that follows is an adaptation of recent work by Evans and Song [35].

A GENERALIZED APPROACH TO THE REACTION OF A POROUS SOLID IN A PACKED BED WITH A GAS UNDER ISOTHERMAL CONDITIONS

First of all, let us consider the reaction of a porous solid with a reactant gas, such as discussed in Chapter 4. By assuming isothermal conditions, first-order kinetics, and the absence of bulk flow effects, the governing equations may be written as

$$\nabla^2 \psi - \sigma^2 \psi \xi^{F_g - 1} = 0 \tag{7.4.5}$$

where

$$\sigma = \frac{F_p V_p}{A_p} \sqrt{\frac{(1 - \varepsilon)k A_g}{D_e V_g}\left(1 + \frac{1}{K_E}\right)} \tag{7.4.6}$$

ψ is the dimensionless concentration driving force for reaction, and ξ the dimensionless position of the reaction front within the grain; the quantities appearing in Eqs. (7.4.5) and (7.4.6) have been defined in Chapter 4.

The rate of reaction of the individual grains may be expressed as

$$\partial \xi / \partial t^* = - \psi \tag{7.4.7}$$

where

$$t^* = \frac{bk(C_{A0} - (C_{C0}/K_E))A_g}{\rho_s F_g V_g} t \tag{7.4.8}$$

i.e., the previously defined dimensionless time.

Equations (7.4.5)–(7.4.8) were originally derived for a constant reactant concentration in the bulk or, more precisely, at the outer surface of the pellet. However, since the system is linear these equations may be generalized in the following manner for a variable bulk concentration driving force

$$\left(C_{A0} - \frac{C_{C0}}{K_E}\right) = \theta\left(\bar{C}_{A0} - \frac{\bar{C}_{C0}}{K_E}\right)$$

where $\bar{C}_{A0}$ and $\bar{C}_{C0}$ are constants.

$$\partial\xi/\partial t^* = -\psi\theta \tag{7.4.9}$$

where now the concentrations appearing in the definition of t^* are $\bar{C}_{A0}$ and $\bar{C}_{C0}$. θ with $0 \le \theta \le 1$, which is not known at this stage, will be used to allow for the variation in the bulk concentration (surface concentration) of the reactant gas during its passage through the bed.

The overall extent of reaction X for the individual pellets may be written as

$$X = \frac{\int_0^1 \eta^{F_p - 1}(1 - \xi^{F_g})\, d\eta}{\int_0^1 \eta^{(F_p - 1)}\, d\eta} \tag{7.4.10}$$

which may be differentiated to give

$$\frac{dX}{dt^*} = \theta\left[\frac{\int_0^1 \eta^{(F_p-1)} F_g \xi^{(F_g-1)}\psi\, d\eta}{\int_0^1 \eta^{(F_p-1)}\, d\eta}\right], \tag{7.4.11}$$

For a nonconstant value of C_{As} we may write in general

$$dX/dt^* = \beta\theta \tag{7.4.12}$$

where $\beta = [\quad]$ on the r.h.s. of Eq. (7.4.11) and may be regarded as akin to the effectiveness factor (η) encountered in heterogeneous catalysis except that β will depend on the extent of reaction of the solid (e.g., [36]).

We note that in general β is a function of X, $\hat{\sigma}$, F_g, and F_p, but this functional relationship may be readily evaluated by numerical means, using the information provided in Chapter 4. We note, furthermore, that in Eq. (7.4.12) β depends on θ only through the dependence of X on θ. This relationship is available and was in fact given by writing the macroscopic balance over the bed in Eq. (7.4.3). We may proceed by generating tabulated values of β versus X with $\hat{\sigma}$, F_g, and F_p as parameters; these values may then be used in the integration of the component balance equation. Tabulated values of β against X with $\hat{\sigma}$ as a parameter are given in Table 7.1.

A less rigorous but perhaps more convenient way to proceed would be to make use of the approximate relationship given in Chapter 4 between dX/dt^* and the other system parameters:

$$\frac{dX}{dt^*} = \left[g'_{F_g}(X) + \hat{\sigma}^2\left\{p_{F_p}(X) + \frac{2}{N'_{Sh}}\right\}\right]^{-1} \tag{7.4.13}$$

where $\hat{\sigma} = \sigma/\sqrt{2F_g F_p}$.

TABLE 7.1

$\beta = dX/dt^*$ (at $\theta = 1$)[a]

Extent of reaction X	$F_g = 3$					
	$\hat{\sigma}$					
	0.3	0.5	0.8	1.2	1.8	3.0
0.00	2.72	2.37	1.88	1.42	1.02	0.653
0.05	2.63	2.30	1.82	1.38	0.984	0.614
0.10	2.55	2.23	1.77	1.33	0.942	0.575
0.15	2.46	2.16	1.71	1.28	0.898	0.533
0.20	2.37	2.08	1.65	1.23	0.853	0.490
0.25	2.28	2.01	1.59	1.18	0.806	0.444
0.30	2.18	1.93	1.53	1.13	0.758	0.396
0.35	2.08	1.85	1.47	1.07	0.707	0.347
0.40	1.98	1.76	1.40	1.02	0.655	0.297
0.45	1.88	1.68	1.33	0.959	0.600	0.251
0.50	1.77	1.58	1.26	0.898	0.542	0.213
0.55	1.65	1.49	1.19	0.833	0.483	0.182
0.60	1.54	1.39	1.11	0.765	0.422	0.155
0.65	1.41	1.28	1.02	0.693	0.361	0.132
0.70	1.28	1.17	0.930	0.617	0.306	0.112
0.75	1.14	1.04	0.832	0.535	0.257	0.0941
0.80	0.986	0.909	0.724	0.448	0.212	0.0779
0.85	0.819	0.760	0.603	0.357	0.170	0.0626
0.90	0.629	0.587	0.461	0.265	0.129	0.0477
0.95	0.399	0.374	0.284	0.166	0.0843	0.0319
0.99	0.137	0.124	0.0877	0.0581	0.0340	0.0144

Extent of reaction X	$F_g = 2$					
	$\hat{\sigma}$					
	0.3	0.5	0.8	1.2	1.8	3.0
0.00	1.87	1.69	1.40	1.10	0.808	0.522
0.05	1.82	1.65	1.37	1.07	0.787	0.503
0.10	1.78	1.61	1.34	1.05	0.764	0.481
0.15	1.73	1.57	1.31	1.02	0.740	0.458
0.20	1.68	1.53	1.28	0.993	0.715	0.434
0.25	1.63	1.49	1.24	0.964	0.689	0.406
0.30	1.58	1.44	1.20	0.933	0.660	0.375
0.35	1.52	1.39	1.17	0.901	0.630	0.339
0.40	1.47	1.35	1.13	0.867	0.596	0.297
0.45	1.41	1.29	1.08	0.831	0.560	0.251
0.50	1.35	1.24	1.04	0.791	0.519	0.212
0.55	1.28	1.18	0.992	0.749	0.474	0.181
0.60	1.21	1.12	0.941	0.703	0.420	0.155
0.65	1.13	1.05	0.885	0.651	0.359	0.132
0.70	1.05	0.979	0.823	0.592	0.304	0.112
0.75	0.964	0.899	0.754	0.524	0.255	0.0937
0.80	0.865	0.809	0.675	0.440	0.210	0.0779
0.85	0.752	0.705	0.581	0.348	0.168	0.0625
0.90	0.616	0.578	0.460	0.259	0.127	0.0477
0.95	0.437	0.406	0.285	0.162	0.0828	0.0319
0.99	0.193	0.151	0.0943	0.0588	0.0334	0.0142

TABLE 7.1 (continued)

| Extent of reaction X | $F_g = 1$ | | | | | |
| | $\hat{\sigma}$ | | | | | |
	0.3	0.5	0.8	1.2	1.8	3.0
0.00	0.966	0.912	0.812	0.679	0.526	0.353
0.05	0.966	0.912	0.812	0.679	0.526	0.353
0.10	0.966	0.912	0.812	0.679	0.526	0.353
0.15	0.966	0.912	0.812	0.679	0.526	0.353
0.20	0.966	0.912	0.812	0.679	0.526	0.353
0.25	0.966	0.912	0.812	0.679	0.526	0.353
0.30	0.966	0.912	0.812	0.679	0.526	0.353
0.35	0.966	0.912	0.812	0.679	0.526	0.353
0.40	0.966	0.912	0.812	0.679	0.526	0.298
0.45	0.966	0.912	0.812	0.679	0.526	0.243
0.50	0.966	0.912	0.812	0.679	0.526	0.228
0.55	0.966	0.912	0.812	0.679	0.515	0.187
0.60	0.966	0.912	0.812	0.679	0.414	0.156
0.65	0.966	0.912	0.812	0.679	0.352	0.133
0.70	0.966	0.912	0.812	0.654	0.289	0.116
0.75	0.966	0.912	0.812	0.530	0.247	0.0949
0.80	0.966	0.912	0.812	0.453	0.216	0.0775
0.85	0.966	0.912	0.672	0.354	0.167	0.0612
0.90	0.966	0.912	0.504	0.264	0.129	0.0470
0.95	0.966	0.658	0.319	0.168	0.0834	0.0317
0.99	0.488	0.243	0.118	0.0689	0.0357	0.0151

[a] Reprinted with permission from Evans and Song, *Ind. Eng. Chem. Process Design Develop.* **13**, 146 (1974). Copyright by the American Chemical Society.

Thus for the variable gas concentration we may write

$$dX/dt^* = \beta'\theta \tag{7.4.14}$$

where

$$\beta' \equiv \left\{ g'_{F_g}(X) + \hat{\sigma}^2 \left[p'_{F_p}(X) + \frac{2}{N'_{Sh}} \right] \right\}^{-1} \tag{7.4.15}$$

and $g'_{F_g} \equiv d(g_{F_g})/dX$; $p'_{F_p} \equiv d(p_{F_p})/dX$. While this definition of β' is only approximate, the procedure is appealing since Eq. (7.4.15) is an analytical expression in which allowance is made for gas–solid mass transfer. We now proceed by substituting the relationship

$$\frac{(1 - \varepsilon_v)}{b} \frac{\partial C_s}{\partial t} = \frac{(1 - \varepsilon_v)(1 - \varepsilon)}{b} \rho_s \frac{\partial X}{\partial t}$$

in (7.4.2) and then substituting in (7.4.1) which gives a relationship between $\partial C_g/\partial z$ and $\partial X/\partial t$. A similar equation may be derived for the gaseous product. This second equation is divided by K_E and subtracted from the first. The resultant equation is

$$-U_0 \left(\bar{C}_{A0} - \frac{\bar{C}_{C0}}{K_E} \right) \frac{\partial \theta}{\partial y} = \frac{(1 - \varepsilon_v)(1 - \varepsilon)}{b} \rho_s \left(1 + \frac{1}{K_E} \right) \frac{\partial X}{\partial t} \tag{7.4.16}$$

where ρ_G is the molar density of the gas, ε the pore volume in the pellet, ρ_s the molar density of the solid and

$$\theta = \theta(y) = \left(\bar{C}_{A0} - \frac{C_{C0}}{K_E}\right)\bigg/\left(\bar{C}_{A0} - \frac{\bar{C}_{C0}}{K_E}\right)$$

which is a dimensionless reactant concentration driving force. On defining

$$y^* = \frac{kA_g}{U_0 F_g V_g}[1 - \varepsilon_v]\left(1 + \frac{1}{K_E}\right)[1 - \varepsilon]y \qquad (7.4.17)$$

the dimensionless form of Eq. (7.4.16) is written as

$$-\frac{\partial \theta}{\partial y^*} = \frac{dX}{dt^*} = \theta \beta' \qquad (7.4.18)$$

with

$$\theta = 1, \qquad y^* = 0, \qquad t^* \geq 0 \qquad (7.4.19)$$

and

$$X = 0; \qquad t^* = 0 \qquad (7.4.20)$$

Equation (7.4.18) may be expressed in the following integral form:

$$\frac{dX}{dt^*} = \beta' \exp\left[-\int_0^{y^*} \beta' \, d\lambda^*\right] \qquad (7.4.21)$$

where λ is a dummy variable.

Since β' is known as a function of X [viz. Eq. (7.4.15)], Equation (7.4.21) is readily integrated, e.g., using Simpson's rule and a Runge–Kutta procedure.

An alternative, perhaps more accurate procedure would be to use the original definition of β and work through Eq. (7.4.11), although the approximate method described here should work quite well except for the very initial stages.

On performing the indicated integration, we obtain the desired results, viz. plots of X (the extent of reaction) against y^* (the dimensionless distance) with t^* as a parameter. Typical computed results are given in Fig. 7.16 for large N_{Sh}. Here the abscissa and parameter have been so chosen as to result in a nearly identical set of curves irrespective of the value of σ. Consequently, Fig. 7.16 can be used for any value of σ.

Note that the reaction front assumes a sigmoid shape after the initial period and that this sigmoid shape travels through the bed essentially unchanged. While this unchanging profile was to some extent imposed on the system by neglecting the time derivative of the concentration in the gas phase, it can be shown that for systems at atmospheric pressure the behavior seen is a natural property of the system.

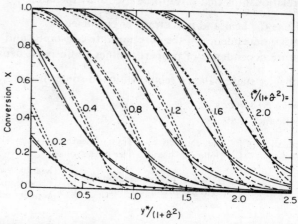

FIG. 7.16. Plot of conversion against a normalized distance down the bed for a packed bed of spherical pellets made up of spherical grains. $\cdots$, $\hat{\sigma} = 0.03$; $- \cdot -$, $\hat{\sigma} = 0.3$; ———, $\hat{\sigma} = 1.0$; $- - -$, $\hat{\sigma} = 3.0$; $- - -$, $\hat{\sigma} = 10.0$

THE GENERAL CASE OF NONISOTHERMAL SYSTEMS

The dimensionless representation of the reaction between a gas and a porous solid in a packed bed arrangement under isothermal conditions is helpful because it provides a convenient means for estimating the behavior of many reaction systems. However, the majority of systems encountered in practice are likely to be a great deal more complex in that neither the assumption of isothermal conditions nor the postulate of equimolar counterdiffusion (i.e., no change in the molar flow rate) are likely to be acceptable.

Since it would be impracticable to develop a general dimensionless formulation that would encompass all the possible cases, we shall discuss the considerations that must enter the mathematical representation of these systems and then present some illustrative examples.

Formulation for nonisothermal systems

The formulation of nonisothermal systems where there may be a significant change in the molar flow rate of the reactant gas will in general consist of the statement of:

heat balance on the gas,
heat balance on the solids
mass balance on the solids,
component balances on the gas, and
overall mass balance on the gas.

The typical relationships that have to be used may include:

correlations for heat and mass transfer coefficients,
temperature dependence of the reaction rate constants, and
temperature dependence of property values (diffusivity, density, etc.).

Let us consider these balance relationships in somewhat more detail:

heat balance on the gas
will express the fact that

$$\begin{bmatrix} \text{rate of accumulation of} \\ \text{heat within the gas} \end{bmatrix} = \begin{bmatrix} \text{net convective} \\ \text{inflow of heat} \end{bmatrix} - \begin{bmatrix} \text{rate of heat trans-} \\ \text{fer to the solids} \end{bmatrix}$$

$$+ \begin{bmatrix} \text{rate of heat generation due} \\ \text{to gas phase reactions} \end{bmatrix} + \begin{bmatrix} \text{heat inflow due to axial} \\ \text{dispersion and} \\ \text{conduction in the gas} \end{bmatrix}$$

(7.4.22)

We note that in many cases we may neglect the accumulation term [i.e., the l.h.s. of Eq. (7.4.22)]; furthermore the effect of axial dispersion and conduction in the gas may also be neglected, particularly when the Peclet number is small.

heat balance on the solids
may be written as

$$\begin{bmatrix} \text{rate of accumulation} \\ \text{of heat in the solids} \end{bmatrix} = \begin{bmatrix} \text{rate of heat transfer to} \\ \text{the solids from the gas} \end{bmatrix} + \begin{bmatrix} \text{rate of heat generation due} \\ \text{to heterogeneous reactions} \end{bmatrix}$$

$$+ \begin{bmatrix} \text{heat inflow due to} \\ \text{solid–solid conduction} \end{bmatrix}$$

(7.4.23)

Since the heat capacity/unit volume of the solids is much greater than that of the gas, the accumulation term cannot be neglected in this case.

component balance on the gas

$$\begin{bmatrix} \text{rate of accumulation} \\ \text{of reactant product} \\ \text{of component } i \end{bmatrix} = \begin{bmatrix} \text{convective inflow} \\ \text{of component } i \end{bmatrix} + \begin{bmatrix} \text{inflow of component } i \\ \text{due to axial dispersion} \end{bmatrix}$$

$$+ \begin{bmatrix} \text{net production of com-} \\ \text{ponent } i \text{ due to chemical} \\ \text{reaction in the gas phase} \end{bmatrix} + \begin{bmatrix} \text{net transfer of component} \\ \text{from solid to the gas phase} \end{bmatrix}$$

(7.4.24)

component balance on the solids

$$\begin{bmatrix} \text{rate of accumulation} \\ \text{of component } j \end{bmatrix} = \begin{bmatrix} \text{rate of production or consumption of} \\ \text{component } j \text{ due to chemical reaction} \end{bmatrix} \quad (7.4.25)$$

It is noted that the accumulation term and the axial dispersion term may often be neglected. It may be helpful to illustrate the nature of the last two terms on the r.h.s. of Eq. (7.4.24). If we were to establish a balance on water vapor in the gas phase, then drying of a porous solid would correspond to the physical transfer of the component in question from the solid to the gas. In contrast, the gas phase oxidation of hydrogen to water vapor would have to be represented by a chemical reaction term.

Another problem may arise when we wish to establish a component balance for a species that is both generated by a heterogeneous surface reaction and consumed by a homogeneous gas phase reaction; viz. a carbon monoxide balance in sintering. Here the transfer of CO from the solid carbon would be described by the last term on the r.h.s. of Eq. (7.4.24), whereas the consumption of CO due to oxidation would be represented by the third term.

Finally, *the overall mass balance* on the gas has to express the fact that the mass flow rate of the gas through the bed may change due to the *net* generation or consumption of gaseous reactants and products: Thus we have:

$$\begin{bmatrix} \text{net change in the mass} \\ \text{flow rate with distance} \end{bmatrix} = \begin{bmatrix} \text{net rate of production} \\ \text{of gaseous species} \end{bmatrix} \qquad (7.4.26)$$

Let us illustrate these concepts through two concrete examples.

THE MATHEMATICAL MODELING OF SINTER BED OPERATIONS

In a recent paper Muchi and Higuchi [37] proposed the following simplified model for the development of the gas and solid temperature distribution in sinter beds.

In a "real" sinter bed operation a mixture of fine ore particles, return sinter, BOF dust,† fluxing materials, and coke breeze is fed onto a continuous grate as sketched in Fig. 7.10. The mixture is ignited at the top and combustion of the coke partially melts the other feed components which results in a strong but porous *sinter*; this sinter is an important feed to the iron blast furnace.

The construction of a complete mathematical model for the sinter bed process would involve a formidable task because of the large number of interrelated chemical and physical processes that take place simultaneously. However, the principal feature of the process, namely the passage of a combustion front and a temperature "wave" through the bed, may be adequately represented.

In developing their simplified model the authors made the following assumptions:

(a) axial dispersion and lateral inhomogeneities were neglected,
(b) radiation heat transfer effects were neglected,

---

† Fine dust particles emitted from the basic oxygen furnace and recovered in electrostatic precipitators or venturi scrubbers [38].

(c) the only chemical reaction considered was the combustion of the coke and this was considered to proceed fully to CO_2 with all the heat generated occurring on the surface of the solid particles.

While all these assumptions represent an oversimplification, in particular assumption (c), the results that may be derived from them will provide a first approximate representation for the system.

On the basis of these assumptions Muchi and Higuchi represented their system by the following equations:

$$G \frac{\partial}{\partial z}(c_{p_G} T_G) - ha(T_G - T_s) = \varepsilon_v \rho_G \frac{\partial}{\partial t}(c_{p_G} T_G) \qquad (7.4.27)$$

(heat balance on the gas)

$$-G \frac{\partial}{\partial z}\left(\frac{C_{O_2}}{\rho_G}\right) - \left(1 + 12 \times 10^{-3} \frac{C_{O_2}}{\rho_G}\right) R_o^* = \varepsilon_v \rho_G \frac{\partial}{\partial t}\left(\frac{C_{O_2}}{\rho_G}\right) \qquad (7.4.28)$$

(oxygen balance)

$$-G \frac{\partial}{\partial z}\left(\frac{C_{CO_2}}{\rho_G}\right) + \left(1 - 12 \times 10^{-3} \frac{C_{CO_2}}{\rho_G}\right) R_c^* = \varepsilon_v \rho_G \frac{\partial}{\partial t}\left(\frac{C_{CO_2}}{\rho_G}\right) \qquad (7.4.29)$$

(CO_2 balance)

$$-\frac{\partial G}{\partial z} + 12 \times 10^{-3} R_c^* = \frac{\partial}{\partial t}(\varepsilon_v \rho_G) \qquad \text{(overall mass balance)} \qquad (7.4.30)$$

$$C_{CO_2} + C_{O_2} + C_{N_2} = 4.46 \times 10^{-5}(273/T_G) \qquad (7.4.31)$$

(overall mass balance)

$$\rho_G = (32 C_{O_2} + 44 C_{CO_2} + 28 C_{N_2}) \times 10^{-3}(273/T_G) \qquad (7.4.32)$$

(equation of state for the gas)

$$ha(T_G - T_s) + (\Delta H_c) R_c^* = (1 - \varepsilon)\rho_s \frac{\partial}{\partial t}(C_{p_s} T_s) \qquad (7.4.33)$$

(heat balance on the solids)

where G is the mass velocity of the gas (mass units), z the coordinate in the direction of flow, R_c^* the overall reaction rate of coke (g-atom C/min cm^3), ρ_G the gas density, h the heat transfer coefficient, a the surface area/unit volume of bed ΔH_c the heat of reaction, ε_v the void fraction, and t the time.

We note that Muchi and Higuchi used an empirical expression for relating R_c^* to the temperature, the carbon content of the bed, and the oxygen concentration within the gas.

The reader may find it of interest to compare the formulation given here with the general procedures described in the preceding section. We note that the authors used a mixture of molar and mass units in their formulation, which gave rise to the factors 12, 28, and 44 in their Eqs. (7.4.30) and (7.4.32). The use of concentration units for describing the gas composition required that the density be evaluated at each position for each time step. While no computational difficulties were reported by the authors, usually care has to be taken in such calculations in order to avoid numerical instability. The use of molar units and mole fractions (mole ratios) might have led to somewhat tidier equations.

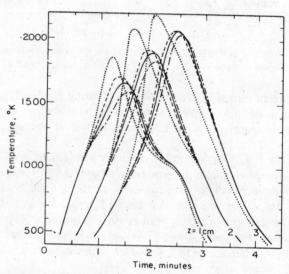

FIG. 7.17. Calculated effect of oxygen concentration in the first ignition furnace on the temperature distributions of solid particles in a sinter bed [37]. Values of $C_{O_2,0}$: $\cdots$, 1.126×10^{-5}; $---$, 0.752×10^{-5}; ———, 0.563×10^{-5}; $-\cdot-$, 0.376×10^{-5}.

Some computed results are presented in Figs. 7.17, 7.18 and 7.19, showing the effect of the oxygen concentration, the diameters of the solid carbon particles, and the fraction of coke in the charge.

Example 7.4.1 The Modeling of Incinerators [39]. Let us consider the combustion and pyrolysis of cellulose particles in contact with air in a packed arrangement such as that sketched in Fig. 7.20. Such a system corresponds to a rather idealized model of packed bed incinerators where the solid

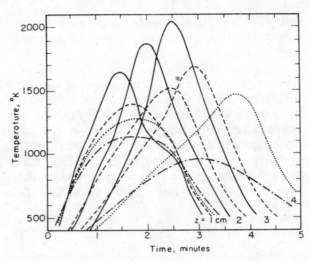

FIG. 7.18. Effect of diameter of solid particles on the temperature distribution in a sinter bed [37]. Values of γ_0: ———, 0.15; – – –, 0.20; ···, 0.24; – · –, 0.28.

feed is made up of numerous components including metals, glass, and cellulose (in the form of paper, grass cuttings, etc.). Nonetheless, the combustion of cellulose-type materials is an important aspect of the operation of incinerators.

Referring to Fig. 7.20, the solid particles are ignited and the resultant flame front is made to move progressively downward in the z direction. As the solid particles are brought into contact with the hot gas, both combustion and pyrolysis may occur; the vapor products of pyrolysis may be burned in the gas space if the gas with which it is in contact is hot enough and contains a sufficient amount of oxygen. Alternatively, the pyrolysis products may escape from the system unburned, which would be undesirable in a real system, because it would cause air pollution.

One principal objective in the modeling of such systems is to define the criteria under which the products of pyrolysis are completely burned and to understand how these criteria are affected by the other process variables. In constructing such a model the governing equations may be derived by establishing a heat and mass balance over an infinitesimal section of the bed. The following assumptions are made:

(1) The reactor is adiabatic and there is no radial transport of heat.
(2) Axial dispersion is neglected.

(3) The velocity is assumed to be uniform over the cross section and length, and temperature effects on the velocity are neglected.

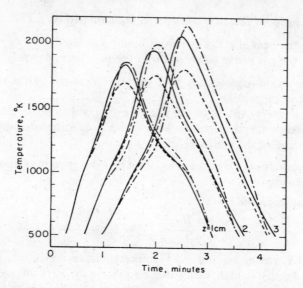

FIG. 7.19. Effect of coke mass fraction in the burden on the temperature distribution in a sinter bed [37]. Values of m_c: – · –, 0.04; ———, 0.03; – – –, 0.02.

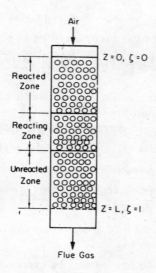

FIG. 7.20. Representation of a packed-bed incinerator.

(4)   The heats of reaction considered are confined to:

(a) . the heat effect of pyrolysis and
(b)   the gas phase combustion of the products of pyrolysis.

This assumption neglects any heat generation caused by the combustion of the solid carbon residue.

(5)   The property values of the gas were assumed to be identical to those of the air except for the specific heat which was calculated by taking a weighted average of the constituents.

Within the framework of these assumptions the governing equations will take the following form:

Heat balance on the solids

$$\frac{\partial T_s}{\partial t} = -\frac{T_s}{\rho_s}\frac{\partial \rho_s}{\partial t} - \frac{3hr_p^2}{r_0^3 c_{ps}\rho_s}(T_s - T_g) - \frac{r_1}{c_{ps}\rho_s}\Delta H_1 \frac{r_p^3}{r_0^3} \qquad (7.4.34)$$

where T_g and T_s are the gas and solid temperatures, respectively; t is the time; ρ_s the solid density (which varies due to pyrolysis); $r_p(t)$ the radius of the cellulose particles, the initial value of which is r_0; c_p the specific heat; r_1 the rate of pyrolysis; and ΔH_1 the heat of pyrolysis, considered to correspond to the heat of reaction for

$$(C_6H_{10}O_5)_x \longrightarrow 5xCH_2O + xC; \qquad \Delta H_1$$

The heat balance on the gas may be written as

$$\frac{\partial T_g}{\partial t} = \frac{3(1-\varepsilon)}{c_{pg}\rho_g}h(T_s - T_g)\frac{r_p^2}{r_0^3} - \frac{1}{\rho_g \varepsilon}\frac{\partial(GT_g)}{\partial z} - \frac{r_2 \Delta H_2}{c_{pg}\rho_g} \qquad (7.4.35)$$

where G is the mass flow rate of the gas, r_2 the rate of gas phase combustion, and ΔH_2 the heat of combustion.

Here ΔH_2 is considered to correspond to the heat of reaction for

$$CH_2O + O_2 \longrightarrow CO_2 + H_2O$$

The mass balance on the gaseous intermediates is written as

$$\frac{\rho_g \varepsilon}{\overline{M}}\frac{\partial x_A}{\partial t} = \frac{-x_A}{\overline{M}}\frac{\partial G}{\partial z} - \frac{G}{\overline{M}}\frac{\partial x_A}{\partial z} + r_1 \frac{5}{m_{ce}}(1-\varepsilon)\frac{r_p^3}{(r_0)^3} - r_2\varepsilon \qquad (7.4.36)$$

where x_A is the mole fraction of the gaseous intermediate and m_{ce} is the formula weight of cellulose: 162.

The mass balance on the oxygen may be expressed as

$$\frac{\rho_g \varepsilon}{\overline{M}}\frac{\partial x_B}{\partial t} = -\frac{x_B}{\overline{M}}\frac{\partial G}{\partial z} - \frac{G}{\overline{M}}\frac{\partial x_B}{\partial z} - r_2\varepsilon \qquad (7.4.37)$$

where x_B is the mole fraction of the oxygen.

The rate of pyrolysis was calculated using the following expression proposed by Roberts [40]:

$$\frac{\partial \rho_s}{\partial t} = -(\rho_s - \rho_f)K_{10} \exp\left(\frac{-E_1}{RT_s}\right) = r_1 \qquad (7.4.38)$$

where ρ_f is the final density of the solid on the completion of pyrolysis,

$$\rho_f/\rho_s \simeq 0.2.$$

The equation of continuity for the gas is given as

$$\frac{\partial G}{\partial z} = r_1(1 - \varepsilon)\frac{r_p^3}{r_Q^3} - \frac{\partial(\varepsilon\rho_g)}{\partial t} \qquad (7.4.39)$$

Finally, the rate expression for the combustion of the gaseous intermediates was expressed as

$$r_2 = K_{20}\exp(-E_2/RT_g)\,\rho_g^2 x_A x_B \qquad (7.4.40)$$

The subsidiary equations required to complete the statement of the problem include relationships for the temperature dependence of the density and viscosity of the gas, correlations for the heat transfer coefficient, and an expression relating the density of the solid particles to the progress of the pyrolysis.

It was decided that the most convenient way of representing this information was the use of empirical expressions that were deduced from experimental data reported by Essenhigh and coworkers [41, 42]. The following two relationships were used:

In the region $\rho_s > \rho_f$, i.e., during pyrolysis

$$r_p = r_0\left[\frac{0.98}{1.98 - (\rho_s/\rho_{s0})}\right]^{1/3} \qquad (7.4.41)$$

and during burnout, i.e., when $\rho_s = \rho_f$

$$r_p = r_0\left[\frac{0.68}{0.58} - \frac{0.68}{0.58}\frac{t}{t_b}\right]^{1/2} \qquad (7.4.42)$$

t_b is the time required to burn out the particle and was estimated as 2.5 times the time that a particle takes to reach its final density. It may be noted here that although both Eqs. (7.4.41) and (7.4.42) are approximate, nonetheless they express the experimental observations that during pyrolysis the particle size does not change a great deal, whereas during burnout there is a parabolic relationship between time and the diameter of the residual particle.

The initial and boundary conditions for this system must specify that the initial temperatures and concentrations correspond to ambient conditions, and that ambient conditions be maintained at the inlet.

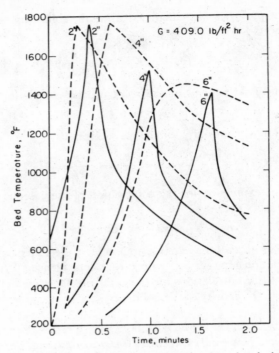

FIG. 7.21. Comparison between experimental (– – –) and calculated (——) temperature profiles in an incinerator.

The system of Eqs. (7.4.34)–(7.4.42) was put in the form of ordinary differential equations using the method of characteristics (e.g., [43]) and the finite difference form of the resultant expressions was integrated numerically using the Runge–Kutta method. Some computed results are shown in Figs. 7.21–7.23 together with actual experimental measurements.

Figure 7.21 shows the propagation of the flame front in a model incinerator about $3\frac{11}{16}$ in. in diameter and some 12 in. long packed with cellulose particles in the 4–8-mesh size range. It is seen that the agreement between measurements and predictions is quite poor, which would indicate that the model described is rather oversimplified. However, in Fig. 7.22 the agreement between measurements and predictions is rather better with regard to the propagation of the flame front; furthermore, inspection of Fig. 7.23 shows that the model appears to provide a reasonable prediction for the relationship between the net emission of the vapors (produced by pyrolysis) and the mass flow rate of the air.

This example is an indication of the complexities involved in the modeling of incinerator systems; if we wished to obtain a faithful description of the

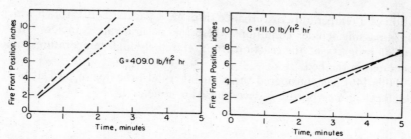

FIG. 7.22. Flame front positions in an incinerator as a function of time. ——————, theoretical; – – –, experimental.

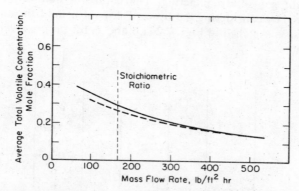

FIG. 7.23. Effect of gas flow rate on volatile leaving the exhaust of an incinerator. ——————, theoretical; – – –, experimental; feed O_2 concentration, 0.21.

transient temperature profiles within the system, much more reliable information would have to be fed into the model concerning the kinetics of pyrolysis and the possible collapse of the bed upon pyrolysis. (This collapse of the bed would make the assumption of stationary solids invalid.) However, the model appears to provide a reasonable representation for the net emission of volatiles and can serve at least as a first approximation in this regard.

With the growing interest in environmental problems incineration is becoming an important branch of gas–solid reaction systems. For this reason we shall present a brief survey of data on incineration kinetics in Chapter 8.

Further examples of packed solids undergoing reaction with a gas under nonisothermal conditions are provided by underground coal gasification and by packed-bed regeneration of "coked" catalyst particles. A mathematical model describing the former has been developed by Sherwood [44] and one describing the latter by Johnson et al. [45]. In both cases the formation of a high-temperature reaction zone which travels through the bed was predicted. Johnson et al. were able to show that there exists a natural rate of propagation of a temperature disturbance through a packed bed through which a gas is passing. Under conditions where this natural rate coincides with the rate of progress of the reaction front, extremely high temperatures can be generated in the bed.

Hills [46] has considered the nonisothermal behavior of packed beds in which calcium carbonate is decomposed to lime.

7.5   Fluidization

Fluidized systems represent a potentially attractive alternative to fixed-bed operations for gas–solid contacting. The nature of fluidization is perhaps best illustrated by considering the upward flow of a fluid through a packed bed of solids, as sketched in Figs. 7.24a,b and c, for progressively increasing fluid flow rates.

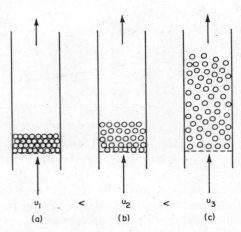

$$u_1 \quad < \quad u_2 \quad < \quad u_3$$

(a)   (b)   (c)

FIG. 7.24. (a) Fixed bed, (b) fixed bed of maximum voidage, (c) fluidized bed.

It is seen that at low fluid velocities the system behaves like a *packed bed*; thus the relationship between the pressure drop across the bed and the flow rate is given by the Ergun equation, which was described in Section 7.4. As the fluid velocity is increased this will cause a corresponding increase in the pressure drop across the bed until a state is reached when the pressure drop across the bed equals the weight of the bed. At this point, illustrated in Fig. 7.24b, the particles in the bed will become rearranged so as to offer less resistance to the flow and the *bed expands* to attain the loosest possible packing. At higher fluid velocities the bed expands further, the particles become freely suspended in the gas stream, and the *bed attains the fluidized state*.

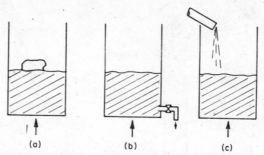

FIG. 7.25. Liquidlike properties of fluidized beds. (a) Light bodies are buoyant in fluidized systems. (b) Fluidized beds may be drained. (c) Fludized bed feeders.

From a macroscopic point of view, gas–solid fluidized beds may be regarded as well-stirred, boiling liquids; the liquid-like properties of fluidized beds are readily shown in Fig. 7.25.

This behavior of fluidized systems represents the major attractiveness of these systems for effecting gas–solid reactions. The good solids mixing helps to minimize temperature variations and renders these systems attractive for carrying out highly exothermic or endothermic chemical reactions. The liquid-like properties of fluidized beds make them very attractive from the viewpoint of solids handling.

The principal drawbacks of fluidized systems are their inability to handle sticky materials, which could lead to agglomeration and eventually to the blockage of the whole system, and the power dissipation entailed in keeping the bed fluidized. Fluidized beds had been used extensively in the chemical industry for carrying out heterogeneous catalytic reactions (catalytic cracking, manufacture of naphthahlene, etc.) especially where close temperature control and frequent catalyst regeneration was necessary.

In addition to catalytic reactors, fluidized beds have been used extensively for carrying out gas–solid reactions; some examples of industrial use are given in Figs. 7.26–7.29.

7.5.1 Some General Properties of Fluidized Systems

THE FLUIDIZATION CURVE

The onset of fluidization described previously (Fig. 7.22) may be conveniently represented through the use of the *fluidization curve* shown in Fig. 7.30. In this graph the logarithm of the pressure drop across the bed is plotted against gas velocity on a logarithmic scale; here the line AB corresponds to the pressure drop across the fixed bed before fluidization takes place; the region BC represents the rearrangement of the bed to provide the minimum

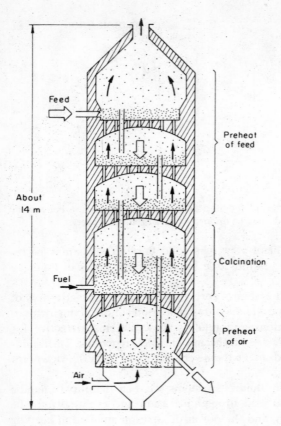

FIG. 7.26. Commercial lime-stone calciner [47].

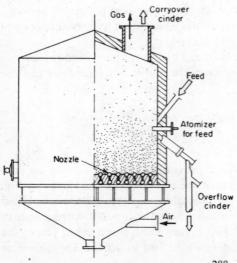

FIG. 7.27. Fluidized bed for roasting sulfides [47].

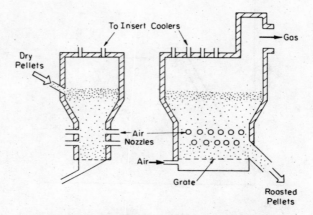

FIG. 7.28. Another design of sulfide roaster [47].

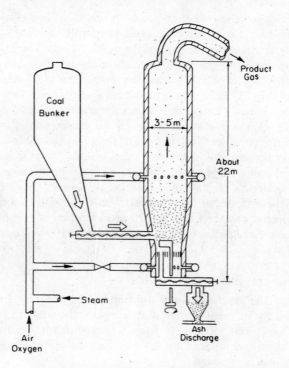

FIG. 7.29. Fluidized bed coal gasification unit [47].

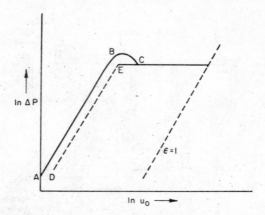

FIG. 7.30. Variation of pressure drop with fluid flow rate in a fluidized bed.

fluidization voidage, when the pressure drop across the bed equals the weight of the bed; the bed is considered fluidized at point C and the corresponding voidage ε_{mf} is the minimum voidage of the fluidized bed. The broken line DE shows the pressure drop–gas velocity relationship for a packed bed with an initial voidage of ε_{mf}. The ABC curve represents the ideal behavior of gas–solid fluidized beds. There are many systems that do not follow this pattern exactly. Two commonly found departures are *channeling* and *slugging*. In a physical sense, channeling corresponds to the preferential flow of the gas through certain vertical sections of the bed; thus under these conditions a part of the bed may become fluidized while the remainder stays in a packed state. Slugging would mean the presence of rather large gas bubbles, occupying most of the cross section of the bed; the periodic collapse of the bed, as particles fall through these large bubbles, would then cause the fluctuations in the pressure drop across the bed. For a further discussion of slugging and other nonideal behavior of fluidized systems the reader is referred to the specialized literature [47, 48].

THE MINIMUM FLUIDIZATION VELOCITY

As noted earlier, at the onset of fluidization the pressure drop across the bed multiplied by its cross-sectional area equals the weight of the bed; this relationship may then be used in conjunction with the Ergun equation to calculate the minimum fluidization velocity, i.e., the minimum gas velocity at which fluidization is initiated.

Upon considering unit cross-sectional area of the bed, we have

$$\underset{\text{pressure drop}}{\Delta P} = \underset{\text{weight of the bed}}{L_{mf}(1 - \varepsilon_{mf})g(\bar{\rho}_s - \bar{\rho}_g)} \tag{7.5.1}$$

where L_{mf} is the bed height of the minimum fluidization velocity, and $\bar{\rho}_s$ and $\bar{\rho}_g$ are the densities of the solid and gas, respectively. Upon substituting for $\Delta P/L_{mf}$ from the Ergun equation, i.e., Eq. (7.3.2), after some algebra we obtain

$$\frac{1.75}{\varphi_s \varepsilon_{mf}^3}\left(\frac{d_p u_{mf} \bar{\rho}_g}{\mu}\right)^2 + \frac{150(1-\varepsilon_{mf})}{\varphi_s^2 \varepsilon_{mf}^3}\left(\frac{d_p u_{mf} \bar{\rho}_g}{\mu}\right) = \frac{d_p^3 \bar{\rho}_g(\bar{\rho}_s - \bar{\rho}_g)g}{\mu^2} \qquad (7.5.2a)$$

For a given particle size and ε_{mf}, Eq. (7.5.2a) provides a quadratic equation for determining the minimum fluidization velocity.

We note that as a practical matter the minimum fluidization velocity depends quite strongly on ε_{mf}, the void fraction at the onset of fluidization; the value of ε_{mf} will in general depend on both the nature of the material and on the particle size, as illustrated in Fig. 7.31.

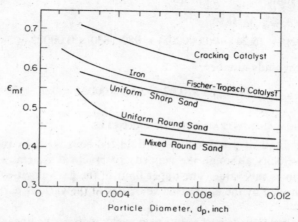

FIG. 7.31. Void fraction at the onset of fluidization for various materials.

When neither ε_{mf} nor φ_s is known, the following approximate formula [49] may be used for estimation:

$$1/\varphi_s \, \varepsilon_{mf}^3 \simeq 14 \qquad (7.5.2b)$$

and

$$(1 - \varepsilon_{mg})/\varphi_s^2 \varepsilon_{mf}^3 \simeq 11 \qquad (7.5.2c)$$

Substituting these formulas in (7.5.2a), we obtain

$$d_p u_{mf} \bar{\rho}_g/\mu = [(33.7)^2 + 0.0408 \, d_p^3 \bar{\rho}_g(\bar{\rho}_s - \bar{\rho}_g)g/\mu^2]^{1/2} - 33.7 \qquad (7.5.3)$$

which can be approximated by

$$u_{mf} = d_p^2(\bar{\rho}_s - \bar{\rho}_g)g/1650\mu \qquad \text{for} \quad N_{Re_p} < 20 \qquad (7.5.4a)$$

and

$$u_{mf}^2 = d_p(\bar{\rho}_s - \bar{\rho}_g)g/24.5\rho_g \quad \text{for} \quad N_{Re_p} > 1000 \quad (7.5.4b)$$

Example 7.5.1 Calculate the minimum fluidization velocity for hematite iron oxide particles, 200 μm in diameter, with hydrogen, at 900°C.

SOLUTION From a handbook we can easily find that the density of hematite, $\bar{\rho}_s$, is 5.24 g/cm³ and the viscosity of hydrogen, μ, at 900°C is 0.00022 g/cm sec. Supposing the pressure is 1 atm we can calculate the density of hydrogen, $\bar{\rho}_g$, at 900°C by employing the ideal gas law. The value is 0.0000205 g/cm³.

Now that we have the values of $\bar{d}_p$, $\bar{\rho}_s$, $\bar{\rho}_g$, and μ, we may use these values in Eq. (7.5.4a) or (7.5.4b) to obtain the minimum fluidization velocity. Let us assume $N_{Re_p} < 20$ first:

$$u_{mf} = d_p{}^2(\bar{\rho}_s - \bar{\rho}_g)g/1650\mu \quad (7.5.4a)$$
$$= (0.02)^2 \times (5.24 - 0.0000205) \times 980/1650 \times 0.00022 = 5.65 \text{ cm/sec}$$

Check the Reynolds number:

$$N_{Re_p} = 0.0000205 \times 0.02 \times 5.65/0.00022 = 0.01 < 20$$

ELUTRIATION AND ENTRAINMENT OF PARTICLES

The lower limit of gas velocities in fluidized beds is set by the minimum fluidization velocity although the majority of practical systems are operated at least 3–5 times this value. The upper limit of the gas velocities in fluidized beds is determined by the elutriation velocities of the various particles within the system.

As a first approximation the elutriation velocity (i.e., the linear gas velocity at which a particle of given size is swept out of the bed) may be estimated from the terminal falling velocity.

The terminal falling velocity u_t of single particles may be calculated by equating the weight of the particle with the drag force of the fluid that acts on it in free fall. Thus we have

$$u_t = \begin{cases} \dfrac{d_p{}^2 g(\bar{\rho}_s - \bar{\rho}_g)}{18\mu} & \text{for} \quad N_{Re_p} < 0.4 \quad (7.5.5) \\[2ex] \left[0.0178 \dfrac{g^2(\bar{\rho}_s - \bar{\rho}_g)^2}{\bar{\rho}_g \mu}\right]^{1/3} d_p & \text{for} \quad 0.4 \le N_{Re_p} \le 500 \quad (7.5.6) \\[2ex] \left[\dfrac{3.1 d_p\, g(\bar{\rho}_s - \bar{\rho}_g)}{\bar{\rho}_g}\right]^{1/2} & \text{for} \quad 500 \le N_{Re_p} \le 2 \times 10^5 \quad (7.5.7) \end{cases}$$

The ratio u_t/u_{mf} has been shown to vary from about 90 for small particles to about 10 for large particles, which represents the theoretical limits of operation. In practice, these limits may be even narrower if elutriation is to be avoided because most practical fluidized beds contain a range of particle sizes. Equations (7.5.5)–(7.5.7) provide the theoretical limits of the superficial gas velocity at which elutriation can take place. Practical experience has shown, however, that in fluidized beds, the particles whose terminal falling velocity is smaller than the superficial gas velocity are not entrained (or elutriated) immediately but rather the elutriation occurs at a finite rate.

The rate at which these particles are removed from a fluidized bed may be expressed as

$$d\mathscr{E}_i/dt = -k_e \mathscr{E}_i \qquad (7.5.8)$$

where $\mathscr{E}_i$ is the weight of particles of size $d_{p,i}$ in the bed, k_e is the elutriation constant, thus $d\mathscr{E}_i/dt$ is the rate of elutriation from the bed of particles $d_{p,i}$ in, for example, kg/sec.

Figure 7.32 shows a plot obtained by Yagi and Aoichi (see Kunii and Levenspiel [47]). On the ordinate of this plot the quantity $\mathscr{E}_i/A_B$ denotes the weight of the particles of given size $(d_{p,i})$ per unit cross-sectional area of the bed, whereas u_0 designates the superficial gas velocity.

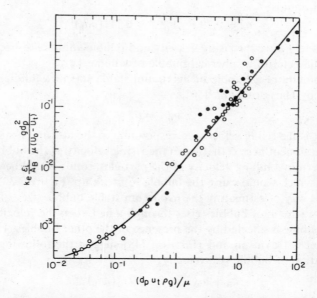

FIG. 7.32. Correlation for the elutriation constant (S. Yagi and T. Aoichi, quoted by Kunii and Levenspiel [47]): ○, continuing operation, ●, batch operation.

Consideration of the quantities appearing on the coordinate axes, together with the shape of the experimental curve, indicates the following:

(1) The higher the difference between the terminal falling velocity and the actual linear gas velocity in the bed, the larger the elutriation constant.

(2) The deeper the bed, i.e., the larger the quantity $\mathscr{E}_i/A_B$, the smaller the value of the elutriation constant; it follows that for the same quantity of fines, deeper beds would tend to reduce the rate of elutriation.

BUBBLES IN FLUIDIZED BEDS

It has been shown that in most gas–solid fluidized systems of practical interest a portion of the gas, equivalent to the amount required to maintain the bed in a fluidized state, is more or less uniformly distributed in the bed, whereas the remainder passes through in the form of *gas bubbles*. An alternative way of stating this would be to say that the amount of gas in excess of that corresponding to the minimum fluidization velocity will pass through the system as gas bubbles.

Davidson and Harrison [48] have studied bubbling phenomena in fluidized beds extensively and have shown that the rising velocity of gas bubbles in fluidized beds may be estimated with the aid of the Davies and Taylor equation, which was originally developed for the rise of large (spherical cap) gas bubbles in liquids:

$$u_B = 0.79 g^{1/2} V_B^{1/6} = 0.711 (g d_B)^{1/2} \qquad (7.5.9a)$$

where u_B and V_B are the rising velocity and bubble volume, respectively, and d_B is the diameter of a spherical bubble of volume V_B.

The velocity of a bubble of maximum stable size in a fluidized bed was estimated by Harrison *et al.* [50] as

$$u_{B,S} = u_t \qquad (7.5.10)$$

where $u_{B,S}$ is the rising velocity corresponding to the maximum stable bubble size; the argument here is that when the rising velocity of the bubble is larger than the terminal falling velocity of the particles, solids would be drawn into the wake, which would cause the bubble to break up. Figure 7.33 provides a convenient way of estimating the maximum stable bubble size.

When a stream of bubbles rises through a bed, then the velocity of an individual bubble is affected by the presence of the other bubbles. Under these circumstances Davidson and Harrison [48] suggest the following equations for estimation of bubble rise velocity:

$$u_B = u_0 - u_{mf} + 0.711 (g d_B)^{1/2} \qquad (7.5.9b)$$

Gas bubbles have a major influence on the operation of fluidized systems; their presence is responsible for the good solids mixing, which is a desirable

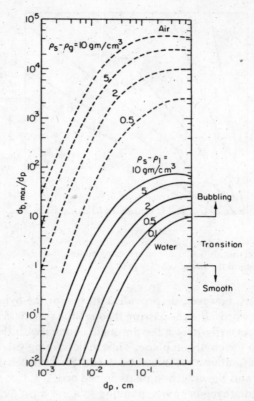

FIG. 7.33. Maximum stable bubble size and conditions for bubbling and smooth fluidization as estimated by Davidson and Harrison [48].

feature. On the other hand, the portion of the gas passing through in the form of bubbles is not brought into intimate contact with the bed, as is the gas contained in the continuous, i.e., emulsion, phase of the system. Kunii and Levenspiel [47] developed a comprehensive model for representing the interaction between gas bubbles (the discontinuous phase) and the emulsion phase of gas–solid fluidized beds.

They considered three groupings of bubbling systems as depicted in Fig. 7.34:

(1) small bubbles, where $u_B \simeq u_{mf}$,
(2) large bubbles, where $u_B > 5u_{mf}$,
(3) an intermediate region, where $u_{mf} < u_B < 5u_{mf}$.

Unless the beds are very shallow, small bubbles do not play an important role in modifying the transfer processes between the gas and the solids.

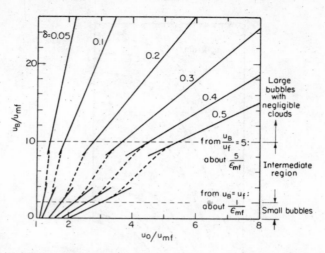

FIG. 7.34. Relationship of bubble velocity, bubble fraction of bed, superficial gas velocity, and minimum fluidizing velocity [47].

Large bubbles, however, do provide a means for the bypassing of the bed by gas; in other words, the gas passing through the system in the form of large bubbles will necessarily have a less intimate contact with the solids than the gas contained in the emulsion phase. Thus in describing gas–solids contacting in these systems, allowance must be made for transfer between the gas (bubble) phase and the emulsion phase in the bed.

The intermediate region, corresponding to $u_{mf} < u_B < 5u_{mf}$, is particularly complex, because here the bubble is surrounded by a *circulating cloud of particles*, whose presence contributes an additional resistance to transfer between the gas bubbles and the emulsion phase.

Kunii and Levenspiel proposed the following procedure describing the exchange between the gas bubble, the surrounding cloud, and the emulsion. Consider the transfer of species A from the bubble phase through the cloud surrounding the bubble to the emulsion phase, as shown in Fig. 7.35. Denoting the concentration of the transferred species in the bubble, cloud, and emulsion by $C_{A,B}$, $C_{A,C}$, and $C_{A,E}$, respectively, an overall mass balance may be written:

$$u_B \, dC_{A,B}/dz = \tilde{k}_{BC}(C_{A,B} - C_{A,C}) \simeq \tilde{k}_{CE}(C_{A,C} - C_{A,E}) = \tilde{k}_{BE}(C_{A,B} - C_{A,E})$$

change of concentration in the bubble during its rise

(7.5.11)

where $\tilde{k}_{BC}$, $\tilde{k}_{CE}$, and $\tilde{k}_{BE}$ denote the transfer coefficients between the bubble-cloud, cloud–emulsion, and the overall bubble–emulsion phases, respectively.

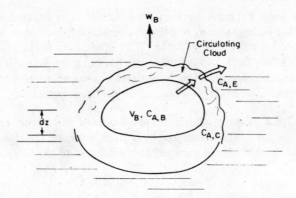

FIG. 7.35. Schematic representation of the transfer process between the bubble cloud and the emulsion.

We note here that all these coefficients have the dimension of time^{-1}. It is readily shown that the overall transfer coefficient $\tilde{k}_{BE}$ is related to $\tilde{k}_{BC}$ and $\tilde{k}_{CE}$ in the following manner:

$$1/\tilde{k}_{BE} \simeq 1/\tilde{k}_{BC} + 1/\tilde{k}_{CE} \qquad (7.5.12)$$

An alternative way of representing the exchange of mass between the bubble and the emulsion phases is to define the cross flow ratio π:

$$\pi = \tilde{k}_{BE}/(u_B/L) \qquad (7.5.13)$$

where L is the total depth of the bed. In physical terms, the *cross flow ratio* corresponds to the number of times the contents of the bubble are replaced during its passage through the bed.

In order to calculate the cross flow ratio π or the overall exchange coefficient between the bubble and the emulsion, $\tilde{k}_{BC}$ and $\tilde{k}_{CE}$ must be estimated.

On the basis of some empiricism and some fluid dynamics, Kunii and Levenspiel proposed the following formulas for estimating these quantities:

$$\tilde{k}_{BC} = (4.5u_{mf}/d_B) + (5.85D^{1/2}g^{1/4}/d_B^{5/4}) \qquad (7.5.14)$$

and

$$\tilde{k}_{CE} \simeq 6.8(\varepsilon_{mf} Du_B/d_B^3)^{1/2} \qquad (7.5.15)$$

where d_B is the bubble diameter and D the gas phase diffusivity. The procedure outlined through Eqs. (7.5.11)–(7.5.15) for estimating the overall exchange coefficient between gas bubbles and the emulsion phase is necessarily an approximation because of the oversimplifications involved in both the model and the formulas used. Perhaps the greatest difficulty is associated with the fact that the average bubble size d_B is difficult to estimate. Indeed, it has been suggested that one may use Eqs. (7.5.11)–(7.5.15) to estimate the average bubble size from cross flow data. Nonetheless, the bubbling model outlined here and described in much more detail by Kunii and Levenspiel provides a means for estimating conversion in "real" industrial fluidized systems.

Example 7.5.2   Zinc concentrates are fluidized with air at 1500°F. If the bed diameter is 5 m and the bed height is 15 m, estimate the cross flow ratio, assuming that the bubbles in the system correspond to the largest stable size.

DATA

   mean particle diameter: 200 μm
   solid density: 3.8 g/cm^3
   gas density: 3.0×10^{-4} g/cm^3
   gas viscosity: 4.0×10^{-4} g/cm sec
   gas diffusivity: 1.25 cm^2/sec
   $u_0 = 10$ cm/sec

SOLUTION

$$\bar{\rho}_s - \bar{\rho}_g = 3.8 \text{ g/cm}^3, \qquad d_p = 2 \times 10^{-2} \text{ cm}$$

From Fig. 7.33,

$$d_{b,\text{max}}/d_p = 2 \times 10^3$$

Therefore,

$$d_{b,\text{max}} = 2 \times 10^3 d_p = 40 \text{ cm}$$

From Eq. (7.5.4a),

$$u_{mf} = \frac{d_p^2(\bar{\rho}_s - \bar{\rho}_g)g}{1650\mu} \qquad \text{for} \quad N_{Re_p} < 20$$

$$= 2.26 \text{ cm/sec}$$

Checking the Reynolds number, we see that

$$N_{Re_p} = d_p u_{mf} \bar{\rho}_g/\mu = 3.39 \times 10^{-2} < 20$$

From Eq. (7.5.2b), $1/\varepsilon_{mf}^3 = 14$; therefore $\varepsilon_{mf} = 0.41$, $u_0 = 10$ cm/sec. From Eq. (7.5.9b),

$$u_B = 10 - 2.26 + 0.711(980 \times 40)^{1/2} = 148.5 \text{ cm/sec}$$

From Eq. (7.5.15),

$$(\tilde{k}_{CE}) \simeq 6.8(\varepsilon_{mf}\, Du_B/d_B^3)^{1/2}, \qquad D = 1.25 \text{ cm}^2/\text{sec}$$

$$(\tilde{k}_{CE})_B \simeq 6.8(0.41 \times 1.25 \times 148.5/40^3)^{1/2} = 0.234/\text{sec}$$

$$(\tilde{k}_{BC})_B = 4.5(u_{mf}/d_B) + 5.85(D^{1/2}g^{1/4}/d_B^{5/4}$$

$$= 4.5\left(\frac{2.26}{40}\right) + 5.85\left(\frac{(1.25)^{1/2}(980)^{1/4}}{(40)^{5/4}}\right) = 0.618/\text{sec}$$

$$1/(\tilde{k}_{BE})_B \cong 1/(k_{BC})_B + 1/(\tilde{k}_{CE})_B$$

$$(\tilde{k}_{BE})_B = (k_{BC})_B(k_{CE})_B/[(k_{BC})_B + (k_{CE})_B] = 0.170/\text{sec}$$

$$\pi = \frac{k_{BE}}{u_B/L} = 0.00114 \times L = 1.71$$

7.5.2 Flow Patterns in Gas–Solid Fluidized Beds

FLOW PATTERN OF THE SOLIDS

In fluidized beds with height-to-diameter ratios not exceeding about 3 : 1 it is reasonable to assume that the solids are well mixed. Thus for a bed such as shown in Fig. 7.36, it may be assumed that the distribution of the solid residence times will correspond to a completely mixed CSTR system, such as was shown in Fig. 7.4.

The wide spread of residence times inherent in completely mixed continuous reactors may be a disadvantage if complete conversion of the solids is desired [48]. For this reason sometimes stagewise contacting is arranged

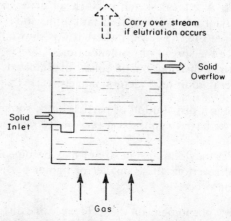

FIG. 7.36. Diagram of a fluidized bed with continuous solids feed and removal.

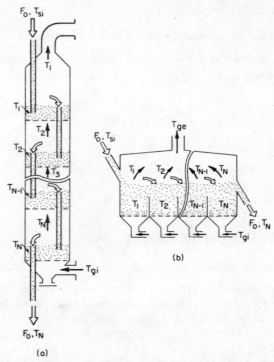

FIG. 7.37. Heat exchange in multiple fluidized beds. (a) Countercurrent contacting, (b) crosscurrent contacting [47].

in fluidized systems. A typical sketch of such an arrangement is given in Fig. 7.37. The behavior of several stirred tank units in series has been discussed earlier (e.g., see Fig. 7.9 *et seq.*) where it was shown that this arrangement tends to reduce the spread of residence times.

The residence time distribution of solids in fluidized systems may be complicated by several additional factors, namely:

(a)   When there is a distribution of particle sizes, some of the fines may be elutriated; thus we have to consider a *carryover stream* in addition to the overflow (Fig. 7.37).

(b)   The particles may diminish in size or may become less dense as a result of the chemical reaction; these smaller or lighter particles may then be elutriated as reaction proceeds.

The effect of these phenomena on the residence time distribution of solids in fluidized beds has been extensively treated by Kunii and Levenspiel [47].

FLOW PATTERN OF THE GAS

In spite of extensive investigations using tracers over the past two decades, our understanding of the gas flow patterns in fluidized systems is still incomplete. The usual method of measuring residence time distribution through the use of tracers gives ambiguous results because of the previously discussed cross flow effects or material transfer between the gas bubbles and the emulsion phase.

While attempts have been made to assign *axial dispersion coefficients* to the gas flow in fluidized beds or to represent the flow field in terms of a series of well-mixed vessels, none of these methods is really satisfactory. Perhaps the most promising approach to the problem of gas flow patterns is provided by a bubbling model [47].

Here we consider that the gas passing through the bed may be divided into two streams: the bubble phase, which travels through the system with linear velocity

$$u_B = (u_0 - u_{mf})/\delta \qquad (7.5.16)$$

where

$$1 - \delta = L_{mf}/L \qquad (7.5.17)$$

and the gas associated with the emulsion phase, with linear velocity

$$u_E = \frac{u_{mf}}{\varepsilon_{mf}} - \left[\frac{\alpha u_0}{1 - \delta - \alpha\delta} - \alpha u_{mf} \right] \qquad (7.5.18)$$

where α is a constant which has a value of about 0.4. Note that when u_0 is large, more particularly, when $u_0/u_{mf} > 6$–10, u_E becomes negative, which means that the gas associated with the emulsion is actually moving downward.

The experimental findings of gas flow patterns in fluidized beds (in the interpretation of which it is impossible to separate the effects of the emulsion and the bubble phases) indicate that the overall flow pattern lies between plug flow and complete mixing. Yoshida and Kunii [51] showed that experimental measurements of tracer dispersion may be adequately interpreted in terms of the bubbling bed model through the use of the exchange coefficients between the emulsion and the bubble phases.

As we shall discuss subsequently in describing the behavior of vigorously bubbling beds (which correspond to the majority of industrial applications) it is reasonable to assume as a first approximation that

(a) *the solids and the gas associated with the emulsion phase are completely mixed,* while

(b) *the bubble phase moves through the system in plug flow,* and

(c) The interchange between the emulsion and the bubble phases may then be described with the aid of Eqs. (7.5.11)–(7.5.15).

7.5.3   Heat and Mass Transfer in Fluidized Beds

Heat and mass transfer phenomena in fluidized beds may be divided into the following categories:

heat and mass transfer between the gas in the emulsion phase and the solids,

heat and mass transfer between the bubble phase and the emulsion, and

heat transfer between the bed and the walls.

GAS–SOLID TRANSFER

In the majority of the early investigations into heat or mass transfer between the solids and the gas, care was taken to operate the beds close to the minimum fluidization velocity; thus the formation of bubbles was avoided or at least minimzed [52–55].

These early workers found that the heat and mass transfer rates were quite rapid, so that the region of transfer was confined to a thin layer at the bottom of the bed and the remainder of the bed was in equilibrium with the gas. The thickness of this layer, to which effective concentration or temperature gradients were confined, was of the order of 6–10 particle diameters.

Figure 7.38 shows experimentally measured gas temperature profiles by Ayers[54], in whose experiments a hot gas was contacted with solids in a cross flow arrangement. The rapid approach by the gas temperature to the temperature of the solids is readily seen. Similar findings were reported in the area of mass transfer[55]; indeed, in all these studies considerable care had to be exercised so that meaningful measurements could be taken, so rapid was the approach to equilibrium. The experimental results were usually represented on plots of the particle Nusselt or Sherwood number against the particle Reynolds number.

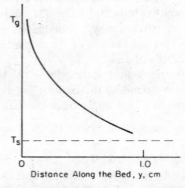

FIG. 7.38. Sketch of gas temperature profile in a fluidized bed.

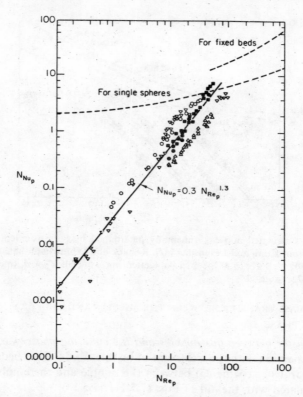

FIG. 7.39. Correlation of data for gas–solid heat transfer in fluidized beds [47]. ○, Richardson and Ayers; ▽, Donnadieu; ■, Heertjes and McKibbins; ●, Walton *et al.*; △, Kettenring *et al.*

Some typical compilations of experimental data are shown in Figs. 7.39 and 7.40 for heat and mass transfer, respectively. These data indicate that for low values of the Reynolds number, the Sherwood or Nusselt number tends to fall below the limiting value of 2 established for single particles or for packed-bed systems [viz. Eqs. (7.3.18), (7.3.23), etc.]. It was suggested by Kunii and Levenspiel [47] that this apparent anomaly may be resolved by considering that even in these systems some fraction of the gas passed through in the form of bubbles. Thus an incorrect driving force has been used in the calculation of the transfer coefficients.

The fact remains, however, that the study of gas–solid heat or mass transfer in fluidized beds is largely an academic matter, because the transfer rate is quite rapid. The principal transfer problem is the exchange of matter or thermal energy between the bubble and the emulsion phases.

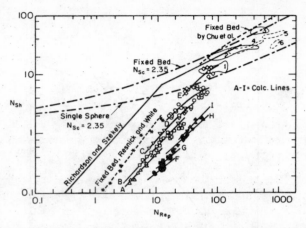

FIG. 7.40. Calculated and experimentally determined mass transfer correlations in fluidized beds from Kunii and Levenspiel [47]. Resnick and White, naphthalene, d_p (mm): △, 0.28; □, 0.40; ▽, 0.47; ○, 0.74; ◇, 1.06. Kettenring *et al.*, silica gel, d_p (mm): ●, 0.36; ▨, 0.50; ▼, 0.72; ▲, 1.00.

HEAT OR MASS TRANSFER BETWEEN THE BUBBLES AND THE EMULSION PHASE

Mass transfer between gas bubbles and the emulsion phase was discussed earlier in connection with the treatment of bubbles in fluidized beds; thus the overall mass transfer coefficient between the bubble and the emulsion phase may be calculated with the aid of Eqs. (7.5.11)–(7.5.15).

Heat transfer between the gas bubbles and the emulsion phase may be calculated in a somewhat similar manner. However, in the establishment of a heat balance about the bubble we shall assume that the rate of transfer between the cloud surrounding the bubble and the emulsion is fast enough so that this process does not contribute a resistance. However, in contrast to the formulation of the mass transfer problem, here too we have to take into consideration the heat capacity of the solids contained in the bubbles.

Thus, the heat balance on the system may be written

$$\begin{bmatrix} \text{heat transferred out} \\ \text{of the gas contained} \\ \text{in the bubble} \end{bmatrix} = \begin{bmatrix} \text{heat taken up} \\ \text{by the solids} \\ \text{in the bubble} \end{bmatrix} + \begin{bmatrix} \text{heat transferred to} \\ \text{the cloud (and hence} \\ \text{to the emulsion)} \end{bmatrix} \quad (7.5.19)$$

Thus we have

$$-\rho_G c_{P,G} u_B \frac{dT_{G,B}}{dz} = \delta_B a' \eta_h h_p [T_{G,B} - T_S] + h_{B,c}[T_{G,B} - T_S] \quad (7.5.20)$$

where $T_{G,B}$ and T_S are the temperatures in the bubble and of the solids, respectively; a' the external surface area per unit solid volume; δ_B the ratio (volume of solids in the bubble)/(volume of the bubble); η_h an effectiveness factor for heat transfer; and h_p the heat transfer coefficient between the fluid and the particles, such as appearing in Fig. 7.39. The quantity $h_{B,c}$ may be estimated from Eq. (7.5.14) using the analogy between heat and mass transfer:

$$h_{B,c} \simeq 4.5 \frac{u_{mf} \bar{\rho}_G c_{p,G}}{d_B} + 5.85 \frac{(k_G \bar{\rho}_G c_{p,G})^{1/2} g^{1/4}}{d^{5/4}} \qquad (7.5.21)$$

The principal practical difficulty in using Eqs. (7.5.19)–(7.5.21) for doing actual heat transfer calculations is associated with the uncertainties involved in the estimation of δ_B, η_h, and even h_p.

Example 7.5.3 A fluidized bed using 200-μm sand particles is being used to preheat hydrogen used in a direct reduction unit. The solids and the gas are in a cross flow arrangement; the solids may be regarded as completely mixed and at a temperature of 1000°C. If the linear gas velocity is 10 times the minimum fluidization velocity and the bubbles are half the maximum stable size, estimate the bed depth required to preheat the gas to 900°C from 25°C inlet temperature.

Assume that in Eq. (7.5.20) $\eta_h \delta_B \simeq 10^{-3}$ and use Fig. 7.39 to estimate h_p.

DATA $\bar{\rho}_s = 1.5 \text{ g/cm}^3$.

SOLUTION The gas temperature varies from 25 to 900°C within the bed. Consequently property values such as gas thermal conductivity and viscosity will vary throughout the bed. However, in this approximate analysis we shall take the properties of the hydrogen as those at a mean gas temperature (462°C):

$$\mu = 1.6 \times 10^{-4} \text{ g/cm sec}, \qquad \bar{\rho}_G = 0.395 \times 10^{-3} \text{ g/cm}^3$$
$$k_G = 0.62 \times 10^{-3} \text{ cal/cm sec °C}, \qquad c_{pG} = 3.5 \text{ cal/g °C}$$

We may now proceed to calculate u_{mf} from Eq. (7.5.4a):

$$u_{mf} = \frac{d_p^2 (\bar{\rho}_s - \bar{\rho}_g) g}{1650 \mu}$$

$$= \frac{(2 \times 10^{-2})^2 (1.5 - 0.395 \times 10^{-3}) \times 980}{1650 \times 1.6 \times 10^{-4}} = 2.23 \text{ cm/sec}$$

Checking the Reynolds number, we have

$$N_{Re_p} = \frac{2 \times 10^{-2} \times 2.23 \times 0.395 \times 10^{-3}}{1.6 \times 10^{-4}} = 0.088 < 20$$

$$u_0 = 10 \mu_{mf} = 22.3 \text{ cm/sec}$$

From Fig. 7.33 for $\bar{\rho}_s - \bar{\rho}_g \simeq 1.5$ g/cm^3, $d_p = 2 \times 10^{-2}$ cm, it was found that

$$d_{B,\,max}/d_p = 5 \times 10^2$$

$$d_{B,\,max} = 5 \times 10^2 \times 2 \times 10^{-2} = 10 \text{ cm}, \qquad d_B = \tfrac{1}{2}d_{B,\,max} = 5 \text{ cm}$$

From Eq. (7.5.9b),

$$u_B = u_0 - u_{mf} + 0.711(gd_B)^{1/2}$$
$$= 22.3 - 2.23 + 0.711(980 \times 5)^{1/2} = 69.84 \text{ cm/sec}$$

Using Eq. (7.5.21), we obtain

$$h_{B,\,C} = 4.5 \frac{u_{mf}\,\rho_G\,c_{p.\,G}}{d_B} + 5.85 \frac{(k_G\,\rho_G\,c_{p,G})^{1/2}g^{1/4}}{d_B^{5/4}}$$

$$= 4.5 \frac{2.23 \times 0.395 \times 10^{-3} \times 3.5}{5}$$

$$+ 5.85 \frac{(0.00062 \times 0.395 \times 10^{-3} \times 3.5)^{1/2}(980)^{1/4}}{(5)^{5/4}}$$

$$= 6.828 \times 10^{-3}$$

$$N_{Re_p} = \frac{d_p u_0 \bar{\rho}_g}{\mu} = \frac{2 \times 10^{-2} \times 22.3 \times 0.395 \times 10^{-3}}{1.6 \times 10^{-4}} = 0.881$$

From Fig. 7.39, it was found that

$$N_{Nu_p} = 0.02 = h_p\,d_p/k_G \qquad h_p = 0.02 \times 0.00062/2 \times 10^{-2} = 0.00062$$

The surface area per unit volume of particle $a' = 6/d_p = 300$ cm^{-1}, and so

$$-\rho_G c_{p.\,G} u_B\, dT_{G,\,B}/dz = (\delta_B{}' a' \eta_h)h_p[T_{G,\,B} - T_s] + h_{B,\,C}[T_{G,\,B} - T_s]$$
$$dT_{G,\,B}/T_{G,\,B} - T_s = -[a'\delta_B \eta_h h_p + h_{B,\,C}/\rho_G c_{p,\,G} u_B]\, dz$$

Integrating over the bed height, we obtain

$$\ln\!\left(\frac{T_s - T_{G,\,out}}{T_s - T_{G,\,in}}\right) = -\frac{a'\delta_B \eta_n h_p + h_{B,\,C}}{\bar{\rho}_G c_{p,\,G} u_B}\,L$$

Substituting numerical values, we have

$$\ln\!\left(\frac{10000 - 900}{1000 - 25}\right) = -\frac{300 \times 10^{-3} \times 0.00062 + 6.828 \times 10^{-3}}{0.395 \times 10^{-3} \times 3.5 \times 69.84}\,L$$

Therefore, $L = 0.6$ cm.

Note that this analysis slightly overestimates the size of the bed required, since on leaving the bed at 900°C the gas carried through in the form of bubbles will be mixed with hotter gas that passed through the emulsion phase. However, when $u_0 = 10u_{mf}$ this latter portion of the gas flow is negligibly small compared with that portion due to the bubbles.

HEAT TRANSFER BETWEEN THE WALLS AND THE EMULSION PHASE

A major attractiveness of fluidized beds for many applications is the high heat transfer rate that may be attained between the emulsion phase and the wall; typical heat transfer coefficients that have been reported in the literature range from about 60 to 500 Btu/hr ft^2 °F (8×10^{-3}–6.5×10^{-2} cal/cm^2 sec^1 °C^1) and similar values have also been found for the heat transfer coefficients between the emulsion and heating or cooling elements (coils, etc.) immersed in the bed.

Numerous models have been proposed to explain the high heat transfer rates attainable in fluidized bed operation. Mickley et al. [56] and coworkers proposed that heat transfer between the bed and a surface takes place in un-steady pulses during a brief residence time of gas–solid pockets in the vicinity of the wall. In many ways this model is analogous to the "penetration theory" for mass transfer.

Figure 7.41 shows typical experimentally measured local heat transfer co-efficients for a system where a small electric heater was immersed into a fluidized bed. The pulselike behavior of the local heat transfer rates is readily apparent here; the troughs or very low values of the local transfer coefficients, also seen in the graph, correspond to times when the heater was in contact with gas bubbles.

While the Mickley model is a useful qualitative concept for the mechanism of heat transfer, it is not helpful for the actual prediction of heat transfer

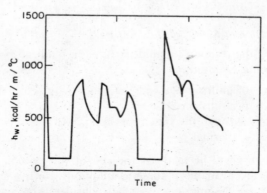

FIG. 7.41. Fluctuations of local heat transfer coefficient from wall to fluidized beds [42]. Glass beads, $u_0 = 18.6$ cm/sec.

coefficients because one cannot estimate either the size of the gas–solid pockets or their residence time near the walls.

A more quantitative model for bed-to-wall heat transfer was proposed by Botterill and coworkers [57, 58], who considered that heat transfer took place by unsteady state conduction through a stagnant gas layer surrounding the particles and through the particles themselves during their brief residence time in the vicinity of the walls. These authors have provided a rigorous formulation for this heat transfer problem and through the solution of the resultant partial differential equations were able to relate the point value of the heat transfer coefficient to the particle size, the properties of the solid particles, and the residence time. This model was found to be consistent with a broad range of experimental observations, although in "real" fluidized beds (as opposed to mechanically agitated systems), the particle residence time cannot be readily determined. Even Botterill's elegant model cannot be used for predicting bed-to-wall heat transfer coefficients.

Thus in estimating the heat transfer coefficients between the bed and the wall or immersed elements, empirical correlations have to be used. A detailed survey of these is available in the text by Kunii and Levenspiel, so here we shall restrict ourselves to quoting a correlation proposed by Leva and Wen [59] that is claimed to represent most of the experimental data in the literature within an accuracy of about 50%:

$$\frac{h_w d_p}{k_G} = 0.16 N_{Re_p}^{0.76} N_{Pr}^{0.4} \left(\frac{\bar{\rho}_S c_{p,s}}{\bar{\rho}_G c_{p,G}}\right)^{0.4} \left(\frac{u_0^2}{g d_p}\right)^{-0.2} \left(\frac{u_0 - u_{mf}}{u_0} \frac{L_{mf}}{L}\right)^{0.36} \quad (7.5.22)$$

The preceding discussion was concerned with convective heat transfer between the bed and the wall. At high temperatures thermal radiation may also become an important mechanism for bed-to-wall heat transfer; it has been shown, however, that thermal radiation is unlikely to play a significant role at temperatures below about 1000°C [60].

7.5.4   The Reaction of Solids in Fluidized Beds

In describing the reaction of solid particles in fluidized systems we shall make the following assumptions:

The solids are of uniform size and elutriation may be neglected.
The solids are perfectly mixed.
The gas passes through the system in the form of plug flow.
The system is isothermal.

Some comments will be made at the end of this section regarding the procedures that are available if these assumptions are relaxed. Within the framework of these assumptions we shall consider three particular cases, namely:

systems where there is no change in gas composition,

systems where $u_0 \ll u_{mf}$ (thus most of the gas passes through in the form of bubbles), and

systems where $u_0 \simeq u_{mf}$ (thus the presence of bubbles may be neglected).

SYSTEMS WHERE THERE IS NO CHANGE IN THE GAS COMPOSITION

When the reactant gas is supplied in large excess we may neglect the existence of an axial concentration gradient in the gas phase; under these conditions the calculation is greatly simplified.

Recall from Chapter 4, Eq. (4.3.81), that the conversion of a single particle in contact with a reactant gas present at a concentration level C_{A0} in the bulk is given by

$$t^* = g_{F_g}(X) + \hat{\sigma}^2\{p_{Fp}(X) + (2X/N'_{Sh})\} \qquad (7.5.23)$$

The Sherwood number appearing in Eq. (7.5.23) can be evaluated with the aid of Fig. 7.40. Then we may proceed by recasting Eq. (7.5.23) so that X is given as an explicit function of t^*; in general, this has to be done numerically so that the desired relationship is obtained in tabular form. Thus we have

$$X = f(\hat{\sigma}, t^*) \qquad (7.5.24)$$

Upon recalling that the solids are considered perfectly mixed, then for a continuous fluidized bed reactor the age distribution of the solids in the bed at time t^* is given by

$$E(t^*) = (1/\bar{t}_R^*) \exp(-t^*/\bar{t}_R^*) \qquad (7.5.25)$$

where $\bar{t}_R^*$ is the nominal retention time or residence time of the solids in the bed, expressed in a dimensionless form, just as t^*. Then by combining Eqs. (7.5.24) and (7.5.25) the mean conversion of the solids leaving the system is given by

$$\overline{X}(\hat{\sigma}, t_R^*) = \int_0^\infty E(t^*)f(\hat{\sigma}, t^*)\, dt^* \qquad (7.5.26)$$

The integral appearing on the r.h.s. of Eq. (7.5.26) may be evaluated numerically.

Example 7.5.4 Calculate the nominal holding time $\bar{t}_R^*$ required for a completely mixed fluidized bed to obtain a mean conversion of 0.95. $\hat{\sigma}$ may be taken as 1.5 and $N'_{Sh} = 3.0$ if the gas composition is uniform.

SOLUTION We start by setting up Table 7.2 with t^* as a function of X to enable us to determine X given t^* by interpolation:

$$t^* = g_{F_g}(X) + \hat{\sigma}^2(p_{F_p}(X) + (2X/N'_{Sh}))$$
$$= g_{F_g}(X) + 2.25p_{F_p}(X) + 1.5X$$

TABLE 7.2

X	t^*	X	t^*
0	0	0.6	1.5486
0.1	0.1924	0.7	1.9557
0.2	0.4047	0.8	2.4567
0.3	0.6406	0.9	3.1316
0.4	0.9048	1.0	4.75
0.5	1.2040		

where

$$g_{F_g}(X) \equiv 1 - (1 - X)^{1/3}, \qquad p_{F_p}(X) = 1 - 3(1 - X)^{2/3} + 2(1 - X)$$

$$\bar{X} = \int_0^\infty E(t^*)f(\hat{\sigma}, t^*)\, dt^* = \int_0^\infty (1/t_R^*)\exp(-t^*/\overline{t_R^*})f(\hat{\sigma}, t^*)\, dt^*$$

$$= \int_0^\infty \exp[-(t^*/\overline{t_R^*})]f[\hat{\sigma}, t^*]\, d(t^*/\overline{t_R^*}) = \int_0^\infty e^{-y}f(\hat{\sigma}, t_R^* y)\, dy$$

where $y = t^*/t_R^*$. This integral may be evaluated using Laguerre's integration formula

$$\int_0^T e^{-y}f(\hat{\sigma}, t_R^* y)\, dy \simeq \sum_{i=1}^n w_i f[\sigma, \bar{t}_R^*)y_i]$$

where the abscissas (y_i) and weight factors (w_i) have been tabulated for various values of n [61].

Because we are seeking a value of t_R^* for which $\bar{X}$ is 0.95, it is necessary to proceed by trial and error. Let us first guess $t_R^* = 20$ and use $n = 15$ (Table 7.3).

Obviously this first guess is too low and we make a second guess of $t_R^* = 30$, using $n = 15$ and the same abscissas and weight factors as before (Table 7.4). This second guess is too high and a third guess is made at some value of t_R^* between 20 and 30 (e.g., by interpolation between the above two guesses). Proceeding in this way we arrive at a final answer of $t_R^* = 24.5$, provided the assumption of uniform gas composition can be made. An alternative assumption is that only the gas in the bubble phase is of uniform composition and that the gas flow directly through the emulsion is negligible compared with that passing through the bed in the form of bubbles, and that the amount of solids in the bubbles and associated clouds is small compared with that in the bed. Under these conditions we allow for the additional mass transfer step from the bubble to the emulsion by means of the relationship

$$(N_{Sh}')^{-1} = (N_{Sh_p})^{-1} + [(1 - \varepsilon_f)6D/\bar{k}_{BE}\,\delta d_p^2 \varphi_s] \qquad (7.5.27)$$

TABLE 7.3

$y_i{}^a$	w_i	$f_i{}^b$	$w_i f_i$
0.09331	$(-1)2.18235$	0.678	1.47963×10^{-1}
0.49269	$(-1)3.42210$	1.0	3.42210×10^{-1}
1.21560	$(-1)2.63027$	1.0	2.63027×10^{-1}
2.26995	$(-1)1.26426$	1.0	1.26426×10^{-1}
3.66762	$(-2)4.02069$	1.0	4.02069×10^{-2}
5.42534	$(-3)8.56388$	1.0	8.56388×10^{-3}
7.56592	$(-3)1.21244$	1.0	1.21244×10^{-3}
10.12022	$(-4)1.11674$	1.0	1.11674×10^{-4}
$\vdots$	$\vdots$	$\vdots$	$\vdots{}^c$
			$\bar{X} = 0.93$

[a] From Abromavitz and Stegun [61].
[b] By interpolation using Table 7.2.
[c] Remaining terms negligible.

TABLE 7.4

$y_i{}^a$	w_i	$f_i{}^b$	$w_i f_i$
0.09331	$(-1)2.18235$	0.857	1.87027×10^{-1}
0.49269	$(-1)3.42210$	1.0	3.42210×10^{-1}
1.21560	$(-1)2.63027$	1.0	2.63027×10^{-1}
2.26995	$(-1)1.26426$	1.0	1.26426×10^{-2}
3.66762	$(-2)4.02069$	1.0	4.02069×10^{-2}
5.42534	$(-3)8.56388$	1.0	8.56388×10^{-3}
7.56592	$(-3)1.21244$	1.0	1.21244×10^{-3}
10.12022	$(-4)1.11674$	1.0	1.11674×10^{-4}
$\vdots$	$\vdots$	$\vdots$	$\vdots{}^c$
			$\bar{X} = 0.969$

[a] From Abromavitz and Stegun [61].
[b] By interpolation using Table 7.2.
[c] Remaining terms negligible.

where N'_{Sh} is the Sherwood number corrected to allow for the additional mass transfer resistance; $\delta = (u_0 - u_{mf}/u_B)$; ε_f is the void fraction at actual fluidizing conditions; $\tilde{k}_{BE}$ the mass transfer coefficient between the bubble and the emulsion, given in Eq. (7.5.12); and $\dot{N}'_{Sh_p}$ the Sherwood number for a single particle, the value of which may be estimated with the aid of the Froessling equation (Chapter 2) using u_{mf}/ε_{mf} as the relative velocity of the gas.

SYSTEMS WHERE MOST OF THE GAS PASSES THROUGH IN THE
FORM OF BUBBLES—VARIABLE GAS COMPOSITION

This case may be handled by applying a slight modification to the procedure discussed in the preceding section. Consider that the reactant gas concentrations at the inlet and exit to the bed are given by C_{A0} and C_{Aex},
respectively.

Upon assuming that the gas in the bubble phase passes through the system
in plug flow, the appropriate effective concentration is the logarithmic mean
between C_{A0} and C_{Aex}, i.e.,

$$(C_{A0})_{lm} = \frac{(C_{A0} - C_{Aex})}{\ln(C_{A0}/C_{Aex})} \tag{7.5.28}$$

The actual mean conversion of the solids may now be calculated with the aid
of Eq. (7.5.26) together with the associated Eqs. (7.5.23)–(7.5.25), and (7.5.27)
with the provision that in computing t^* the quantity $(C_{A0})_{lm}$ has to be used
for irreversible reaction.

C_{Aex} is not known explicitly at this time but may be obtained from an
overall balance as follows:

$$\underset{\substack{\text{molar rate at which} \\ \text{solid product is} \\ \text{generated}}}{G_s \overline{X}(\sigma, t_R{}^*)} = \underset{\substack{\text{molar rate at which gaseous} \\ \text{reactant is consumed}}}{bG_G[(C_{A0} - C_{Aex})/C_{A0}]} \tag{7.5.29}$$

Thus the calculation involves a trial and error procedure:

(1)   For fixed gas and solids flow rates and bed geometry, we estimate
the conversion.

(2)   Then we calculate $\overline{X}$ through the use of Eqs. (7.5.23)–(7.5.28).

(3)   We compare the calculated value of $\overline{X}$ with the estimate and iterate
until convergence is reached.

We note that the actual bed dimensions and geometry enter these calculations
indirectly in the evaluation of k_B and hence $(N'_{Sh})_B$.

Avedesian and Davidson [62] have recently described an investigation in
which carbon particles were burned in a bed of ash particles fluidized by air.
The conclusion was reached that under the conditions of the experiments,
chemical kinetics played no part in controlling the rate of combustion. The
rate was determined by two diffusional steps:

(1)   mass transfer from the bubble phase into the emulsion, and

(2)   diffusion and dispersion within the emulsion phase around each
carbon particle.

The former provided the dominating resistance when there were large
amounts of carbon particles in the bed.

The dependence of the time for complete combustion on such parameters as mass of carbon charged, carbon particle diameter, and carbon density was of the form predicted by the bubbling bed model.

Note that for gas–solid fluidized systems we may estimate the conversion of the gaseous and solid reactants by using much simpler procedures than required for packed-bed systems. The main reason for this difference is due to the assumption made in the representation of fluidized beds, that the bubble phase passed through in plug flow whereas the solids were perfectly mixed.

It must be stressed however, that while the bubbling model provides a convenient means for estimating the behavior of fluidized beds, the predictions based on this model are not very reliable. It follows that calculations using the bubbling model have to be regarded as approximations or perhaps as representing limiting cases, using the largest stable bubble size.

For a review of models describing the behavior of fluidized bed reactors the reader is referred to recent publications by Grace [63] and Pyle [64].

7.6 Gas–Solid Contacting in Kilns, Moving Beds, and Cyclones

In this section we shall present a brief description of gas–solid contacting in kilns, moving beds, cyclones, and transfer line reactors. A substantial number of gas–solid reactions are carried out in equipment of this type, such as ironmaking in the blast furnace (moving bed), the manufacture of cement (kilns), and certain direct reduction processes of iron oxide (kilns or moving beds).

In spite of the major economic importance of these operations, the fundamental aspects of flow patterns and heat and mass transfer in these systems have received very much less attention than for the case of beds and fluidized beds which were discussed in the preceding sections. This scarcity of information necessarily limits the material that can be presented here; indeed the treatment of gas–solid reaction systems in kilns, moving beds, and cyclones will be based on the real (or assumed) analogy of these operations to packed- or fluidized-bed processing.

ROTARY KILNS

Figure 7.42 is a diagram of gas–solid contacting in rotary kilns. It is seen that solids and gases may be contacted in counterflow, concurrent flow, and crossflow arrangements. The figure also indicates the principal symbols used for describing the behavior of kilns.

Let us designate the length and the diameter of the kiln by L' and D', respectively, and let α be the angle formed by the kiln with the horizontal plane, ω the number of times the kiln rotates per unit time, and φ the angle

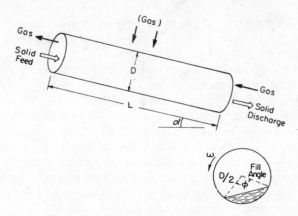

FIG. 7.42. Diagram of gas–solid contacting in a kiln.

of fill. Knowledge of the fill angle enables one to calculate the surface area of the solids exposed to the gas together with the fraction of the cross-sectional area occupied by the solids using simple trigonometric relationships. It has been suggested that for materials having a dynamic angle of repose in the range 35–45°, the following relationship holds for *the angle of fill*:

$$\sin (\varphi/\pi) \simeq 0.19 \qquad (7.6.1)$$

The mean residence time $\bar{t}_R$ of the solids in the kiln may be estimated by using the expression [65]

$$\bar{t}_R = 0.19 L'/(\omega D' \sin \alpha) \qquad (7.6.2)$$

A survey of residence time distribution measurements by Taylor and Jenkins [66] indicates that for most free flowing materials processed in long kilns, it is reasonable to assume that *the solid flow pattern approaches plug flow*. It has also been shown that the solids are well mixed in any horizontal section.

HEAT TRANSFER

Heat transfer in kilns is a complex process, the individual elements of which would include:

(1) direct heat transfer between the solids and the gas by convection and radiation,

(2) heat transfer between the gas and the inside wall of the kiln by radiation and convection, and

(3) heat transfer between the wall and the solids by conduction and radiation.

The direct convective heat transfer between the gas and the solids may be estimated by using conventional turbulent heat transfer correlations between a surface and a moving gas stream; radiative transfer between the solids and the gas may also be estimated by assigning an effective emissivity and absorptivity to the gas space [3], p. 289. This latter problem may be complicated by the presence of luminous solid particles, which would then increase the effective emissivity and absorptivity.

Although heat transfer between the walls and the solids is more difficult to assess because it is a periodic process, an elegant model for this process has been recently proposed by Pearce [67].

For a more detailed discussion of heat transfer in kilns the reader is referred to the excellent paper by Pearce; some further empirical correlations and additional references may also be found in an article by Sass [68].

We stress to the reader that heat (and mass) transfer models or correlations for kilns may only be used for estimating the transfer rate. If kilns are to be used for gas–solids contacting for a new process for which no industrial data are available, pilot plant work would be highly advisable.

MOVING-BED SYSTEMS

In moving-bed systems such as shown in Fig. 7.43 solid and gas streams are contacted in a counterflow arrangement. Perhaps the most economically significant moving-bed operation is the iron blast furnace, a detailed description of which is readily available in the literature (Strassburger [69] and McGannon [70]). In typical moving-bed operations, the flow pattern of both the solids and of the gas may be approximated by assuming plug flow behavior, although as shown in Fig. 7.44a,b this is an oversimplification, as there exist preferential flows in both the gas and the solid streams.

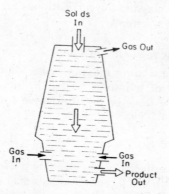

FIG. 7.43. A moving-bed reactor.

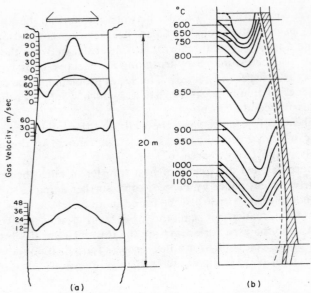

FIG. 7.44. (a) Measured gas velocity profiles in an iron blast furnace; (b) measured isotherms in an iron blast furnace.

This tendency for preferential gas flows (near the walls) could cause serious problems in systems where the reaction is either strongly exothermic or endothermic, because the uneven temperature distribution may aggravate these bypass phenomena [71, 72]. On viewing moving-bed systems within the framework of a fixed coordinate system, it is readily apparent that the linear velocity of the solids is much smaller than that of the gas. It follows that in predicting pressure drop–gas flow relationships, heat and mass transfer coefficients, and the like, we may use the correlations that were previously given in Section 7.3 for fixed-bed systems.

MISCELLANEOUS CONTACTING ARRANGEMENTS

While packed beds (including moving beds), fluidized beds, and kilns account for the majority of gas–solid contacting operations, it may be of interest to mention briefly some of the alternative contacting arrangements that are available.

Cyclones

Cyclones are used extensively for gas cleaning purposes [73, 74], but their application as reactors is more recent. Cyclones may be used for the drying of granular materials [75], the combustion of solid fuel, and the smelting and reduction of ores [76].

While some work has been done on heat transfer cyclones [78–80], no really satisfactory predictive correlations are available at the present.

Conveyed systems

When solid particles are contacted with a gas stream that has a higher linear velocity than the terminal falling velocity of the solid particles, elutriation or conveying of the solids will take place.

A detailed description of such systems has been presented by Gauvin and Gravel [81]. In these operations, the solids are present in relatively small concentration *or in a dilute phase*; furthermore, the *actual contact time* that is available is necessarily limited to small values by the actual time that is taken by the solids to travel through the reactor.

Vibrated beds, annular kilns, etc.

A brief description of the operation of some more exotic gas–solid contacting arrangements such as vibrated beds and mechanically stirred beds is given by Taylor and Jenkins [66].

FORMULATION OF HEAT AND MASS BALANCES

The formulation of heat and mass balances for the contacting arrangements described in this section follows a similar pattern to that given in Sections 7.4 and 7.5.5 dealing with fixed and fluidized beds, respectively.

In the operation of kilns and moving-bed systems the gas and the solid streams may be contacted in a concurrent or in a countercurrent arrangement; the formulation for each of these modes of contacting is quite straightforward and involves the following steps:

(1) Overall heat and mass balances are established for the gas and solid streams.

(2) By writing the equation of continuity, allowance is made for changes in the molar flow rates of the gases and the solids.

(3) The overall balances are coupled to rate expressions describing the reaction of the solid particles.

This procedure will be illustrated for a simple isothermal system, together with the presentation of a general scheme for nonisothermal systems.

ISOTHERMAL SYSTEMS

Let us consider a moving-bed (or a kiln) in which a gas and a solid stream are contacted in a concurrent or countercurrent arrangement, as shown in Fig. 7.45. We shall assume that the reaction scheme is

$$A(g) + bB(s) \rightarrow \text{products} \tag{7.6.3}$$

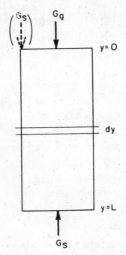

FIG. 7.45. Countercurrent or (concurrent) contacting of a gas and solid stream.

and that there is no change in the total molar flow rate of gas on reaction. Attention will be restricted to steady state behavior. On denoting the molar flow rate of the gaseous reactant and solid stream per unit area of reactor cross section by G_G and G_S, respectively, we can perform a differential material balance to give

$$-G_G \frac{d}{dy}\left(\frac{C_A}{C_{A0}}\right) = \frac{G_S}{b}\frac{dX}{dy} \qquad (7.6.4)$$

A similar equation can be written for the gaseous product if the reaction is reversible:

$$dC_A/dy = f(X, C_A, k, D_e, h_D, etc.) \qquad (7.6.5)$$

The precise functional form will depend on whether or not the solid is initially dense, the geometry of the solid, etc.

The boundary conditions for Eqs. (7.6.4) and (7.6.5) may be written

$$X = 0 \quad \text{at} \quad y = 0 \quad \text{(concurrent case)}$$
$$X = 0 \quad \text{at} \quad y = L \quad \text{(countercurrent case)} \qquad (7.6.6)$$

and

$$C_A = C_{A0} \quad \text{at} \quad y = 0 \qquad (7.6.7)$$

Equations (7.6.4)–(7.6.8) may be solved analytically or numerically, depending on the nature of the functional relationship appearing on the r.h.s. of Eq.

(7.6.5). However, this relatively simple system is ideally suited for solution by the technique described in Sections 7.4 and 7.5 [35].

Let us define the dimensionless concentration θ, the dimensionless length y^*, and the quantity β in the following manner:

$$\theta = C_A/C_{A0} \tag{7.6.8}$$

$$y^* = (k A_g C_{A0}/G_G F_g V_g)(1 - \varepsilon_v)(1 - \varepsilon)y \tag{7.6.9}$$

while

$$\beta = \frac{\int_0^1 \eta^{(F_p - 1)} F_g \varsigma^{(F_g - 1)} \psi \, d\eta}{\int_0^1 \eta^{(F_p - 1)} \, d\eta} \tag{7.6.10}$$

and is tabulated as Table 7.1 (page 272). β may also be approximated by

$$\beta \simeq (g'_{F_g}(X) + \hat{\sigma}^2 [F'_{F_p}(X) + 2/N'_{S_h}])^{-1} \tag{7.6.11}$$

where

$$g'_{F_g}(X) = (d/dx)(g_{F_g}(X)) \qquad \text{etc.}$$

We may proceed by writing

$$d\theta/dy^* = -\theta\beta \tag{7.6.12}$$

[replacing Eq. (7.6.5)] and

$$dX/dy^* = (G_G/G_s)\theta\beta \tag{7.6.13}$$

from Eqs. (7.6.4) and (7.6.13). The boundary conditions may thus be written

$$\theta = 1 \qquad \text{at} \quad y^* = 0 \tag{7.6.14}$$

and

$$X = 0 \qquad \text{at} \quad y^* = 0 \qquad \text{(concurrent flow)} \tag{7.6.15}$$

$$X = 0 \qquad \text{at} \quad y^* = L^* \qquad \text{(countercurrent flow)} \tag{7.6.16}$$

where

$$L^* = [k A_g C_{A0}(1 - \varepsilon)(1 - \varepsilon_v)/G_G F_g V_g]L \tag{7.6.17}$$

is the dimensionless reactor length for countercurrent flow.

Let us define

$$\mathscr{R} = bG_G/G_s \tag{7.6.18}$$

Thus, by combining Eqs. (7.6.13), (7.6.14), and (7.6.18), upon integration we have

$$X - X_0 = -\mathscr{R}(\theta - 1) \tag{7.6.19}$$

where X_0 is the value of X at $y^* = 0$, i.e., X_0 corresponds to the solid conversion at the exit for countercurrent flow. On substituting Eq. (7.6.20) into Eq. (7.6.14), subsequent integration yields

$$y^* = \int_0^X \frac{d\chi}{(\mathcal{R} - \chi)\beta(\chi)} \qquad \text{(concurrent operation)} \qquad (7.4.20)$$

where χ is a dummy variable. For countercurrent flow we have

$$L^* = \int_0^{X_0} \frac{d\chi}{(\mathcal{R} - \chi + X_0)\beta(\chi)} \qquad (7.6.21)$$

Equation (7.6.20) provides the desired relationship between the conversion of the solids and the dimensionless position in the bed for concurrent operation; Eq. (7.6.21) relates the length of the bed to the solids conversion for countercurrent operation. Since β is known as a function of χ the integrals appearing in these expressions are readily evaluated, either through the use of Simpson's rule or by some other numerical method.

Once X is known as a function of y^*, the corresponding $\theta = f(y^*)$ profiles are, of course, readily obtainable by using Eq. (7.6.19).

Figures 7.46–7.49 show a set of computed X and θ values for both concurrent and countercurrent flow. In the computation of these curves, we have assumed that external mass transfer could be neglected, i.e., $N'_{Sh} \to \infty$. Curves with finite values of N'_{Sh} may, of course, be readily constructed by integrating Eq. (7.4.21) or (7.4.22). Note, moreover, that analytical solutions are readily generated for the special case where the overall rate is mass transfer controlled [82].

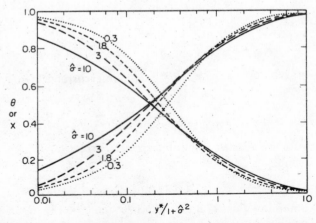

FIG. 7.46. Plot of solids conversion X and fraction of gas remaining unreacted θ against position in reactor, concurrent operation. $F_p = 3$, $F_g = 3$, $\mathcal{R} = 1$.

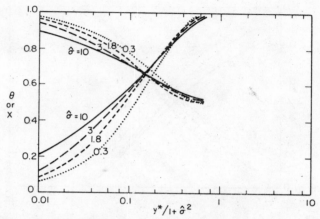

FIG. 7.47. Plot of solids conversion X and fraction of gas remaining unreacted θ against position in reactor, concurrent operation. $F_p = 3$, $F_c = 3$, $\mathscr{R} = 2$.

The abscissas in Figs. 7.46–7.49 have been chosen so that curves for different values of $\hat{\sigma}$ lie very close to one another. Consequently the figures may be used for any value of $\hat{\sigma}$.

It is a relatively straightforward matter to show that the same curves may be applied to the case of a reversible reaction (equilibrium constant K_E) provided the dimensionless variables are redefined as follows:

$$\hat{\sigma} = \frac{V_p}{A_p} \left[\frac{(1 - \varepsilon)k}{2D_E} \frac{F_p}{F_g} \frac{A_g}{V_g} \left(1 + \frac{1}{K_E} \right) \right]^{1/2} \tag{7.6.22}$$

$$y^* = \frac{kA_g(1 - \varepsilon)(1 - \varepsilon_v)(1 + 1/K_E)y}{u_0 F_g V_g} \tag{7.6.23}$$

$$L^* = \frac{kA_g(1 - \varepsilon)(1 - \varepsilon_v)(1 + 1/K_E)L}{u_0 F_g V_g} \tag{7.6.24}$$

$$\mathscr{R} = \frac{bu_0(\bar{C}_{A0} - \bar{C}_{C0}/K_E)}{G_s(1 + 1/K_E)} \tag{7.6.25}$$

where $\bar{C}_{C0}$ is the concentration of gaseous product at the reactor inlet (usually zero).

It is thought that the curves shown in Figs. 7.46–7.49 may be helpful for a preliminary estimation of reactor sizes required to attain a given conversion, or alternatively these plots may also be used to estimate the extent of reaction that may be attained for a given throughput and a specified equipment size.

The application of this general dimensionless approach described here will be illustrated by the following example.

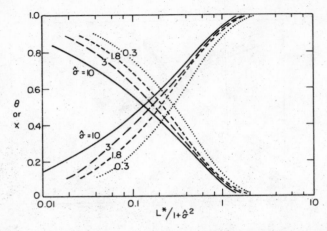

FIG. 7.48. Plot of solids conversion X and fraction of gas remaining unreacted θ at reactor exits against length of reactor, countercurrent operation. $F_p = 3$, $F_c = 3$, $\mathscr{R} = -1$.

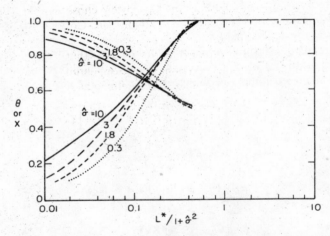

FIG. 7.49. Plot of solids conversion X and fraction of gas remaining unreacted θ at reactor exits against length of reactor, countercurrent operation. $F_p = 3$, $F_g + 3$, $\mathscr{R} = -2$.

Example 7.6.1   It is proposed to reduce wüstite (FeO) pellets with hydrogen in a counterflow arrangement. The wüstite pellets are 1 cm in diameter, have an internal porosity of 0.3, and are fed at a rate of 3×10^{-2} kg-mole/m² sec. The hydrogen flow rate is 4 times the stoichiometric value and it may be assumed that equilibrium at the pellet surface would correspond to 50% H_2O and 50% H_2. Assume isothermal behavior at 900°C.

ADDITIONAL DATA

$k = 10^{-1}$ cm/sec, $D_e = 1.2$ cm^2/sec, $F_g = 3$, $F_p = 3$, $V_g/A_g = 10^{-4}$ cm, $\varepsilon_v = 0.5$, $C_{A0} = 0.0104$ g-mole/liter

Estimate the bed height required to obtain a 99 % conversion of the solids.

SOLUTION The reduction reaction is

$$\text{FeO(s)} + \text{H}_2\text{(g)} \overset{k_1}{\underset{k_2}{\rightleftarrows}} \text{Fe(s)} + \text{H}_2\text{O(g)}$$

The equilibrium constant is

$$\frac{k_1}{k_2} = \left(\frac{C_{\text{H}_2\text{O}}}{C_{\text{H}_2}}\right) = 1$$

$k_1 = k_2 = k = 10^{-1}$ cm/sec

From Eq. (7.6.23),

$$\hat{\sigma} = \frac{V_p}{A_p}\left[\frac{(1-\varepsilon)kF_pA_g}{2D_EF_gV_g}\left(1 + \frac{1}{K_E}\right)\right]^{1/2}$$

$$= \frac{1}{6}\left[\frac{(1-0.3) \times 10^{-1} \times 3 \times 1}{2 \times 1.2 \times 3 \times 10^{-4}}(1+1)\right]^{1/2} = 4.0$$

The input gas is pure hydrogen; hence

$$u_0(C_{A0} - \bar{C}_{C0}/K_E) = G_g = -4 \times G_s$$

From Eq. (7.6.26),

$$\mathscr{R} = \frac{bu_0(\bar{C}_{A0} - \bar{C}_{C0}/K_E)}{G_s(1 + 1/K_E)} = \frac{-4}{1 + 1/K_E} = -2$$

The required conversion $X = 0.99$. From Fig. 7.49 for $\hat{\sigma} = 4.0$ and $X = 0.99$,

$$L^*/(1 + \sigma^2) = 0.5, \qquad L^* = 8.5$$

Using Eq. (7.6.24),

$$L^* = \frac{\bar{C}_{A0}kA_g(1-\varepsilon)(1-\varepsilon_v)(1 + 1/K_E)L}{G_gF_gV_g}$$

$$L = \frac{L^* \times 4G_sF_gV_g}{\bar{C}_{A0}kA_g(1-\varepsilon)(1-\varepsilon_v)(1 + 1/K_E)}$$

$C_{A0} = 0.0104 \times 10^{-3}$ g-mole/cm^3

$G_s = 3 \times 10^{-2}$ kg-moles/m^2 sec $= 3 \times 10^{-3}$ g-mole/cm^2 sec

$$L = \frac{8.5 \times 4 \times 3 \times 10^{-3} \times 3 \times 10^{-4}}{0.0104 \times 10^{-3} \times 10^{-1} \times (1-0.3) \times (1-0.5) \times 2} = 42 \text{ cm}$$

The general case of nonisothermal systems

The dimensionless formulation given previously was helpful because it provided a convenient means for estimating the behavior of reaction systems. In the majority of real problems (which would include the reduction of iron oxide with hydrogen) the assumption of isothermal behavior is not appropriate; moreover, there are many systems where there is a change in molarity upon reaction.

It would be impracticable to present general, all encompassing formulas for the description of these systems; for this reason we shall restrict our attention to a brief review of the considerations that must enter such a general formulation.

The differential balance equations must express the following:

heat balance on the gas,
heat balance on the solids,
component balance(s) on the gas,
mass balance of the solids,
overall mass balance on the gas (if there is a change in molarity).

Thus we have (at steady state):

heat balance on the gas

$$\begin{bmatrix} \text{net convective} \\ \text{inflow of heat} \end{bmatrix} - \begin{bmatrix} \text{rate of heat transfer to} \\ \text{the solids from the gas} \end{bmatrix} + \begin{bmatrix} \text{rate of heat generative} \\ \text{due to gas phase reaction} \end{bmatrix} \quad (7.6.26)$$
$$= \begin{bmatrix} \text{net heat outflow due} \\ \text{to axial dispersion} \end{bmatrix}$$

heat balance on the solids

$$\begin{bmatrix} \text{rate of heat generation} \\ \text{due to heterogeneous} \\ \text{reactions} \end{bmatrix} + \begin{bmatrix} \text{rate of heat transfer} \\ \text{to the solids from} \\ \text{the gas} \end{bmatrix} = \begin{bmatrix} \text{net convective} \\ \text{outflow of heat} \\ \text{in solid stream} \end{bmatrix} \quad (7.6.27)$$

component balance on the gas

$$\begin{bmatrix} \text{net convective inflow} \\ \text{of component } i \end{bmatrix} + \begin{bmatrix} \text{net component inflow} \\ \text{due to axial dispension} \end{bmatrix} + \begin{bmatrix} \text{net production of component} \\ i \text{ due to chemical reaction} \end{bmatrix}$$
$$= \begin{bmatrix} \text{net transfer of component } i \\ \text{from the gas to the solid} \end{bmatrix} \quad (7.6.28)$$

mass balance on the solids

$$\begin{bmatrix} \text{net convective inflow} \\ \text{of solid reactant} \end{bmatrix} = \begin{bmatrix} \text{rate of consumption} \\ \text{of solid reactant} \end{bmatrix} \quad (7.6.29)$$

overall mass balance on the gas

$$\begin{bmatrix} \text{net change in the mass or molar} \\ \text{flow rate of the gas stream} \end{bmatrix} = \begin{bmatrix} \text{net rate of production of} \\ \text{the sum of the species} \end{bmatrix} \quad (7.6.30)$$

Note that this formulation would apply to both concurrent and counter-current systems; thus it should form the logical starting point of the description of kilns, moving beds, etc.

Good illustrative examples of such formulations may be found in the literature [83–85].

7.7   Reactors with a Distribution of Solid Particle Size

In our previous discussions of the behavior of reactors we, assumed that the solid charged to the reactor consisted of pellets or particles all of which were the same size and shape. In some industrial reactors the solid feed consists of particles of widely varying sizes. This is particularly true if the feed material is produced by comminution of larger pieces of solid. From the considerations outlined in earlier chapters it follows that the larger particles in the feed will usually take longer to react than the smaller particles, and this effect must be taken into account when calculating the overall conversion of the solids leaving the reactor.

For a reactor in which the gas conversion is small, the reaction of an individual particle is not affected by the presence of the other particles and a fairly straightforward mathematical development [86] can be employed.

Let us suppose that the particle size distribution can be described by the function $y(R, R_*)$, where $y\,\delta R$ is the weight fraction of particles with radii between R and $R + \delta R$, and R_* is the size of the largest particle in the feed. This implies (assuming other parameters are independent of particle size) that the weight fraction of particles with $\hat{\sigma}$ lying between $\hat{\sigma}$ and $\hat{\sigma} + \delta\hat{\sigma}$ is $y(\hat{\sigma}, \hat{\sigma}_*)\,\delta\hat{\sigma}$ where $\hat{\sigma}_*$ is $\hat{\sigma}$ calculated for the largest particle.

Using the approximate equation developed in Chapter 4,

$$t_x/t_*(1.0) = [g_{F_G}(X) + \hat{\sigma}^2 p_{F_p}(X)]/(1 + \hat{\sigma}_*{}^2) \qquad (7.7.1)$$

where t is the time required to react a particle of reaction modulus σ to extent of reaction X and $t_*(1.0)$ the time required to completely react the largest particle. This can be rewritten formally

$$X = f[\sigma, \sigma_*, t/t_*(1.0)] \qquad (7.7.2)$$

The determination of X, given t, would normally require the solution of (7.7.1) by trial and error, linear interpolation, or some other elementary numerical technique.

For a batch or plug flow reactor all the solid particles have the same residence time and consequently the overall conversion of the solids leaving the the reactor is given by

$$X = \int_0^{\hat{\sigma}_*} y f(\hat{\sigma}, \hat{\sigma}_*, t_R/t_*(1.0))\,d\hat{\sigma} \qquad (7.7.3)$$

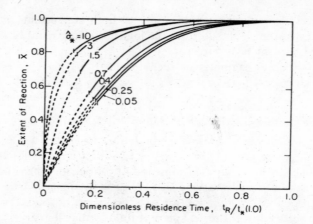

FIG. 7.50. Plot of extent of reaction against dimensionless reactor residence time for spherical particles made up of spherical grains in a batch or plug flow reactor. Particle size distribution given by the Gates–Gaudin–Schumann equation. $F_g = 3$.

where t_R is the reactor residence time. This integral may be readily evaluated using techniques such as Simpson's rule.

Figure 7.50 gives the extent of reaction of spherical particles made up of spherical grains for the particle size distribution given by the Gates–Gaudin–Schumann size distribution (which frequently results when solids are ground):

$$y = m(\hat{\sigma}^{m-1}/\hat{\sigma}_*{}^m) \qquad (7.7.4)$$

with $m = 1.0$. Curves for $\hat{\sigma}_* < 0.05$ lie quite close to the curve for $\hat{\sigma}_* = 0.05$ since for small particles (small $\hat{\sigma}$) the time to achieve complete reaction becomes independent of particle size. The curves for $\hat{\sigma}_* > 10$ lie close to that for $\hat{\sigma}_* = 10$. Between these two extremes we see that the effect of a particle size distribution is to increase the overall rate initially, compared with the rate of reaction of the largest particle, but to leave the time for complete reaction unchanged.

For continuous reactors in which the solid is perfectly mixed (e.g., fluidized bed) then there is a distribution of residence times, and consequently the overall conversion of the solid leaving the reactor is given by

$$\bar{X} = \int_0^{\sigma_*} y \int_0^{\infty} E(t) f(\hat{\sigma}, \hat{\sigma}_*, t_X/t_*(1.0)) \, dt \, d\hat{\sigma} \qquad (7.7.5)$$

where $E(t)$ is the residence time distribution for a reactor of nominal residence time $\bar{t}_R$ and is given by

$$E(tX/t_*(1.0)) = [t_*(1.0)/\bar{t}_R] \exp(-t/\bar{t}_R) \qquad (7.7.6)$$

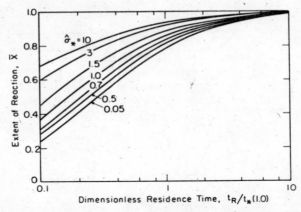

FIG. 7.51. Extent of reaction against reactor residence time for spherical particles made up of spherical grains in a perfectly mixed reactor. Particle size distribution given by the Gates–Gaudin–Schumann equation. $F_g = 3$.

This double integral may be evaluated using numerical techniques such as Simpson's rule and Laguerre integration. Figure 7.51 gives the calculated overall extent of reaction of spherical solid particles leaving a perfectly mixed reactor with the particle size distribution given by Eq. (7.7.4) with $m = 1.0$. Curves calculated for $\hat{\sigma}_* < 0.05$ almost coincide with that for $\hat{\sigma}_* = 0.05$, while those for $\hat{\sigma}_* > 10$ almost coincide with that for $\hat{\sigma}_* = 10$.

These curves for a reactor in which the solid is perfectly mixed would also be applicable to reactors in which gas conversion is not small. In this case $\hat{\sigma}_*$ is calculated using the spatial mean gas concentration in the bed calculated as described at the end of Section 7.5.5.

McIlvried and Massoth [87] have considered an assembly of particles reacting according to the shrinking-core model at low gas conversions. The particle size distributions examined were the normal and log normal distributions.

We shall conclude this chapter with a few general remarks concerning experimental and analytical work involved in the design and development of gas–solid reaction systems.

7.8 Concluding Remarks

In the preceding sections we outlined how the kinetic information obtained for the behavior of single particles may be used for predicting the performance of particle assemblies, packed beds, moving beds, kilns, etc.).

Two general conclusions must emerge from this discussion:

(a) While the description of single-particle behavior may involve some uncertainties, it is even more difficult to represent particle assemblies because of the numerous additional factors (and uncertainties) involved.

TABLE 7.5

Type of reactor	Solids behavior		
	Hold-up (fraction of packed bed in reaction zone)	Mean residence time for inerts	Distribution of residence times
Rotary kilns	8–15%	$t_R \propto L$ under suitable conditions	$j \approx 5$; $D_e \approx 0.7$ cm²/sec
Flighted kilns	8–15%	$t_R \propto L$	$j \approx 5$
Stirred beds	Typically 20%; possible to 40–50%	$t_R \propto L$ (see Fig. 2)	$j = 12 \to 50$; $D_e = 1.22 \to$ cm²/sec
With solids–solids mixing			$j \approx 5$
Vertical moving beds (i) Unvibrated	~100%	$t_R \propto L$	$j \approx 10^2$
(ii) Vibrated	Limited in height		$j = 10^2 \to 10^3$
Vibrated trays (i) Simple	Up to 40%	t_R approx. $\propto L$	$j = 20 \to 60$
(ii) Gas-permeated		(see Fig. 6)	$t_R = 20 \to 60$
Series fluidized beds (i) Vertical	Liable to instability	Liable to instability	j per bed ≈ 1.0
(ii) Horizontal baffled	~50% with freeboard	$t_R \propto$ number of beds	j per bed $\propto 1.0$ Also Ref. 22
Single fluidized bed	~50% with freeboard	Usually suitable for high solids conversion on continuous flow	

(b)   While it is necessary to have information on the kinetic behavior of the gas–solid reaction system under consideration, in many instances the principle factors may be other than those associated with chemical kinetics; viz, heat transfer, mass transfer, possible bypassing of the gas, and short circuiting of the solids.

We shall conclude this chapter with a brief outline of the necessary steps involved in the design and development of a gas–solid reaction system.

Continuous Flow Behavior in Various Types of Tubular Reactors

Gas behavior

Pressure drop	Distribution of residence times	Features determining applications	Typical industrial application
	Reasonably limited in absence of major pulsations	Handle partial fusions; Suffer from accretions on walls	Cement burning
Low	Probably similar to that in stirred beds	Difficulties with partial fusions	Power drying and calcining
Low	$j \propto 10 \to 50$; $D_e \propto 20$ cm^2/sec	Limited in temperature range, adaptable in solids handling; Suited to profiling of temperature and chemical compositions	Powder drying; Uranium processing
Tends to be high	Much work reported packed beds	Aggregated solids feed usual; Poor heat transfer for exothermic reactions	(i) Shaft kilns; Blast furnaces (ii) None known
(i) Low	$j \approx 10$	Limited in size; Adaptable in solids handling	(i) Uranium processing; Catalyst calcining
(ii) Intermediate	$j \approx 3$		(ii) None known
Medium to high; $\propto$ wt. of beds	Dependent on circuitry, near plug-flow possible	(i) Discontinuous often unavoidable Amenable to pressure operation; Solids feed preparation often required	(i) Direct reduction of iron ores, lime burning (ii) None known
Medium to high	Split into lean and dense phase flows for particles $<\frac{1}{3}$ mm.	Gas reaction times below 2 sec desirable	Hydrogen processing

Thermodynamics

The first step in designing a reaction system is to establish the thermodynamic equilibrium relationship between the reactants and products and to calculate the heat effects of the reaction. This step is vital because the equilibrium constants and heat effects, which may be readily calculated [88, 89], will immediately tell us whether the reaction is in fact feasible thermodynamically and whether some special precautions are needed to remove the products and to operate at high or low temperatures, pressures etc. Knowledge of the heat effects, whether the reaction is exothermic or endothermic, again is very important in the selection of the contacting arrangements.

Kinetics

In contrast to the criteria for thermodynamic equilibria, the kinetic parameters of gas–solid reaction systems cannot be calculated; it is to be stressed, furthermore, that because of the possible diversities in experimental conditions, experimental arrangements, and possible differences in the physical state of the solid reactants, great caution must be exercised in the use of kinetic data reported in the literature for design purposes. Perhaps the only exception here would be systems that were found to be mass transfer controlled.

It is generally desirable to perform kinetic measurements of the type described in Chapter 6 so as to obtain values of the kinetic parameters. The depth and scope of such kinetic study will have to be determined by the nature of the system (the extent of prior information available), the possible complications expected, possible variations in feed material, and so forth. The results of these measurements may then be interpreted with the aid of models described in Chapters 3–5.

Selection of the equipment of the gas–solid reaction system

Having obtained information on the thermodynamics of the system and on the kinetic parameters, we must select or narrow down the selection of the actual contacting procedure and equipment to be chosen.

A list of available equipment for the continuous contacting of solids with a gas is given in Table 7.5 taken from an article by Taylor and Jenkins [66]. Highlighting the comments in this table we note that fluidized beds are very attractive from the viewpoint of solids handling and for producing a relatively uniform bed temperature, even for highly exothermic or endothermic reactions. However, fluidized beds are unsuitable for systems where partial fusion of even softening of the material may occur, because the resultant agglomeration may choke the bed. Another possible drawback of continuous fluidized beds is the fact that the solids are perfectly mixed, so that a high conversion of solids would require a long residence time.

Rotary kilns make a rather less efficient use of space, but could handle partial fusion; moving-bed systems are attractive because the solids approach plug flow behavior; however, bypass phenomena and the occurrence of "hot spots" could be a drawback of moving-bed operations.

Pilot-scale studies

If a new process is to be developed or even if the modification of an existing process is contemplated using drastically different raw materials, it is desirable to conduct pilot-scale studies, i.e., to construct a pilot plant with the specific objective of examining the possible difficulties that might arise in the trans-action of single-particle data to multiparticle assemblies. Pilot plant operation should provide information on the handling characteristics of the material (agglomeration, softening, attrition, etc.) for which allowance may then be made in the design on the full-scale equipment. In many instances, it may be desirable to perform the scale-up of the process in several stages. Pilot plant studies and scale-up constitute a very interesting field; the reader is referred to the specialist literature for its more detailed study [90, 91]. A good illustra-tion of these procedures is provided by Stone and Rigg [92], who have pre-sented a readable account of the development through the pilot plant stage of a large-scale reactor for a gas–solid reaction step in a new iron-making process.

Notation

A_B	cross-sectional area of bed
$A(g)$	material A in the gaseous phase
A_g, A_p	surface area of grain and pellet, respectively
a	surface area per unit volume of bed
a'	external surface area per unit solid volume
$B(s)$	material B in the solid phase
b	stoichiometric factor
$C_{A,B}; C_{A,C}; C_{A,E}$	concentration of transferred species A in the bubble, cloud, and emulsion, respectively
C_{Aex}	exit concentration
C_{A0}	inlet concentration or initial concentration
C_g	gaseous reactant concentration
C_{gs}	reactant concentration at the surface of the particle
C_I	concentration of the Ith component
C_s	concentration of solid reactant
c	concentration of the tracer in the exit stream
c_i	tracer concentration at the inlet
c_p	heat capacity
c_t	tracer concentration
D	gas phase diffusivity
D'	diameter of kiln
D_e	effective diffusivity
d_c	diameter of column

d_B	bubble diameter
d_p	particle diameter
$\bar{d}_p$	surface area mean diameter
$\mathscr{E}_i$	weight of particles of size d_{p_i}
$E(r)$	residence time distribution for reactor
$E(t^*)$	age distribution of the solids in bed at time t^*.
E_z, E_r	axial and radial eddy diffusivities
$E(\theta)$	dimensionless tracer concentration at exit in C diagrams
F'_{Fr}	friction term
F_g, F_p	shape factor for grain and pellet, respectively
$F(\theta)$	dimensionless tracer concentration at exit in F diagrams C/c_i
G_G	molar flow rates of gas
G_s	molar flow rates of solid
G_0	mass velocity
G_r	mass velocity in the r direction
G_x, G_z	mass velocity in the x and z directions, respectively
g	gravity constant
$g_F(X), p_{F_p}(X)$	conversion function for small and large $\hat{\sigma}$, respectively
h	heat transfer coefficient
h_{BC}.	heat transfer coefficient between cloud and bubble
h_D	mass transfer coefficient
h_c	convective heat transfer coefficient
h_p	heat transfer coefficient between fluid and particles
h_w	heat transfer coefficient between bed and the wall
j_D	Chilton–Colburn factor
K_E	equilibrium constant
k	reaction rate constant
$\tilde{k}_{BC}, \tilde{k}_{CE}, \tilde{k}_{BE}$	transfer coefficients between the bubble–cloud, cloud–emulsion, and bubble–emulsion phases, respectively
k_e	elutriation constant
k_{eff}	effective thermal conductivity
k_f	thermal conductivity of the fluid
k_s.	thermal conductivity of solids
L	depth of the bed
L'	length of kiln
L^*	dimensionless reactor length
L_{mf}	bed height at minimum fluidization
m	molecular weight
$\bar{m}$	average molecular weight
m_e	mass fraction
N_{Nu}	Nusslet number ($h\,d_p/k$)
N_{Pe}	Peclet number (du/E)

N_{Pr}	Prandtl number $(c_p \mu / k)$
N_{Re_p}	Reynolds number $(\rho u \, d_p / \mu)$
N_{Sc}	Schmidt number $(\mu / \rho D)$
N_{Sh}	Sherwood number $(h \, d_p / D)$
N'_{Sh}	modified Sherwood Number
n	number of completely mixed cells
P	pressure
Q	mass of tracer added
R'	universal gas constant
R	radius of a particle
R_*	size of the largest particle in the feed
$\mathscr{R}$	ratio of molar gas to solid rates defined by Eq. (7.6.19)
r, z, θ	cylindrical coordinate system
r_1	rate of reaction I
r_0	initial radius of cellulose particle
$r_p(t)$	radius of cellulose particle at time t
S^0	dimensionless position of the reaction front
T	temperature
T_G	temperature of the gas
$T_{G, B}$	temperature in bubble
T_s	temperature of the solid
t	time
$t_{X, \sigma}$	time required to react a particle of reaction modulus σ to extent of reaction X
t^*	dimensionless time
$t^*_{(X = 1.0)}$	time required to completely react largest particle
$\bar{t}_R$	time for reaction front to travel the depth of the bed
$\bar{t}_R$	nominal holding time or residence time of material in vessel
t_R^*	residence time of the solids in the bed
U	linear velocity
U_0	linear velocity in empty column
U_S	linear velocity of solids
U_z	axial velocity component
u	velocity
u_B	rising velocity of bubble
u_E	emulsion phase linear velocity
u_{mf}	minimum fluidization velocity
u_t	terminal falling velocity of a particle
V	volume of the vessel
V_B	bubble volume
V_d	dead zone or volume of reactor
V_m	completely mixed region or volume of reactor

V_p	plus flow region or volume of reactor
v	volumetric flow rate of the stream
$\bar{v}$	specific volume
V_g, V_p	volume of grain and pellet, respectively
w	length of bed
X	conversion or overall extent of reaction
X_A	mole fraction of gas A
X_n	weight fraction of solids of diameter d_{pn}
X_0	value of x at $y^* = 0$
x, y, z	cartesian coordinate system
$Y(R, R^*)$	particle size distribution function
y^*	dimensionless length defined in Eq. (7.4.17)
$y\,\delta R$	weight fraction of particles between R and $R + \delta R$
z	dimensionless distance x/L, where L is reactor length
α	angle formed with the horizontal plane
β	function variable defined in Eq. (7.4.11)
β'	defined by Eq. (7.4.15)
δ	dimensionless length defined by Eq. (7.5.17)
δ_B	volume of solids in bubble per volume of the bubble
ε_v	void fraction
$\bar{\varepsilon}$	emissivity
ε	pore volume in pellet
ε_{mf}	minimum voidage of fluidized bed
ε_f	void fraction at actual fluidized conditions
η	dimensionless position in the pellet
η_H	effectiveness factor for heat transfer
θ	dimensionless time t/E_R
θ	dimensionless reactant concluded driving force along reactor
λ	dummy variable
μ	viscosity
ξ	dimensionless position in the pellet
π	cross flow ratio
ρ_f	density of the fluid
ρ_s	molar density of the solid
ρ_G	molar density of gas
$\bar{\rho}_s$	density of solid
$\bar{\rho}_g$	density of gas
$\bar{\sigma}$	gas–solid reaction modulus based on R_A
$\hat{\sigma}$	gas–solid reaction modulus
φ_s	shape factor
φ	angle of fill

ψ dimensionless concentration
ω rotation per unit time
Δp pressure drop
ΔH_c heat of reaction

References

1. P. V. Danckwerts, *Chem. Eng. Sci.* **2**, 1 (1953).
2. O. Levenspiel, "Chemical Reaction Engineering," 2nd ed., Chapter 9. Wiley, New York, 1972.
3. J. Szekely and N. J. Themelis, "Rate Phenomena in Process Metallurgy," Chapter 15. Wiley, New York, 1971.
4. K. G. Denbigh, "Chemical Reactor Theory." Cambridge Univ. Press, London and New York, 1965.
5. J. Happel and H. Brenner, "Low Reynolds Number Hydrodynamics," Chapter 8. Prentice Hall, Englewood Cliffs, New Jersey, 1965.
6. B. P. LeClair and A. E. Hamielec, *Ind. Eng. Chem. Fundam.* **5**, 542 (1968).
7. B. P. LeClair and A. E. Hamielec, *Ind. Eng. Chem. Fundam.* **9**, 608 (1970).
8. S. Ergun, *Chem. Eng. Prog.* **5**, 90 (1960).
9. J. Szekely and N. J. Themelis, "Rate Phenomena in Process Metallurgy," p. 641. Wiley, New York, 1971.
10. Kunii and Levenspiel, "Fluidization Engineering." Wiley, New York, 1969.
11. E. F. Blick, *Ind. Eng. Chem. Process Design and Develop.* **5**, 90 (1960).
12. J. Szekely and R. G. Carr, *Trans. Met. Soc. AIME* **242**, 918 (1968).
13. R. Newell and N. Standish, *Met. Trans.* **4**, 1851 (1973).
14. V. Stanek and J. Szekely, *Can. J. Chem. Eng.* **50**, 9 (1972).
15. H. Schlichting, "Boundary Layer Theory," 6th ed. (J. Kestin, trans.) McGraw-Hill, New York, 1968.
16. V. Stanek and J. Szekely, *Can. J. Chem. Eng.* **51**, 22 (1973).
17. R. H. Wilhelm, *IUPAC* **5**, 403 (1962).
18. L. Z. Balla and T. W. Weber, *AIChE J.* **15**, 146 (1969).
19. H. A. Deans and L. Lapidus, *AIChE J.* **6**, 656 (1960).
20. N. R. Amundson, *Ind. Eng. Chem.* **48**, 35 (1956).
21. P. N. Rowe, K. T. Claxton, and J. B. Lewis, *Trans. Inst. Chem. Eng.* **43**, T14 (1965).
22. F. W. Dittus and L. M. K. Boelter, *Univ. California Publ. Eng.* **2**, 443 (1930).
23. J. Beek quoted by R. Aris, in "Elementary Chemical Reaction Analysis," p. 285. Prentice Hall, Englewood Cliffs, New Jersey, 1969.
24. R. F. Baddour and C. Y. Yoon, *Chem. Eng. Progr. Symp. Ser. No. 32*, **57**, 35 (1961).
25. J. Downing, *Proc. El. Steelmaking Conf.*, AIME, New York, (1968).
26. E. J. Wilson and C. J. Geankoplis, *Ind. Eng. Chem. Fundam.* **5**, 9 (1966).
27. A. D. Gupta and G. Thodos, *AIChE J.* **5**, 608 (1962).
28. G. F. Malling and G. Thodos, *Int. J. Heat Mass Transfer.* **10**, 489 (1967).
29. R. Pfeffer, *Ind. Eng. Chem. Fundam.* **3**, 380 (1964).
30. R. D. Bradshaw and J. E. Myers, *AIChE J.* **9**, 590 (1963).
31. K. R. Jolls and T. J. Hanratty, *AIChE J.* **15**, 199, (1969).
32. C. C. Furnas, *Trans. Amer. Inst. Chem. Eng.* **24**, 142 (1930).
33. N. R. Amundson, *Ind. Eng. Chem.* **48**, 26 (1956).
34. A. Moriyama, *Trans. Iron Steel Inst. Jpn.* **11**, 177 (1971).
35. J. W. Evans and S. Song, *Ind. Eng. Chem. Process Design Develop.* **13**, 146 (1974).

36. J. M. Smith, "Chemical Engineering Kinetics," Chapter 11. McGraw-Hill, New York, 1970.
37. I. Muchi and J. Higuchi, *Trans. Iron Steel Inst. Jpn.* **12**, 55 (1972).
38. J. Szekely (ed.), "The Steel Industry and the Environment," Marcel Dekker, New York, 1972.
39. J. Szekely and J. H. Chen, *Proc. DECHEMA, Frankfurt, Germany* (1973).
40. A. F. Robert, *Combust. and Flame* **14**, 261 (1970).
41. R. H. Essenhigh and T. J. Kuno, *Proc. Nat. Incinerator Conf.* 261 (1970).
42. R. H. Essenhigh, T. J. Kuo, M. Kuwata, and J. B. Stumbar, *Proc. Nat. Incinerator Conf.* 288 (1970).
43. W. H. Ray and J. Szekely, "Process Optimization," p. 359. Wiley, New York, 1973.
44. A. E. Sherwood, Lawrence Livermore Lab. Rep. UCID-16155 to the U.S.A.E.C., 1972.
45. B. M. Johnson, G. F. Froment, and C. C. Watson, *Chem. Eng. Sci.* **17**, 835 (1962).
46. A. W. D. Hills, *Ind. Chem. Eng. Symp. Ser.* **27**, (1968).
47. D. Kunii and O. Levenspiel, "Fluidization Engineering." Wiley, New York, 1969.
48. J. F. Davidson and D. Harrison (eds.), "Fluidization." Academic Press, New York, 1971.
49. C. Y. Wen and Y. H. Yu, *AIChE J.* **12**, 610 (1966).
50. D. Harrison, J. F. Davidson, and J. W. DeKock, *Trans. Inst. Chem. Eng.* **39**, 202 (1961).
51. K. Yoshida and D. Kunii, *J. Chem. Eng. Japan* **1**, 11 (1968).
52. J. C. Chu, J. Kalil, and W. A. Wetteroth, *Chem. Eng. Progr.* **49**, 141 (1953).
53. C. T. Hsu and M. C. Molstadt, *Ind. Eng. Chem.* **47**, 1550 (1955).
54. J. F. Richardson and J. Ayers, *Trans. Inst. Chem. Eng.* **37**, 314 (1959).
55. J. F. Richardson and J. Szekely, *Trans. Inst. Chem. Eng.* **39**, 212 (1961).
56. H. S. Mickley, D. F. Fairbanks, and R. D. Hawthorn, *Chem. Eng. Progr. Symp. Ser.* **51**, 57 (1961).
57. J. S. M. Botterill and J. R. Williams, *Trans. Inst. Chem. Eng.* **41**, 217 (1963).
58. J. S. M. Botterill, K. A. Redish, D. K. Ross, and J. R. Williams, *Proc. Symp. Interaction between Fluids, Particles*, p. 183. Inst. Chem. Eng. 1963.
59. M. Leva and C. Y. Wen, *AIChE J.*, **2**, 482 (1956).
60. J. Szekely and R. J. Fisher, *Chem. Eng. Sci.* **24**, 833 (1969).
61. M. Abromavitz and I. A. Stegun, "Handbook of Mathematical Functions." Nat. Bur. Stds., Washington, D.C., 1966.
62. M. M. Avedesian and J. F. Davidson, *Trans. Inst. Chem. Eng.* **51**, 121 (1973).
63. J. R. Grace, *AIChE. Symp. Ser.* **67**, 159 (1971).
64. O. L. Pyle, *Advan. Chem. Ser. No. 109, Chem. Reaction Eng.*, Amer. Chem. Soc. p. 106, 1972.
65. J. H. Perry (ed.) "Chemical Engineers Handbook," 3rd ed., Sect. 23. McGraw-Hill, New York, 1950.
66. R. F. Taylor and T. R. Jenkins, *Ind. Chem. Eng. Symp. Ser. No. 27*, p. 101, Inst. Chem. Eng., 1968.
67. K. W. Pearce, A heat transfer model for rotary kilns, *Symp. Flames Ind.*, *4th* jointly by the British Flame Res. Comm. and the Inst. of Fuel, Imperial College, London (1972).
68. A. Sass, *Ind. Eng. Chem., Process Design Develop.* **6**, No. 4, 532 (1967).
69. J. H. Strassburger (ed.), "Blast Furnace Theory and Practices," Vols. I and II. Gordon Breach, New York, 1969.
70. H. E. McGannon (ed.), The Making, Shaping and Treating of Steel, 9th ed. U.S. Steel Corp. Pittsburgh, Pennsylvania, (1971).

71. V. Stanek, and J. Szekely, *Can. J. Chem. Eng.* **50**, 9 (1972).
72. V. Stanek and J. Szekely, *Can. J. Chem. Eng.* **51**, 21 (1973).
73. C. J. Stairmand, *Trans. Inst. Chem. Eng.* **29**, 356 (1951).
74. N. Cherrett, *J. Inst. Fuel* **35**, 245 (1962).
75. H. Seidl, *A.S.M.E. and I.M.E. Joint Conf. Combust. U.S.A. Boston* (June 1955).
76. Y. Nogiwa, U.S. Patent 2, 973, 260 (Feb. 8th, 1961).
77. V. V. Vyshenskii, *Tr. Inst. Energ. Alma-Ata* **2**, 294 (1960).
78. P. Klucovsky, J. Haspra and J. Kukji, *Int. Chem. Eng.* **2**, 279 (1962).
79. W. Solbach, *Tonind-Z. Keram. Rundsch.* **82**, 474 (1958).
80. J. Szekely and R. J. Carr, *Chem. Eng. Sci.* **21**, 119 (1966).
81. W. H. Gauvin and J. J. O. Gravel, *Proc. Int. Symp. Interaction between Fluids Particles*, p. 250. Inst. Chem. Eng., London, (1962).
82. J. F. Elliott, *Trans. Met Soc. A.I.M.E.* **227**, 802 (1963).
83. M. Ishida and C. Y. Wen, *Ind. Eng. Chem. Process Design Develop.* **10**, No. 2, 164 (1971).
84. J. Yagi and I. Muchi, *Trans. Iron Steel Inst. Jpn.* **10**, 5, 392 (1970).
85. A. S. Asai and I. Muchi, *Trans. Iron Steel Inst. Jpn.* **10**, 4, 250 (1970).
86. J. W. Evans and S. Song, *Met. Trans.* **4**, 1701 (1973).
87. H. G. McIlvried and F. E. Massoth, *Ind. Eng. Chem. Fundam.* **12**, 225 (1973).
88. J. M. Smith and Van Ness, "Chemical Engineering Thermodynamics." McGraw-Hill, New York, 1970.
89. K. G. Denbigh, "The Principles of Chemical Equilibrium," 3rd ed. Cambridge Univ. Press, London and New York, 1971.
90. R. E. Johnstone and M. W. Thring, "Pilot Plants, Models and Scale-up Methods in Chemical Engineering." McGraw-Hill, New York, 1957.
91. Pilot Plants in the Iron and Steel Industry, Special Rep. No. 96 of the (British) Iron and Steel Inst., London, 1966.
92. J. N. Stone and Tyson Rigg, *Ind. Chem. Eng. Symp. Ser. No. 27*, p. 208. Inst. of Chem. Eng., London, 1968.
93. V. Stanek and J. Szekely, *AIChE J.* **20**, 974 (1974).
94. J. Szekely and J. J. Poveromo, *AIChE J.* **21**, 769 (1975).

Chapter 8

Gas–Solid Reactions of Industrial Importance

8.1 Introduction

In the preceding chapters we presented a general discussion of gas–solid reaction systems. Specific examples of particular systems were introduced into the discussion only to illustrate certain types of behavior. This was consistent with the objectives of the book to present a general treatment of gas–solid reactions including their experimental study and the application of single-particle data to reactor design.

While an exhaustive survey of measurements on gas–solid reactions and of the technologies currently in use for carrying them out would be well beyond the scope of this monograph, in this chapter we shall present a brief review of some key systems of industrial importance. In these brief reviews emphasis will be placed on providing a perspective, that is, the economic significance, together with some comments on current technology and a listing of important references. We hope that by reading these summaries and by following the references listed, the reader will obtain a good introduction to these particular fields.

It would have been very appealing to state that the experimental measurements reviewed on the following pages may be readily interpreted in terms of the models and theories that were discussed in the earlier chapter and that the technologies used for effecting these gas–solid reactions evolved as a result of extensive theoretical work. Unfortunately, this is not the case. The use of the equations developed in Chapters 3–5 for the interpretation of kinetic measurements is possible only under certain circumstances, e.g., when the overall rate is clearly limited by external mass transfer, by pore diffusion, or perhaps by chemical kinetics. The interpretation of intermediate cases where more than one of these steps plays an important role tends to be more

338

problematic because usually not all the necessary information is available. Moreover, at present the design of equipment for gas–solid reactions is, in general, done on an empirical basis with relatively little input from sophisticated kinetics.

Ultimately, the principal usefulness of the material contained in Chapters 1–7 should be in its application to real, practical problems of the type to be discussed in the following sections. It is hoped that by listing those reaction systems of practical interest we shall draw the reader's attention to very interesting areas where a great deal of work could be usefully done. Regarding the organization of the chapter, Section 8.2 is devoted to iron oxide reduction, Section 8.3 is concerned with the roasting of sulfides, whereas the reaction of SO_2 with various solid sorbents is discussed in Section 8.4. A brief review of coal gasification is given in 8.5, and finally the incineration of urban solid wastes is discussed in Section 8.6.

8.2 Iron Oxide Reduction

PERSPECTIVES

The reduction of iron oxides to yield metallic iron is perhaps the technologically most significant gas–solid reaction system. At present the overwhelming portion of the iron oxides processed is being reduced in the iron blast furnace. Worldwide hot metal production in the blast furnace was 551,000,000 tons in 1971 [1]. The corresponding figure for the United States was 95,000,000 tons [1].

Physically, the blast furnace is a packed bed into which a mixture of coke and specially prepared ore (sinter or pellets) is charged at the top and a hot air blast is blown through tuyeres located at the bottom. The descending iron oxide is made to react with the reducing gases (principally CO) produced by the reaction between the hot blast and the coke, ultimately yielding blast furnace hot metal that is tapped from the lower end of the furnace.

A typical blast furnace installation is shown in Fig. 8.1 and the evolution of blast furnace sizes in Fig. 8.2 [3]. As seen, recent developments have tended to produce larger furnaces, especially those with larger hearth diameters. Modern blast furnaces have working volumes ranging from 60,000–150,000 ft^3 (1080–4500 m^3) and may produce up to 10,000 tons of hot metal per day.

In the more recently developed direct reduction processes iron oxide is made to react with a reducing gas in a packed-bed arrangement, in fluidized beds, or in kilns; here the solid reaction product is *sponge iron*, which then has to be melted in an electric furnace.

A good description of direct reduction processes may be found in Wild [4], Miller [5], Astiér [6], and *Proc. South Africa Int. Ferroalloys. Congr.* [7].

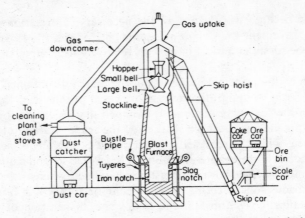

FIG. 8.1. Diagram of a typical iron blast furnace and associated plant.

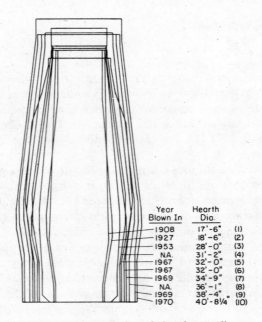

Year Blown In	Hearth Dia.	
1908	17'-6"	(1)
1927	18'-6"	(2)
1953	28'-0"	(3)
N.A.	31'-2"	(4)
1967	32'-0"	(5)
1967	32'-0"	(6)
1969	34'-9"	(7)
N.A.	36'-1"	(8)
1969	38'-4"	(9)
1970	40'-8¼"	(10)

FIG. 8.2. The evolution of blast furnace lines.

(1) U.S. South Works No. 1.
(2) U.S. South Works No. 4.
(3) U.S. Fairless Works No. 1.
(4) Hoesch A. G. Works Phoenix Horde/WG.
(5) Youngstown Sheet and Tube Co.,
    Indiana Harbor No. 4.

(6) Steel Company of Canada 'E'.
(7) Italsider Taranto No. 1.
(8) Russian Krivoi Rog.
(9) Fujii Iron and Steel Co.,
    Nagoya No. 3, Japan.
(10) Mizushima No. 3, Japan.

As discussed by Miller [5], in recent years direct reduction processes have gained wider acceptance, especially in areas where there is a cheap supply of natural gas; however, at present, direct reduction processes account for only a small fraction (less than 5%) of the total iron produced.

EQUILIBRIUM CONSIDERATIONS

The reduction of iron oxides involves one or more of the following steps:

$$
\begin{aligned}
\text{hematite} &\rightarrow \text{magnetite} & Fe_2O_3 &\rightarrow Fe_3O_4 \\
\text{magnetite} &\rightarrow \text{iron} & Fe_3O_4 &\rightarrow Fe \\
\text{magnetite} &\rightarrow \text{wüstite} & Fe_3O_4 &\rightarrow FeO \\
\text{wüstite} &\rightarrow \text{iron} & FeO &\rightarrow Fe
\end{aligned}
$$

(Wüstite is stable only above 570°C)

The thermodynamic equilibria for these reactions are well established for the two major reducing agents used, viz., hydrogen and carbon monoxide [2]. Reduction by carbon monoxide or by hydrogen is reversible. For example, at 1000°C and 1 atm a 27% CO_2, 73% CO mixture or a 40% H_2O, 60% H_2 mixture is in equilibrium with iron and wüstite. Figure 8.3 shows an equilibrium diagram for the reduction of iron oxides with carbon monoxide at various temperatures. A similar diagram for hydrogen is available in Strassburger [2].

In reduction by carbon monoxide the reactions listed above are nearly isoenthalpic (having heats of reaction less than 5 kcal/g-atom oxygen) except for the reduction of hematite to magnetite, which is exothermic with a heat of reaction of approximately 12 kcal/g-atom oxygen at 1000°C.

Reduction by hydrogen is endothermic, at temperatures normally encountered in laboratory experiments and industrial plants, the heat of reaction

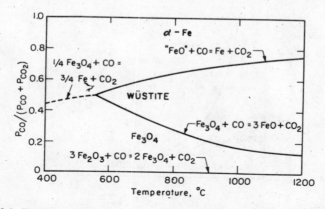

FIG. 8.3. Equilibrium diagram for iron oxides in the presence of carbon monoxide-carbon dioxide gas mixtures.

being approximately 11 kcal/g-atom oxygen for the reduction of magnetite to wüstite. The exception is the reduction of hematite to magnetite, which is slightly exothermic (roughly 2 kcal/g-atom oxygen).

Many early investigators made categorical statements concerning the rate-controlling mechanism in iron ore reduction on the basis of plots of the extent of reaction, measured by laboratory reduction of spherical pellets, against time. We stressed in Chapters 3 and 4 that the difference between such plots is small for the limiting cases of ash layer diffusion control and control by chemical kinetics. It is therefore difficult to ascertain the rate-controlling mechanism merely on the basis of such plots. The difficulty is compounded if the pellets used in the reduction studies are initially porous. The results of some of the earlier investigators must therefore be accepted with caution. A convincing demonstration of a rate-controlling mechanism would employ evidence other than the shape of the reaction extent versus time curve. Such other evidence might be the effect of pellet size on time to reach a given extent of reaction (proportional to diameter squared for ash layer diffusion control, to diameter for chemical control of reaction of dense pellets, and independent of diameter for chemical control of porous pellets). Alternatively, the effect of temperature might be studied; a low apparent activation energy would then be evidence of ash-layer diffusion control.

Figure 8.4 is a plot of the data of von Bogdandy and Janke [9] on the reduction of Malmberget pellets using mixtures of hydrogen and water vapor. One might conclude from these data that ash layer diffusion provides the major resistance to the progress of reaction, for pellet diameters greater than

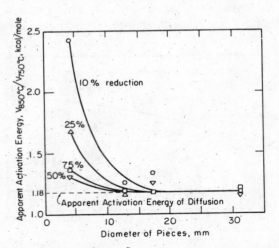

FIG. 8.4. The apparent activation energy as a function of particle size and extent of reaction for Malmberg pellets [9].

15 mm under the conditions of the experiment, except for about the first 10% of reduction.

However, a high activation energy cannot be unambiguously interpreted as resulting from control by the chemical step(s) in the reaction sequence. The reduction of wüstite or magnetite to iron frequently results in a dense iron layer [10]. The transport of matter across such a dense iron layer is by solid state diffusion, a process with a high activation energy.

Figures 8.5 and 8.6 from the work of Kohl and Engell [11] illustrate this point. Figure 8.5 is a plot of the reduction of wüstite at 800°C in hydrogen

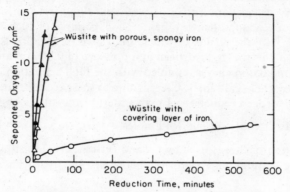

FIG. 8.5. Progress of reduction with the formation of porous spongy iron and with the formation of a covering layer of iron [11].

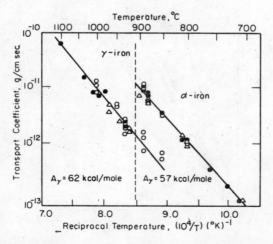

FIG. 8.6. Dependence of the product $C_0 D_0$ on the temperature for the reduction of wüsutite by hydrogen. ●, reduction by H_2/H_2O; ○, reduction by CO/CO_2; △, Kohl and Engell.

containing 2.7% water vapor at a total pressure of 680 Torr. The sharp decrease in the reaction rate brought about by the formation of a dense iron layer is obvious. Figure 8.6 is an Arrhenius plot of the "transport coefficient" $C_0^* D_0$, calculated from rate measurements in experiments where dense iron layers formed. Here C_0^* is the concentration of oxygen in iron which is in equilibrium with wüstite and D_0 is the diffusivity of oxygen in iron. The high activation energy for the diffusion process is readily apparent.

If the ore samples used are of a geometry such that the interfacial area between oxide and metal does not change during the course of reaction (e.g., platelike pellets of small thickness compared with edge length) then valid conclusions may be drawn concerning the rate-controlling mechanism from plots of the extent of reaction against time. A good example of such work is that of Quets *et al.* [12], the results of which are shown in Fig. 8.7. Reduction of the magnetite plates was by hydrogen at a pressure of 0.86 atm and the linear nature of the reduction curves, coupled with the high activation energy, suggests control by chemical kinetics. According to these data the first-order rate constant for hydrogen reduction of magnetite to iron was given by

$$k \simeq 2.0 \times 10^{-2} \exp[-14,600/RT] \text{ moles oxygen/cm}^2 \text{ sec atm hydrogen}$$

A similar approach was used by Lu and Bitsianes [13] to study the reduction of dense natural and synthetic hematites by hydrogen and carbon

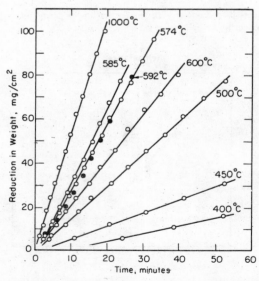

FIG. 8.7. The linear relationship between extent of reaction and time for the reduction of magnetite plates by hydrogen [12].

monoxide. The specimens used had the form of cylinders or rectangular parallelopipeds and an impervious enamel coating was applied to some of the external surfaces (e.g., the curved surface in the case of cylinders). In this way reaction was limited to a receding planar interface of constant area. Diffusion through the ash layer presented a significant resistance in these experiments as evidenced by the curvature of plots of reaction extent against time.

Most investigators have assumed that the multiple reactions taking place during iron ore reduction occur within a narrow region of the pellet. This is probably justified in the case of dense pellets exemplified by the work of McKewan [14], whose results are given in Fig. 8.8. However, Spitzer et al. [15] presented a variant of the shrinking-core model in which allowance was made for the existence of three reaction interfaces within a hematite pellet undergoing reduction to iron.

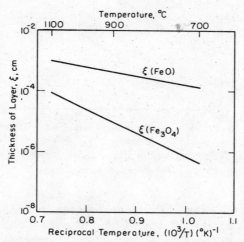

FIG. 8.8. Calculated thickness of layers of FeO and Fe_3O_4 in the reduction of hematite [14].

Two interesting papers relevant to the topic of iron ore reduction are those of Olsson and McKewan [16]. These investigators measured the diffusivity of H_2–H_2O through porous iron formed by the reduction of iron oxides and came to the conclusion that ash layer diffusion had had considerable influence on the progress of reaction in previous reduction studies carried out by McKewan.†

Seth and Ross [17] studied the reduction of ferric oxide composites by hydrogen and used a shrinking-core model to interpret their results. Figure

† Private communication in von Bogdandy and Engell [8].

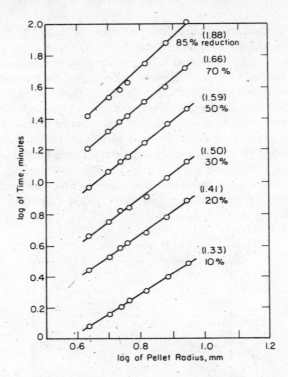

FIG. 8.9. A plot of time required to attain a given degree of reduction against the pellet radius, after Seth and Ross [17]; slopes in parentheses.

8.9 is a plot on a logarithmic scale of the time required to reach a given percentage reduction against pellet radius. Let us recall that the slopes of such plots should be unity for chemical control (for the reduction of dense pellets) and have the value of two for ash layer diffusion control. Clearly, for the conditions of these experiments both chemical kinetics and ash layer diffusion played a role in determining the progress of reaction.

Most of the investigations of iron ore reduction described above have been on natural or synthetic iron oxides of low porosity. Studies of the reduction of porous oxides are becoming increasingly popular as porous pellets and sinter continue to supplant relatively nonporous natural lump ores as blast furnace feed material. As long ago as 1936 Joseph reported [18] that the porosity of an iron ore had some influence on the rate at which it could be reduced to iron. El-Mehairy [19] reported a similar effect. McAdam et al. [20] studied the effect of various parameters, including solid structure, on the rate of reduction of pellets made from ironsand concentrate.

An extensive series of experiments has been carried out by Turkdogan and coworkers [21–23] on the reduction of porous iron ores. In this investigation, which is an excellent example of the contemporary approach to gas–solid reactions, attention was focused on the reduction of natural hematite of roughly 30% porosity by hydrogen and by carbon monoxide. A wide range of temperatures and particle sizes was studied, enabling these workers to explore the diffusion controlled and chemically controlled regimes of hydrogen reduction; "unidirectional" experiments akin to those of Lu and Bitsianes (described above) were also carried out. Widespread use was made of techniques for characterizing the porous reactants and products including metallography, electron microscopy, and mercury porosimetry, and for measuring pore volume, pore surface area, and pore diffusivity. These investigators observed the rate minimum in reduction of hematite by hydrogen in the range 600–700°C. The rate minimum has been noted by other workers [24]. Turkdogan and coworkers observed that the pores became larger as the reduction temperature was increased. Consequently the effective diffusivity of the gaseous products and reactants within the iron layer was found to increase markedly with increasing reduction temperature. Turkdogan used a model akin to that of Ishida and Wen [25] for the interpretation of his experimental results. This model has been discussed in some detail in a previous chapter.

Other papers on iron ore reduction include those of Warner [26], Edmiston and Grace [27], Landler and Komarek [28], Walker and Carpenter [29], and Themelis and Gauvin [30]. A comprehensive text on this subject, covering work published before the mid 1960s, is that of Bogdandy and Engell [8].

8.3   The Roasting of Sulfides

PERSPECTIVES

The roasting of metal sulfides is of considerable technological and commercial interest because the ores of the majority of heavy metals such as copper, nickel, lead, and zinc are in a sulfide form. One important step in the recovery of these metals from their ores involves roasting, that is, their oxidation to metal oxide.

The tonnages involved are illustrated in Tables 8.1 and 8.2, which indicate the annual production of copper, lead, zinc, and pyrites in the world and in the United States, respectively. The actual course of the roasting operations is illustrated by the following example using lead sulfide:

$$PbS + 3O_2 = 2PbO + 2SO_2 \tag{1}$$

$$2PbS + 4O_2 = 2PbSO_4 \tag{2}$$

$$2PbO + 2SO_2 + O_2 = 2PbSO_4 \tag{3}$$

TABLE 8.1

World Production of Major Mineral Commodities[a]

Pyrite	1969	1970	1971
Copper.			
Smelter/thousand metric tons	5817	6096	6117
Lead			
Smelter/thousand metric tons	3218	3310	3213
Zinc			
Smelter/thousand metric tons	4974	4891	4740
Pyrites			
Smelter/thousand metric tons	21,035	22,391	21,541

[a] From "Metal Statistics 1973," Amer. Metal Market Publ., Fairchild Publ., New York (1973).

TABLE 8.2

Mineral Production in the U.S.[a]

	1968	1970	1971
Pyrites, thousand long tons	872	—	808
Metals			
Copper (recoverable content of ores, etc.), short tons	1,204,621	1,719,675	1,984,484
Lead (recoverable content of ores, etc.), short tons	359,156	571,767	578,550
Zinc (recoverable content of ores, etc.), short tons	529,446	534,136	502,543

[a] Bureau of Mines [30].

In general, the formation of the oxide is favored at high temperatures, whereas the formation of the sulfate is favored—at least thermodynamically—at lower temperatures. The complex phase equilibria that have to be considered in roasting operations have been studied in Ingraham [32] and by Kellogg and Basu [33]; in fact, diagrams of the type shown in Fig. 8.10 are frequently called Kellogg diagrams.

Table 8.3 shows the heats of reactions for various roasting systems [34]. It is seen that these reactions are highly exothermic and the heats of reaction are of the same order of magnitude as that for the combustion of carbon. It can be readily calculated that if zinc sulfide were roasted with air the result-ant adiabatic temperature would be of the order of 1700°C, which would

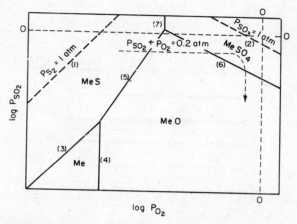

FIG. 8.10. Schematic predominance diagram for the system Me–O–S; $t = $ constant.

TABLE 8.3

Heats of Reaction for Various Roasting Operations [34]

$FeS_2 + 2\frac{3}{4}O_2 = \frac{1}{2}Fe_2O_3 + 2SO_2$;	$\Delta H_{298} = -205$ kcal
$ZnS + \frac{3}{2}O_2 = ZnO + SO_2$;	$\Delta H_{290} = -110$ kcal
$Cu_2S + \frac{3}{2}O_2 = Cu_2O + SO_2$;	$\Delta H_{298} = -95$ kcal

provide a molten product. This has to be avoided for practical reasons, and because of this all the processes developed for roasting operations are designed to allow the ready dissipation of the heat of reaction and to avoid local overheating. One example of an early roasting system is the Hereschoff furnace, a diagram of which is shown in Fig. 8.11. As seen in the figure, the ore enters at the top and drops from hearth to hearth while the sulfide ore particles are reacted through their contact with the rising gases. The vessel is lined with fire brick and is equipped with rotating rabble arms attached to a central shaft. These rabble arms provide a measure of agitation and prevent excessive local overheating. Further developments in roasting technology include the use of a suspension roaster (Fig. 8.12), which consists of a Hereschoff furnace with the hearths removed. Other roasting arrangements include the use of a sinterbed as shown in Fig. 8.13; the sinterbed operation has been discussed earlier in Chapter 7.

Perhaps the most commonly used arrangement for the roasting of sulfide ores is the fluidized-bed system. As discussed in Chapter 7, fluidized beds are ideal arrangements for performing reactions where large volumes of solids

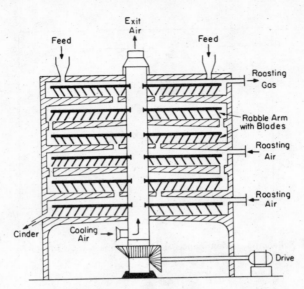

FIG. 8.11. A multiple-hearth furnace for roasting.

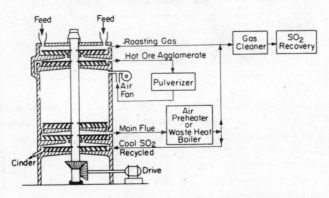

FIG. 8.12. A multiple-hearth roaster and associated plant.

have to be handled and where uniformity of temperature and rapid dissipation of heat are required. A typical fluidized bed roaster is shown in Fig. 8.14. These roasters may be up to 6–10 m (20–30 ft) in diameter and up to 10–13 m (30–40 ft) high.

KINETICS AND THERMODYNAMICS OF ROASTING REACTIONS

As discussed in the preceding section dealing with iron oxide reduction, while a great deal of work has been done on the kinetics of the roasting of sulfides, relatively little of this information is being used in the design of

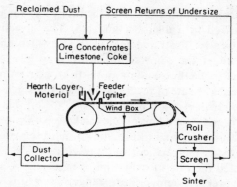

FIG. 8.13. A sintering machine and associated equipment

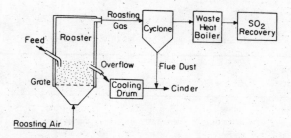

FIG. 8.14. A fluidized-bed roaster and auxiliary plant.

equipment. Moreover, our understanding of the individual rate-controlling steps in roasting systems is rather less complete than that found for iron oxide reduction. In the following, we shall present a very brief survey of the information that is available on the thermodynamics and kinetics of the roasting of zinc sulfide, copper sulfide, and lead sulfide. As in all the sections in this chapter, our principal objective is to provide the reader with a good compilation of the important literature on this subject. For a more complete, recent review of the subject the reader is referred to an article by Gray *et al.* [35].

THE ROASTING OF ZINC SULFIDE

The thermodynamics of zinc sulfide roasting have been studied by numerous investigators. The classical work on this subject is the paper by Ingraham and Kellogg [36]. Since iron tends to be present in the vast majority of naturally occurring zinc sulfide ores, the zinc–iron–oxygen–sulfur system is of considerable practical interest. This problem has been discussed extensively in Gray *et al.* [35] and some useful papers on this subject have been written by Kubaschewski [37] and Benner and Kenworthy [38]. Numerous kinetic studies have been made of oxidation of zinc sulfide; the results of some of the more significant investigations and the scope and range of the variables studied is summarized in Tables 8.4 and 8.5. The diversity of the kinetic measurements is clearly illustrated in Fig. 8.15, which shows that kinetic measurements taken at the same temperature may give overall rates that could differ by a factor of a thousand or more. In an interesting study, Ong and coworkers measured the rate of reaction of sphalerite single crystals in oxygen at pressures ranging from 6–640 Torr at 700–870°C [39]. The rate of oxidation was found to be linearly dependent on the thickness of the oxide layer formed. At higher temperatures thermal instabilities were observed.

In what is perhaps the classical study of zinc sulfide oxidation, Cannon and Denbigh studied the rate of oxidation of naturally occuring, single crystals of sphalerite over the temperature range 680–940°C and with oxygen partial pressures between 0.014 and 0.5 atm [40]. The oxidation was aniso-

TABLE 8.4

Results of Kinetic Studies of Zinc Sulfides

Ref.	Crystal-linity[a]	Shape	Size (mm)	Porosity (%)	Iron (wt %)
38	s	plate	$13 \times 3 \times 1.5$	negligible	—
39	s	irregular	few mms	negligible	0.1–0.3
40	p	sphere	9.5–16.0	variable	—
41	p	sphere	4.0–16.0	2–35	12.9
42	p	plate	$20 \times 1.5 \times 2.3$	3	0.1–0.35

[a] s denotes single crystals, p prepared polycrystalline particles.

TABLE 8.5

System Conditions, Activation Energies, and Transition Temperatures
for Various Zinc Sulfides

Ref.	Temp. range (°C)	Gas composition	E (kcal/g-mole) I	E (kcal/g-mole) II	Transition temp. (°C)
38	700–870	6–640 Torr gO_2 in N_2	60.3	—	845
39	680–940	1.4–50.0% O_2 in N_2	49.9	—	830
40	500–1440	Air	41.0^a	3.0	675
			20^b	—	—
41	740–1020	20–100% O_2 in N_2	—	7	—
42	640–830	Air	62.5	3.5	750

[a] Refers to unsintered pellets.   [b] Refers to sintered pellets.

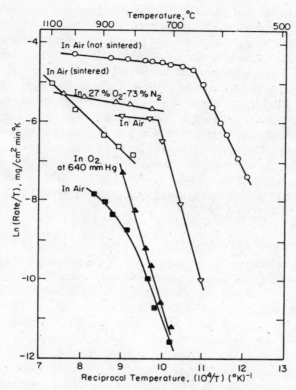

FIG. 8.15. The rate of zinc sulfide oxidation according to numerous investigators. Polycrystalline spheres and plates: ○, Denbigh and Beveridge; △, Natesan and Philbrook; ▽, Gerlach and Stichel. Single crystals: ▲, Ong *et al.*; ■, Cannon and Denbigh.

tropic and all the comparative measurements were made on the 110 fcc plane. These authors found that below 830°C the reaction rate was chemically controlled. Above 830°C the overall rate was found to be markedly affected by the diffusion of oxygen or sulfur dioxide through the reacted product layer. Another study of this system was made by Denbigh and Beveridge [41], who worked with pellets rather than single crystals. In a more recent investigation, Natesan and Philbrook [42] measured the rate of oxidation of zinc sulfide pellets over the temperature range 740–1020°C using a thermogravimetric technique. Their results were interpreted on the assumption that the rate of oxidation was controlled by the transport of gases through the zinc oxide product layer. This conclusion was supported by the low value of the apparent activation energy obtained. Gerlach and Stichel [43] studied the oxidation of sintered briquettes of zinc sulfide over the temperature range 640–

830°C. They found that below $\overline{7}$50°C the reaction was chemically controlled but above this value the overall rate was controlled by product layer diffusion. An interesting recent investigation was reported by Mendoza *et al.* [44], who investigated this reaction within the range 1030–1334°K and interpreted their results in terms of a sophisticated distributed model.

With regard to the application of these kinetic measurements for engineering purposes, that is, the use of single-particle data for the representation of multiparticle systems, the work of Natesan and Philbrook [45] is noteworthy.

In the electrolytic processing of zinc it is desirable to convert the sulfide to a sulfate rather than to an oxide. The problem of zinc sulfate formation and the kinetic and thermodynamic criteria that would favor this process have also been extensively studied [46–51]. The study of zinc sulfide roasting in sinterbeds and in fluidized beds was investigated by Woods and Harris [51], Beveridge [52], and Kellogg [53].

THE ROASTING OF COPPER SULFIDE

The phase relationships of importance in the roasting of copper sulfide have been studied by numerous investigators including Roseboom [54], Ingraham [55], Espelund [56], Floyd and Willis [57], and Blanks [58]. While numerous investigations have been carried out, the actual overall rate-controlling step has not yet been conclusively determined [59–66]. Shah and Khalafalla [62, 63] suggested that the reaction is chemically controlled at temperatures between 260 and 400°C. In contrast, McCabe and Morgan [59] considered that either gaseous or solid state diffusion through the product layer is the rate-limiting factor. It would appear that a great deal of further work would be warranted in this field.

The reader may find it somewhat disturbing that numerous, large-scale industrial processes are operating on systems for which the kinetic information is so very limited, as in the case of the roasting of copper sulfides. As a practical matter, however, the rate at which roasters may be run is controlled by heat balance considerations, i.e., the rate at which heat may be carried away from the system, rather than by the kinetics of the reaction. This may explain why the actual kinetic problems for copper sulfide roasting have not yet been resolved.

ROASTING OF LEAD SULFIDE

The equilibrium diagram for the roasting of lead sulfide was established by Kellogg and Basu [33]. Subsequent work has been done by Esdaile [67] and Rosen and Witting [68] The kinetic information available on the roasting of lead sulfide is rather fragmented and at the present no definite conclusions may be drawn regarding the rate-limiting factor [69–72].

8.4   The Absorption of SO$_2$ by Solids

The abatement of sulfur dioxide emissions is a major problem in most industrialized nations that use fossil fuel for power generation. It has been shown that over 35 million tons of SO$_2$ were emitted into the atmosphere in 1970 in the United States [73]. Electric power generation using fossil fuel was the single largest contributor (55%), while copper and lead smelters processing sulfide ores accounted for about 10% of the total emission.

The pressure from regulatory bodies such as the Environmental Protection Agency in the U.S. has stimulated a great deal of research and development effort aimed at reducing SO$_2$ emissions. The problem of handling the emissions from the stacks of power stations is particularly difficult because of the very large gas volumes involved and the relatively low absolute levels of SO$_2$ content. A typical 1000 MW boiler would produce some 2,100,000 ft^3 (or about 60,000 m^3) of effluent gases, containing a few tenths of one percent of SO$_2$. Smelter emissions are typically at higher concentration levels (a few percent of SO$_2$ in the waste gas). There is scope for increasing the concentration by various plant modifications (e.g., using oxygen rather than air in smelting) and many metallurgical facilities have for a number of years been converting waste SO$_2$ to sulfuric acid or liquefied SO$_2$. For this reason, in this section emphasis will be placed on the removal of sulfur dioxide from power plant stack gases.

Due to recent research efforts numerous methods have been proposed for reducing the SO$_2$ content of stack gases. Good reviews of the available and proposed technologies may be found in Slack [73] Mills and Perry [74], Raben [75], and Burchard et al. [76]. A list of operating and planned full-size lime/limestone desulfurization systems in the United States is given in Table 8.6.

The techniques actually available for SO$_2$ removal from flue gases may be broadly classified into " wet " and " dry " systems. The wet methods employ a slurry or solution of some absorbent for SO$_2$, whereas the dry systems use dry particles of absorbent in such gas–solid contacting devices as fluidized beds, packed beds, and entrainment reactors. It is with these latter methods that we are concerned here, although wet methods appear to be at least as satisfactory as the dry systems.

A further classification can be made on the basis of whether the absorbent is regenerated or "thrown away." The former methods employ a separate step(s) to convert the spent absorbent back into a usable form, at the same time generating a salable sulfur species (sulfuric acid, elemental sulfur, etc.) or one that can readily be converted into a useful product (H$_2$S, a gas stream containing high levels of SO$_2$, etc.). In contrast the throw away processes aim to produce a sulfur-containing compound (e.g., calcium sulfate) that can be

TABLE 8.6

Operating and Planned Full-Size Lime/Limestone Desulfurization Facilities in the U.S.

Utility Company, plant	Absorbent	Megawatts	New or retro.	Scheduled startup	Fuel	Supplier
1. Union Electric Co. (St. Louis), Meramec No. 2	CaO	140	R	September 1968	3.0% S coal	C-E
2. Kansas Power & Light, Lawrence Station No. 4	CaO	125	R	December 1968	3.5% S coal	C-E
3. Kansas Power & Light, Lawrence Station No. 5	CaO	430	N	December 1971	3.5% S coal	C-E
4. Commonwealth Edison, Will County Station No. 1	CaOC$_3$	175	R	February 1972	3.5-% S coal	Babcock & Wilcox
5. City of Key West, Stock Island	CaCO$_3$	37	N	June 1972	2.75% S fuel oil	Zurn
6. Kansas City Power & Light, Hawthorne Station No. 3	CaO.	130	R	Mid-late 1972	3.5% S coal	C-E
7. Kansas City Power & Light, Hawthorne Station No. 4	CaO	140	R	Mid-late 1972	3.5% S coal	C-E
8. Louisville Gas & Electric Co., Paddy's Run Station No. 6	Ca(OH)$_2$	70	R.	Mid-late 1972	3.0% S coal	C-E
9. Kansas City Power & Light, La Cygne Station	CaCO$_3$	840	N	Late 1972	2.3% S coal	Babcock & Wilcox
10. Arizona Public Service Co., Cholla Station	CaCO$_3$	115	R	Early 1973	0.4-1% S coal	Research Cottrell

	Sorbent	Capacity (MW)	N/R	Start-up	Coal	Contractor
11. Duquesne Light Co. (Pittsburgh), Phillips Station	CaO	100	R	May 1973	2.3% S coal	Chemico
12. Detroit Edison Co., St. Clair Station No. 6	CaCO₃	180	R	December 1973	2.5–4.5% S coal	Peabody
13. Ohio Edison, Mansfield Station	CaO	1800 (2 boilers)	N	Late 1974–Early 1975	3% coal	Chemico
14. Tennessee Valley Authority, Widow's Creek Station No. 8	CaCO₃	550	R	April 1975	3.7% S coal	TVA
15. The Montana Power Co., Colstrip Units 1 & 2	Fly Ash +CaO	720 (2 boilers)	N	May 1975	0.8% S coal	CEA
16. Northern States Power Co., (Minnesota), Sherborne County Station, Nos. 1 & 2	CaCO₃	Total 1360	N	May 1976 (No. 1, 700 MW)	0.8% S coal	C-E
17. Northern Indiana Public Service Co., Kankakee 14	CaO or CaCO₃	500	N	June 1975	3.0% S coal	Not selected
18. Mohave-Navajo, Mohave Module (Vertical)[a]	CaCO₃/CaO	160	N	March 1974	0.5–0.8% S coal	Procon/UOP/Bechtel
19. Mohave-Navajo, Mohave Module (Horizontal)[a]	CaCO₃/CaO	160	N	December 1973	0.5–0.8% S coal	South California Edison/Stearns Roger
20. Salt River Project, Navajo Station	CaCO₃	2250 (3 boilers)	N	July 1977	0.5–0.8% S coal	Procon/UOP/Bechtel
21. Southern California Edison, Mohave Station (Horiz/Vert)	CaCO₃/CaO	1500 (2 boilers)	R	December 1976	0.5–0.8% S coal	Not selected

[a] Southern California Edison is program manager.

disposed of as a waste material without serious difficulty. The future market for sulfur and sulfuric acid is uncertain, because it may depend to a large extent on the number of SO_2 removal facilities of the regenerating type that are to be built. Perhaps because of this fact and due to the additional complexities involved in regeneration, the present interest in dry processes appears to center around throw away processes, in particular, the absorption of sulfur dioxide by lime, produced by injecting limestone into a power plant furnace, or the use of a fluidized bed of lime particles as a combustion device for coal or oil.

Limestone is attractive for this purpose because of its low cost (perhaps as low as \$2/ton [73]), widespread availability throughout the U.S., and the inertness of the absorption product (calcium sulfate). Limestone becomes less attractive, however, when it is recognized that the principal impurity is magnesium carbonate. This yields magnesia, which is also capable of absorbing sulfur dioxide in the furnace. However, the absorption product (magnesium sulfate) is soluble in water and water pollution caused by rainwater leaching of the absorption product waste dump may cause problems unless the less common and more expensive limestones containing low magnesium levels are used.

A further difficulty with the limestone injection method of curtailing sulfur dioxide emissions is the short residence time (1 or 2 sec) of the solid at the temperature at which the important reactions

$$CaCO_3 = CaO + CO_2$$
$$CaO + SO_2 + \tfrac{1}{2}O_2 = CaSO_4$$

take place at an appreciable rate. In pilot plant tests to date, sulfur dioxide absorbed has typically been less than 50% of that generated by combustion. As a consequence, limestone injection is likely to find its major use as an "add on" or retrofit device to existing power plant boilers, where the low capital cost (limestone handling, grinding, and injection facilities together with increased dust collecting capacity) and short down time for installation are appealing and the stop gap nature of the method is likely to be adequate for the economic life of the plant. This picture would, of course, change if the kinetics of the absorption reaction could be improved to the point where a greater proportion of the sulfur dioxide were absorbed.

Recently, a considerable amount of development work has been carried out on the fluidized bed combustion of coal and oil [114–117]. Such a bed would consist largely of lime particles fluidized by air or air enriched by oxygen. Coal or oil is injected into the bed. Steam is generated within heat exchanger coils located within the bed and is used to generate electricity in a conventional steam turbine. If the bed is operated at elevated pressures, it is feasible to generate electricity by means of a gas turbine located in the gas stream leaving the bed.

One of the attractions of such a fluidized-bed combustor is that sulfurous gases generated by the fuel combustion are trapped by the lime (which is continuously replenished by the addition of limestone). Since the residence time of the lime is long, in comparison with the 1 or 2 sec of the limestone injected into a conventional furnace, there is hope of absorbing most of the SO$_2$ without using excessive quantities of limestone.

As a final remark on the utility of limestone as an absorbent for sulfur dioxide, the reader is referred to the novel approach of Bartlett and Huang [77] to SO$_2$ absorption in the roasting of sulfide ores. In their suggested process, the sulfur dioxide is trapped at point of generation by lime within the roasting sulfide ore pellet.

The thermodynamics of SO$_2$ absorption by limestone has been described by Reid [78]. On the assumption that conversion of limestone into lime is necessary first step, a temperature somewhat greater than 1400°F (750°C) was calculated as being necessary to bring about a calcination of calcium carbonate in a furnace atmosphere of 14% CO$_2$ (a typical value for a power plant). Figure 8.16 shows the SO$_2$ concentration in a gas containing 2.7% oxygen in equilibrium with magnesia and with calcia at various temperatures. It is ob-

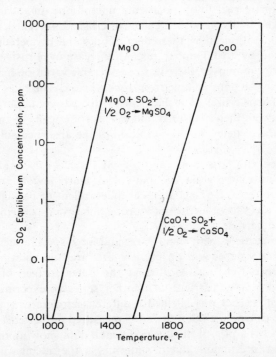

FIG. 8.16. Calculated equilibrium SO$_2$ concentration in flue gas in contact with CaO and MgO [78]. 2.7% O$_2$ in flue gas (unit activity of solids).

vious that at this level of oxygen calcia is an ineffective absorbent of SO_2 above about 1900°F (1000°C) and magnesia becomes ineffective above about 1400°F (750°C). Since gas temperatures in the furnace of a coal fired power plant boiler are typically above 2000°F (1040°C) [72], absorption of SO_2 takes place only after the absorbent has left the furnace and entered the cooler sections of the boiler. Therefore, absorption of SO_2 by the lime must take place in the short span of time that the absorbent spends between 1900 and 1400°F (1000 and 750°C) (where CO_2 absorption presumably provides a competing reaction).

The key role played by gas–solid reaction kinetics in SO_2 removed by limestone should be obvious from the above remarks; particularly important is the rate of the reaction

$$CaO + SO_2 + \tfrac{1}{2}O_2 = CaSO_4$$

A recent paper by Borgwardt and Harvey [79] on the kinetics of this reaction is interesting, not only as a report of an extensive experimental investigation, but also because the conclusions presented are in conformity with the ideas developed in this book. Borgwardt and Harvey characterized eleven diverse types of carbonate rock by a polarizing microscope and scanning electron microscope. The rocks were then calcined, crushed, screened, and subjected to further examination under these microscopes and by mercury penetration porosimeter and by a BET apparatus. An increase in porosity was observed after calcination; moreover, there was considerable variation in the mean pore diameter and in the pore volume from rock to rock. The rates of reaction of each of three particle sizes of each calcine with a gas containing 3000 ppm of SO_2 at 980°C were then measured and microprobe scans of the reacted calcines were made to determine the location of the absorbed sulfur within the particle.

For calcines with large pores and low BET surface areas (which we may interpret here as a large grain size and consequently low $\hat{\sigma}$), the rate of SO_2 absorption (g-mole SO_2/g CaO sec) was independent of particle size and the absorbed SO_2 was shown by the microprobe to be evenly distributed throughout the reacted particle.

For calcine particles with small pores and large specific surface areas, the rate was markedly dependent on particle size, and the microprobe revealed that the absorbed SO_2 was located at the outer surface of the particle. Borgwardt and Harvey carried out another series of experiments in which the same carbonate rock was calcined at different temperatures and the rate of reaction of the calcine with SO_2 at 760°C measured on particles of 0.0096-cm size. Figure 8.17 is a plot of these experimental data showing the decrease of the surface area (presumably due to sintering) as the calcination temperature is raised and the importance of this surface area to solid reactivity. Borgwardt

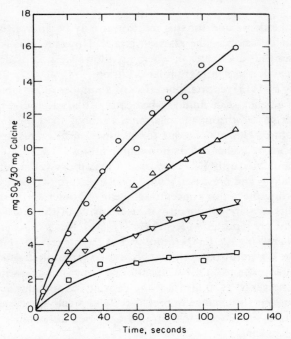

FIG. 8.17. Sorption of SO_2 by limestone as a function of surface area of its calcine [79]. Calcination temperature (°C) and $S_g (\times 10^4 \text{ cm}^2\text{g})$, respectively; ○, 790, 7.3; △, 890, 50; ▽, 980, 2.4; □, 1100, 0.95. Reprinted with permission from *Environm. Sci. Tech.* **6**, 350 (1972). Copyright by the American Chemical Society.

and Harvey suggested that the capacity of a lime (amount of SO_2 absorbed at very long contact times) is determined by closing of the pore mouths as the calcium sulfate (which has a larger specific volume than calcia) is formed within the pores. Thus, the smaller particles and those with lower BET surfaces showed higher capacities because considerable internal reaction had taken place for these systems before the pore mouths became sealed.

Pigford and Sliger [80] have interpreted earlier data by Borgwardt [81] and by Coutant *et al.* [82] on the absorption of SO_2 by lime, using a model similar to the grain model discussed in this book. The principal difference between the Pigford–Sliger model and the grain model is that in the former it was assumed that the interfacial reaction within the grains presents a negligible resistance to the progress of reaction and that the reaction may be limited by intragranular diffusion (solid state or pore diffusion through an ash layer surrounding each grain.) In the case of sulfur dioxide absorption by calcium oxide, the product of reaction (calcium sulfate) has a higher specific volume than the solid reactant. It follows that the ash layer surrounding each grain may indeed be dense and for this reaction it may be appropriate to include intragranular diffusion in the reaction model. However, this is perhaps best done as described in Chapter 4, where an allowance was made for both the interfacial reaction *and* intragranular diffusion.

Archer *et al.* [115] reported the effect of a number of variables on the performance of a bench-scale fluidized-bed combustor, and found that the retention of sulfur passed through a maximum at about 1550°F on increasing the bed temperature. This may be due to the lower surface area and porosity of limes produced by higher calcination temperatures [118].

Ehrlich and McCurdy [117] reported a maximum sulfur retention in the vicinity of 1500°F and presented a microprobe picture of a partly reacted lime particle showing the sulfur concentrated near the outer surface.

The results of some gravimetric measurements of the rates of absorption of SO_2 by lime are reported in Kearns *et al.* [116]. Rates of reaction increased slightly on increasing the temperature within the range 1350–1700°F, but the total capacity for sulfur fell, presumably due to more rapid blocking of pore mouths by sulfate at the higher temperatures. The rate was insensitive to total pressure in the range 1–10 atm but the capacity of the lime increased by a factor of two over this change in pressure. Solid reactivity increased with decreasing particle size. Precalcination in a CO_2 rich atmosphere resulted in a more reactive calcine.

In summary, the absorption of SO_2 by lime is a rather complex gas–solid reaction. Under conditions likely to be encountered in electric power generation the reaction would appear to be controlled by pore diffusion. Considerable complication is introduced by the fact that the pore structure of the lime will depend on the time–temperature history of the lime during (and perhaps subsequent to) calcination. Furthermore, reaction is accompanied by considerable changes in pore structure, which may cut short the reaction before all the lime has reacted [79, 119].

Perhaps the most promising regenerative dry process for the removal of sulfur dioxide from stack gases is the alkalized alumina process, which has undergone extensive development by the U.S. Bureau of Mines and others. Here the absorption step takes place at 300–350°C, which is a relatively low temperature compared with the lime absorption processes discussed previously. The absorption unit, perhaps a fluidized bed, would be located between the steam generation section of the boiler and the stack. The alumina provides a support rather akin to a catalyst support for the absorbent proper, which is Na_2O, at a level of about 20%. The absorption reaction is

$$2NaAlO_2 + SO_2 + \tfrac{1}{2}O_2 = NaSO_4 + Al_2O_3$$

Regeneration is carried out in a reducing atmosphere (using reformed natural gas or other suitable source) at a temperature of about 700°C.

$$Na_2SO_4 + Al_2O_3 + 3H_2 = 2NaAlO_2 + H_2S + 2H_2O$$

The hydrogen sulfide gas formed by regeneration can be converted to elemental sulfur by the well-known Claus process whose major reactions are

$$2H_2S + 3O_2 = 2H_2O + 2SO_2$$
$$SO_2 + 2H_2S = 2H_2O + 3S$$

The major difficulty with the alkalized alumina process appears to be the attrition of the solid as it is recycled from the absorber to the regenerator. A recent report from the Bureau of Mines [83] describes the evaluation of some solid oxides, including alkalized aluminas, for this process. All of the oxides were found to show adequate capacity and reactivity but were subject to excessive attrition, as evaluated by the investigators.

Krishnan and Bartlett [84] have recently published the results of their investigation of the kinetics of sulfation of reduced alunite. Alunite, $K_2SO_4 \cdot 3Al_2(SO_4)_3 \cdot 6H_2O$ occurs in vast deposits in the U.S. and can be converted to $K_2O \cdot 3Al_2O_3$ by reduction with hydrogen. The mixed oxide produced is akin to the alkalized alumina described above and is capable of being used in a regenerative scheme. Below 300°C absorption of SO₂ brought about sulfation of the K_2O component alone but above 300°C a mixed sulfate $KAl(SO_4)_2$ is formed. Below 300°C Krishnan and Bartlett found from microprobe studies that the absorbed sulfur was evenly distributed throughout their porous compacts, indicating that the reaction was under chemical control. This was confirmed by the fact that the rate of reaction between reduced alunite and SO₂ was independent of the size of the compact; Krishnan and Bartlett interpreted their results in terms of a model [85] somewhat similar to the grain model described in Chapter IV with the difference that a distribution of grain sizes, rather than a unique grain size, was used. For certain size distributions (e.g., a log-normal size distribution with a standard deviation of intermediate magnitude) the rate of reaction of the compact was calculated to be proportional to the fraction of solid unreacted, under conditions of chemical control. Krishnan and Bartlett calculated the rate of sulfation of the Al_2O_3 fraction of the reduced alunite above 300°C by subtracting the rate of sulfation of K_2O (calculated by extrapolation of the measured rate below 300°C) from the observed rate of sulfation. As with the sulfation of the K_2O part, below 300°C, the sulfation of the Al_2O_3 component was found to proceed at a rate proportional to the amount of Al_2O_3 remaining. Marked differences in the activation energy and in the order with respect to oxygen partial pressure were observed for the two reactions, however; water vapor was found to enhance sulfur dioxide absorption, particularly at low temperatures. From surface area measurements, Krishnan and Bartlett concluded that the absence of water vapor promoted the blocking of the pores.

A similar investigation was carried out by Krishnan and Bartlett on the kinetics of sulfation of alkalized alumina [86]. Both dry gases and gases containing up to 0.024 atm water vapor were used. Sodium sulfate was the primary reaction product in the temperature range 300–600°C but sodium sulfite could be detected by wet chemical analysis, and at 500°C and above, sodium aluminum sulfate was found to form at the later stages of sulfation. On sulfation under dry conditions, the rate was found to decline sharply after about 20 % of the Na_2O had been sulfated, whereas in the presence of water vapor reaction would proceed to 100 % sulfation of Na_2O at a gradually declining rate. The surface areas of samples sulfated under dry conditions were found to be much lower than those of samples sulfated in the presence of water vapor. Krishnan and Bartlett attributed the difficulty of achieving high levels of sulfation under dry conditions to the sealing of pore mouths by reaction products.

Reaction in the presence of water vapor was found to proceed at a rate independent of oxygen partial pressure and proportional to sulfur dioxide partial pressure. Furthermore, the rate was found to be proportional to the fraction of Na_2O unreacted, in line with Bartlett's findings on alunite and the mathematical model [113] mentioned previously. This proportionality between the rate of reaction and the fraction of Na_2O unreacted $(1 - F_{Na})$ may be deduced from the straight line observed when the fraction unreacted is plotted against time on a semilogarithmic scale as shown in Fig. 8.18.

Microprobe traverses of partly reacted and sectioned particles showed only small gradients in the sulfur level within the particle. The rate of reaction was only weakly dependent on the mesh size of the alkalized alumina particle in the range 8–100 mesh. From these two facts, Krishnan and Bartlett concluded that the reaction was proceeding under chemical control for the conditions of their experiments, despite the low apparent activation energy of 2 kcal/g-mole. Small amounts of nitric oxide were found to increase the rate significantly.

An alternative regeneration scheme is that under development by Showa Denko in Japan. A fluidized or entrainment reactor is used to react magnesia with the SO_2 in a stack gas at a relatively low temperature (150°C). Partly reacted magnesia is then regenerated in a second reactor heated to about 800°C yielding a gas stream with a sufficiently high SO_2 concentration for sulfuric acid manufacture or SO_2 liquefaction.

Another Japanese development is the process employing manganese oxide by Mitsubishi Heavy Industries. This is a dry absorption process utilizing an entrainment reactor at about 150°C. The solids are separated from the gas stream cyclone and an electrostatic precipitator and then subjected to a wet regeneration involving solution of the manganese sulfate in water and precipitation of a hydrated manganese oxide by ammonia and air, which can be recycled to the absorber. The process avoids the problems of absorbent attrition and decline in absorbent reactivity common in systems employing a dry regeneration. The disadvantage of manganese absorption systems is that soluble manganese compounds are produced that may be toxic.

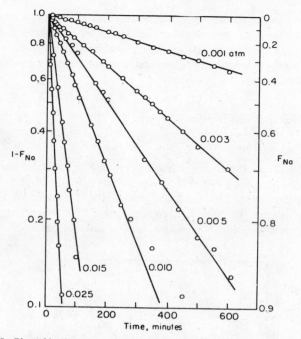

FIG. 8.18. Plot of $\log(1 - F_{Na})$ reacted against time for various concentrations of SO_2 in a study of the kinetics of sulfation of reduced alunite [86].

Finally, Van Hecke and Bartlett [87] investigated the reaction of manganese nodules (accretions of manganese and other naturally occurring oxides on the ocean floor) with atmospheres typical of stack gases, over the temperature range 300–500°C. The rate of reaction was found to be first order with respect to sulfur dioxide, was proportional to the fraction of solid still remaining to be sulfated, and was moreover substantially independent of particle size in the range 14–20 mesh to 100–170 mesh. Further evidence that sulfation of the crushed nodules, which are highly porous, was proceeding under chemical control was provided by microprobe traverses of sectioned partly reacted particles. The sulfur was found to be uniformly distributed through the particles. Part of the interest in the use of manganese nodules for absorption of sulfur dioxide from stack gases stems from the fact that they contain a few tenths of a percent of copper, nickel, and cobalt which may be rendered soluble by sulfation. Van Hecke and Bartlett were able to extract most of the copper, nickel, and cobalt from their sulfated nodules by boiling in water.

In closing this section it may be stated that the reaction of SO_2 with lime, magnesia, alkalized alumina, and other reagents is likely to generate a great deal of interest in the forthcoming years because of the environmental and economic importance of the processes that are based on these reactions. It is thought that both the theoretical and the experimental techniques described in the earlier portions of this text could be relevant to the planning of future studies of these systems.

8.5 Combustion and Gasification of Coal

PERSPECTIVES

Figure 8.19 shows the energy consumption in the United States together with projections beyond 1970. The United States, with about 6–7% of the world's population produces (and consumes) about a third of the world's annual energy output. In 1970 this required some 550 million tons of coal, 700 millions ton of oil, and 500 million tons of natural gas.

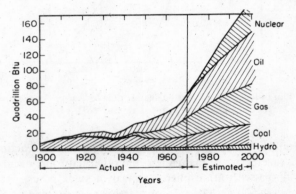

FIG. 8.19. U.S. energy consumption for this century. 1 quadrillion Btu = 965 billion H³ gas, 175 million 661 oil, or 36 million tons coal.

As seen in the figure, the total energy consumption in 1970 was about 70 quadrillion Btu $\simeq 70 \times 10^{15}$ Btu or $\simeq 28 \times 10^{15}$ kcal of which about 20% was being supplied by the combustion and gasification of coal. It is also observed that natural gas and oil appear to have attained a nearly dominant role in the energy supply during the past 30 years. However, the recent realization of the decline of oil and natural gas reserves in the U.S. has stimulated a great deal of interest in the development of technologies, through the use of which coal may be used as an energy source in place of oil and natural gas.

In the classification of gas–solid reaction systems both the combustion and the gasification (reaction to produce a combustible fuel) of coal fall in the category where there is no solid reaction product. The analytical treatment of these systems has been discussed in Chapter 4. For this reason the brief treatment to be presented in this section will be confined to the following topics:

(1) characterization and classification of coal,
(2) technologies available for the combustion of coal,
(3) technologies available for the gasification of coal.

It would be desirable to link the purely descriptive material presented in the following pages to the theoretical analyses that were given in the earlier chapters. However, at the present little work has been done (or is available in the open literature) on real coals (as opposed to kinetic studies on graphite)† to allow this. It is hoped that the intensive research and development work proceeding in the U.S. and also in Western Europe at the present will change this picture.

CHARACTERIZATION AND CLASSIFICATION OF COAL [89]

In the characterization of coal two types of analysis are usually used: "proximate analysis" and "ultimate analysis."

Proximate Analysis

This simple analysis consists of the determination of the *moisture, ash, volatile matter, fixed carbon content*, and possibly the calorific value of the coal.

The moisture content of the coal is usually determined in an "air dried" condition and the ash is a collective term for all the mineral matter, which usually consists of CaO, MgO, iron oxides, and some alkalies. The volatile matter of coal is defined as the weight loss of dry coal when heated to a "standard temperature" in the absence of air, for a "prescribed period of time,"

The solid residue left after deducting the moisture—the volatile matter and the ash—is termed the "fixed carbon." (It is to be noted that fixed carbon does *not* denote the total carbon content of the coal because some carbon is necessarily present in the volatile matter.) Thus the components of the proximate analysis have to add up to 100%.

It is to be stressed to the reader that, as its name implies, the proximate analysis is a relatively crude description of the composition of coal. It is worthwhile to remember this term, however, because it is used quite frequently both in coal technology and in the combustion of refuse.

† For an excellent review on the kinetics of graphite oxidations the reader is referred to Walker *et al.* [121].

368 8 GAS–SOLID REACTIONS OF INDUSTRIAL IMPORTANCE

Ultimate analysis of coal

By ultimate analysis we mean the actual elemental composition of coal, that is, the percentage of carbon, hydrogen, sulfur, carbon, and nitrogen. For the purpose of comparison the ultimate analysis is usually presented on an ash- and moisture-free basis.

Some comment may be made on the principal elements from which coal is constituted.

Carbon, Hydrogen, and Oxygen. These elements make up the true coal substance, and as shown subsequently, their proportion is a good guide to the properties of coal.

Nitrogen. Coals usually contain from about 0.5–2.0 wt % nitrogen; when coal is distilled or carbonized about 15% of the nitrogen is converted into ammonia together with small amounts of cyanogen and pyridine bases.

Sulfur. The sulfur content of coal ranges from about 0.5 to 5%. Sulfur is usually present in three forms:

(1) in combination with iron as pyrite, FeS_2,
(2) organic sulfur compounds,
(3) $CaSO_4$.

The sulfur content is an important, but undesirable property of coals. If the coal is to be used for the production of metallurgical coke, its sulfur content has to be limited because of the deleterious effect of sulfur on the properties of steel. The sulfur content of coal used as fuel is also undesirable because of the resultant SO_2 in the flue gases. At present, low-sulfur coals are at a premium and much research is being done on the desulfurization of coal.

The classification of coal

The properties of coal, as represented by its proximate and ultimate analysis, together with some other factors, form the basis of certain coal classification systems. As with most systems of classification, the subdivisions are somewhat arbitrary, but since these are widely used a brief mention of them should be made here. The following is a listing of the popularly recognized types of coal; some properties and analyses of these coals are also given in Tables 8.7–8.9.

Peat is not a coal, strictly speaking, but it represents the starting point in the coal series. On a dry, ash-free basis peat has a carbon content of about 60% with about 30% associated oxygen.

Lignites mark the transition of peat to coal and form the lowest rank of pure coals. The air-dried material has a moisture content of about 15 to 20%.

TABLE 8.7

Stages in the Transition of Coals

	Wood	Peat	Lignite		Bituminous[a]				Carbon-aceous	Anthracite
			Brown	Black	1	2	3	4		
Air-dried										
Moisture	20	20	18	15	10	3	1	1	1	1
Volatile matter, less moisture	—	50	47	41	35	34	32	30	11	8
Fixed carbon	—	27	28	32	45	58	62	64	84	88
Ash	0.5	3	7	12	10	5	5	5	.4	3
Btu/lb	6400	7700	9,900	10,200	10,700	13,900	14,300	14,400	15,000	15,000
Ash-free dry										
Carbon	50	60	67	74	77	84	85.6	87	92	94
Hydrogen	6.5	6	5.5	5.4	5	5	5	5.3	4	3
Oxygen	43	32	26	19	16	8	5.4	4.7	2	2
Sulfur and nitrogen	0.5	2	1.5	1.6	2	3	4	3	2	1
Btu/lb	8000	10,000	13,000	13,900	13,400	15,100	15,200	15,300	15,800	15,600

[a] 1, Lignitous, long-flame steam and house coal; 2, Para-bituminous, hard steam and house coal; 3, Para-bituminous, gas and coking coal; 4, Ortho-bituminous, coking coal (Durham).

TABLE 8.8

Parr's Classification of Coal

Class of coal	Volatile matter %		Calorific value (Btu/lb)		kcal/kg	
	Low	High	Low	High	Low	High
Anthracite	0	8	15,000	16,500	3,780	4,150
Semianthracite	8	12	15,000	16,000	3,780	4,150
Bituminous A	12	24	15,000	16,500	3,780	4,150
B	24	50	15,000	16,500	3,780	4,150
C	30	55	14,000	15,000	3,530	3,780
D	35	60	12,500	14,500	3,150	3,530
Lignite	35	60	11,000	12,500	2,770	3,150
Peat	55	80	9,000	11,000	2,260	2,770
Cannels	60	80	15,000	16,500	3,780	4,150

Dry, ash free lignites have a carbon content ranging from 60 to 75% together with an oxygen content ranging from 20 to 25%.

Sub-bituminous coals form a group between lignites and bituminous coals. This material has a high moisture and volatile content and *no coking power*. On a dry, ash-free basis the carbon content ranges from 75 to 83% while the oxygen is between 10 and 20%.

Bituminous coal is black and banded in appearance. The term bituminous covers a broad range of properties and this type of coal is usually prefixed with coking, medium, or noncoking. On a dry, ash-free basis the carbon content of bituminous coals is within the range 75–90% and the volatile matter ranges from 20 to 45%.

Semibituminous coal forms a group between bituminous coals and anthracite. The carbon content ranges from 90 to 93% and volatile matter is between about 10 and 20%, with oxygen ranging from 2 to 4%. Semibituminous coals have been used for steam raising (the term "steam coal") and those with high coking power find use in the production of metallurgical coke.

Anthracites form the highest rank of coal with a carbon content over 93%, less than 6% volatiles, and a zero coking power.

The reader may find this traditional, essentially qualitative classification of coal rather out of tune with the rest of the text, where stress has been placed on the quantitative characterization of solid reactants. The state of affairs currently persisting is due to two factors:

(a)   Coal is a very complex substance; the quantitative microscopic characterization is a very difficult task.

TABLE 8.9

Average Analyses of Coal in Cahaba Coal Field, Alabama[a,b]

Bed	Proximate analysis (%)				Ultimate analysis (%)					Btu	Ash softening temp. (°F)	Number of analyses averaged[c]
	Moisture	Volatile matter	Fixed carbon	Ash	Hydrogen	Carbon	Nitrogen	Oxygen	Sulfur %			
Bibb County												
Coke	2.7	34.7	57.4	5.2	5.3	79.7	1.4	7.1	1.3	14,160	2050	1
Clark	3.1	34.7	57.1	5.1	5.3	78.5	1.4	8.7	1.1	13,930	2260	4
Thompson	4.1	34.8	56.0	5.0	5.3	76.6	1.3	10.5	1.1	13,660	2140	2
Jefferson County												
Lower Nunnally	2.3	33.6	57.4	6.7	5.3	77.5	1.7	7.3	1.5	13,860	2160	1
Upper Nunnally	2.2	31.1	54.3	12.4	5.0	73.3	1.6	7.0	0.7	12,980	2430	1
Harkness	2.6	32.9	54.9	9.5	5.2	74.4	1.6	8.1	1.0	13,230	2350	2
Helena	2.0	34.3	56.3	7.4	5.3	77.1	1.7	8.1	0.4	13,660	2460	1
Shelby County												
Clark	2.3	35.0	54.3	8.2	5.1	75.9	1.4	8.6	0.6	13,580	2200	2
Gholson	4.0	34.7	58.2	3.1	5.5	79.3	1.5	10.0	0.7	14,150	2170	3
Thompson	2.8	29.9	54.8	12.4	5.0	71.4	1.2	8.6	1.2	12,620	2200	2
Helena	2.7	32.8	55.0	9.4	5.2	74.6	1.4	8.8	0.4	13,210	2480	2
Montevallo	2.9	37.9	51.9	7.2	5.3	75.4	1.0	10.3	0.7	13,490	2370	2
St. Clair County												
Harkness	2.2	33.7	55.9	8.2	5.1	75.3	1.7	7.5	2.1	13,520	2200	3
Clark	2.5	36.7	55.6	5.2	5.4	78.4	1.4	8.6	1.0	13,940	2230	1
Helena	4.7	33.4	56.9	5.0	5.5	76.4	1.6	10.6	0.9	13,610	1920	1

[a] Analyses by U.S. Bureau of Mines on as-received basis. Rank is high volatile A bituminous. Samples are from mine faces.

[b] After 1973 Keystone Coal Industry Manual.

[c] Each analysis is the averaged composite of three or more samples.

(b)   The actual amount of work done on coal characterization has not been very extensive during the past two decades.

This situation, however, is likely to change with the increasing amount of attention that is being paid at present to coal characterization and coal gasification.

In addition to the above-mentioned classification there exist numerous others including those adopted by the European Economic Community (EEC) and the British Coal Board. For an informative discussion on these the reader should consult Brame and King [89].

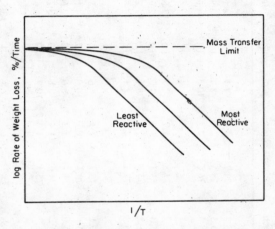

FIG. 8.20. Plot of rate of weight loss against reciprocal temperature for coal particles of identical size but different reactivity when exposed to carbon dioxide.

At this stage it may be instructive to introduce yet another classification of coal, namely, its *reactivity*. The reactivity of various coals is illustrated schematically in Fig. 8.20. It is seen that the more reactive the coal, the lower the temperature at which the transition occurs from chemical control to mass transfer control on the traditional Walker-type representation. In general, it may be stated that the reactivity of coal decreases with increasing rank; thus lignite would tend to be the most reactive whereas anthracite would be the least reactive, and certain pyrolitic graphites would be even less reactive.

It is to be stressed to the reader that even in this context reactivity is a qualitative concept; in general, one would determine the reactivity by evaluating the fractional reaction (i.e., weight loss) of a coal particle for certain "standard" conditions, such as temperature, pressure, and reactant gas composition. A great deal more work has to be done in this area to make the concept of coal reactivity quantitative.

Combustion of coal

In most industrial nations the coal mined is used largely for power generation and for the production of metallurgical coke. Figure 8.21 shows a plot of the fossil-fuel steam-electric generating capacity in the United States [90]. At present these fossil-fuel power generating stations, account for about 80% of the total electric power generated in the United States. Statistics also show that in 1970 some 62% of these power stations used coal as a fuel [90]. It follows that the combustion of coal is responsible for about half the electric power generated in the U.S.; this proportion is somewhat higher in some of the West European countries.

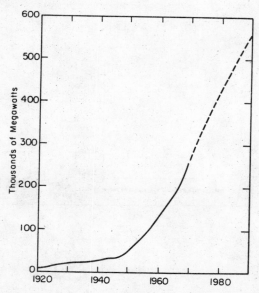

FIG. 8.21. Fossil-fuel electric generating capacity for the contiguous United States.

Modern power stations have a generating capacity of the order of 500–1100 MW; the coal-fired units generally use coal in a pulverized form. One such arrangement, the cyclone furnace, is illustrated in Figs. 8.22 and 8.23 [91]. In addition, a typical boiler layout is shown in Fig. 8.24.

For a good detailed discussion of steam generation through the use of pulverized coal, the reader is referred to [89] and [91]. The use of gas turbines, and more specifically *the combined cycle operation*, offers a potentially attractive alternative route to steam turbines for power generation. Such a combined cycle is shown in Fig. 8.25 where it is seen that the fuel is burned in a gas turbine which generates electric power; moreover, the exhaust gases from

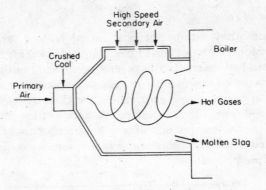

FIG. 8.22. A cyclone furnace.

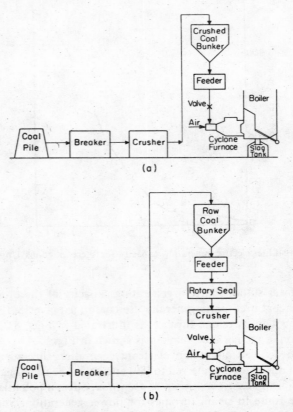

FIG. 8.23. Bin firing (a) and direct firing (b) systems for coal preparation and feeding to a cyclone furnace.

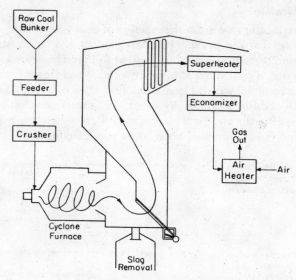

FIG. 8.24. Cyclone furnace and associated boiler.

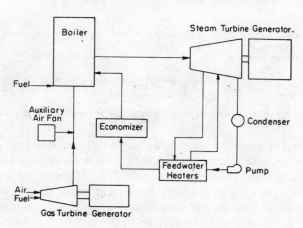

FIG. 8.25. Combined gas turbine–steam turbine cycle.

the gas turbine are fed to a boiler and the steam thus generated is used to drive a steam turbine. In principle, the use of such combined cycles offers a marked increase in overall thermal efficiency. An interesting discussion on the status of power generation through the use of gas turbines has been presented by Hedley [92]. For a comprehensive review of the kinetics of combustion of pulverized coal the reader is referred to [120].

Coal gasification

By coal gasification we mean the reaction of coal with air or oxygen under such conditions that the gaseous reaction product is a combustible fuel. In coal gasification three possible routes may be followed, as shown in Fig. 8.26.

It is seen that if coal is made to react with air and steam (at atmospheric pressures) the resultant gas will have a calorific value of about 120 to 150 Btu/ft³ (1150–1430 kcal/m³) principally because of the large amount of diluent nitrogen present in the air. Such gas with a low calorific value (e.g., producer gas) could find direct industrial application as a gaseous fuel.

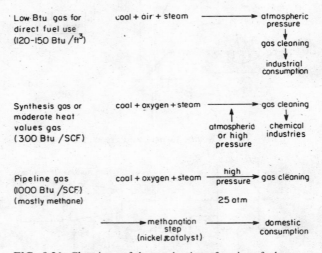

FIG. 8.26. Chemistry of the production of various fuel gases.

When coal is made to react in an oxygen–steam mixture the resultant gaseous product will consist primarily of carbon monoxide and hydrogen (with some CO_2) and will have a calorific value of about 300 Btu/ft³ (2850 kcal/m³). This gas may be used for subsequent processing by the chemical industry (synthesis gas) for the manufacture of ammonia, methanol, etc., or may also form a component of a synthetic natural gas for domestic consumption.

When gasification is carried out with steam and oxygen at high pressures some methane is produced in the gasification step. Additional methane may be generated *through a catalytic methanation step*. The end product is a gas, consisting largely of methane with a calorific value of about 1000 Btu/ft³ (9000 kcal/m³), which is termed *synthetic natural gas* or *pipeline quality gas*.

The principal reactions are the same in all these processes and are listed in the following:

(1) $C + O_2 = CO_2$ ($-$ 169,000   Btu/lb-mole, $-$94,000   kcal/kg-mole)

(2) $C + \frac{1}{2}O_2 = CO$ ($-$ 47,500   Btu/lb-mole   26,400   kcal/kg-mole)

(3) $C + 2H_2 \rightleftharpoons CH_4$ ($-$ 32,100   Btu/lb-mole, $-$17,890 kcal/kg-mole)

(4) $C + H_2O \rightleftharpoons CO + H_2$ ($+$ 75,400   Btu/lb-mole, $+$41,900 kcal/kg-mole)

(5) $C + CO_2 \rightleftharpoons 2CO$ ($+$ 74,200 Btu/lb-mole, $+$41,200 kcal/kg-mole)

(6) $CO + H_2O \rightleftharpoons CO_2 + H_2$ ($+$ 20,100 Btu/lb-mole, $+$11,200 kcal/kg-mole)

(7) $CO + 3H_2 \rightleftharpoons CH_4 + H_2O$ ($-$ 57,790 Btu/lb-mole, $-$38,270 kcal/kg-mole)

It is seen that reactions (1, 2, 3, and 7) are exothermic, whereas the remainder are endothermic. In practice the reactor or gasifier temperature is controlled by the amount of steam that is added. When oxygen rather than air is used, a larger amount of steam may be added, which results in a richer product gas. With regard to coal gasification processes, at present the principal interest is in the production of pipeline quality gas as a means of replacing natural gas as supplied to domestic consumers.

Excellent critical reviews of coal gasification technology are available in two recent reports, one by a select committee appointed by the National Academy of Engineering [93], and the other by Perry [94]. Both reports contain useful bibliographies.

While numerous processes have been proposed and are at the pilot plant stage, the two commercially available coal gasification processes for the production of synthesis gas or pipeline quality gas are the Lurgi process and the Koppers–Totzek process.

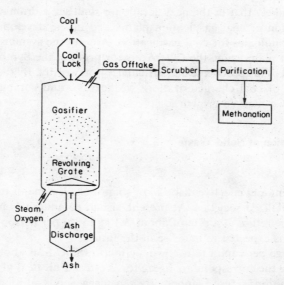

FIG. 8.27. The Lurgi gasifier.

A schematic diagram of the Lurgi process is shown in Fig. 8.27. The system operates at a pressure of about 50 atm, with a steam/oxygen ratio of about 1:3; the resultant gas, after treatment to remove the CO_2, has a calorific value of about 500 Btu/ft^3 (4650 kcal/m^3).

The Koppers–Totzek process, the general layout of which is shown in Fig. 8.28 [95], employs pulverized coal and operates just slightly above atmospheric pressure. The product gas, before purification for CO_2 removal, is stated to have a calorific value of about 270 Btu/ft^3 (2560 kcal/m^3).

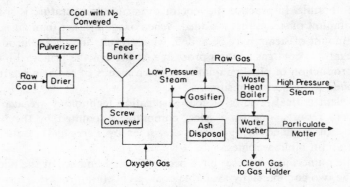

FIG. 8.28. The Koppers–Totzek gasifier.

It is very likely that in the next decade we shall see a dramatic growth in the construction of coal gasification plants and in the development of new improved technologies for coal gasification. Among the systems currently in the large pilot plant stage, the CO_2 Acceptor process, the Synthane process, and the IGT U-Gas process seem particularly noteworthy. By the same token the literature of coal characterization, gasification, and combustion is also likely to show a marked growth.

8.6   Incineration of Solid Waste

PERSPECTIVES

The efficient and effective disposal of solid waste is a major problem faced by all industrialized societies. At present, urban communities in the United States produce refuse at a rate of about 2.2 kg (5 lb) per person per day and this quantity is expected to increase in the future.

The average per capita refuse production in the U.S. is about 2 kg (4 lb) per day, while the corresponding figure for Western Europe is about 1 kg/day [96]. By definition, refuse is inherently a heterogeneous substance that even in a given locality may show quite drastic variations both with regard to composition and physical consistency.

Several methods have been used for the characterization of refuse. Some of these have been derived from the techniques employed for the characterization of solid fuels such as the proximate and ultimate analyses. The reader will recall from the previous section that by proximate analysis we mean the moisture, volatile, fixed carbon, and ash content of the refuse. The ultimate analysis gives the actual chemical composition of the refuse. Still another classification which is frequently employed provides a breakdown of the refuse components according to their origin, viz., glass, leather, paper, etc.

TABLE 8.10

"Typical" Refuse Composition[a]

Category	Weight % (as fired)[b]	Component	Weight %
Metal	8.7	Moisture (H_2O)	28.16
Paper	44.2	Carbon (C)	25.62
Plastics	1.2		
Leather and rubber	1.7	Oxygen (O)	21.21
Textiles	2.3	Hydrogen (H)	3.45
Wood	2.5	Sulfur (S)	0.10
Food waste	16.6	Nitrogen	0.64
Yard waste	12.6	Ash	20.82
Glass	8.5		
Miscellaneous	1.7		

[a] From Wilson [97].

[b] This weight distribution shows the effects of moisture transfer between the categories in the refuse during storage and handling. For example, the food waste tends to lose moisture and the paper absorbs moisture.

A "typical" refuse composition shown in Table 8.10 indicates that in the U.S., paper, yard waste, and food waste are the predominating refuse components. Table 8.11 shows the mean composition of municipal refuse in the U.S.; the standard deviations, also listed, indicate the variations that one might expect.

There are three basic methods available for the disposal of refuse: dumping, sanitary landfill, and incineration.

Dumping either onto designated areas or into the open sea (much of New York City's refuse is being handled this way at the present) is becoming less acceptable because of environmental considerations and because of the scar-

TABLE 8.11

Estimate of National Annual Average Composition of Municipal Refuse[a]

Component	Data[b] samples utilized	Mean wt %	Mean (100% total)	Standard deviation $S(X)$	Confidence limits (95%)
Glass	23	9.7	9.9	4.37	1.89
Metal	23	10.0	10.2	2.18	0.93
Paper	23	50.3	51.6	11.67	5.04
Plastics	9	1.4	1.4	.96	0.74
Leather, rubber	9	1.9	1.9	1.62	1.25
Textiles	17	2.6	2.7	1.80	0.93
Wood	22	2.9	3.0	2.39	1.06
Food wastes	23	18.8	19.3	10.95	4.73
		97.6			

[a] Excluding yard wastes and miscellaneous categories.
[b] Several data sets were not presented in a form suitable for extracting the weight fractions of all of the above refuse components.

city of land for this purpose, thus the use of sanitary landfills and incineration is likely to show appreciable growth in the coming years.

Problems relating to the disposal of solid wastes are quite well documented through the numerous publications sponsored by the Environmental Protection Agency. The volume edited by Wilson [97] provides a useful introduction to these problems.

In the brief treatment to be presented here we shall restrict our attention to incineration, because this is the only operation involving solid waste treatment that entails gas–solid reactions.

THE INCINERATION PROCESS

The primary function of incineration processes is to reduce the volume of the refuse and at the same time to effect its sterilization. This is accomplished by reacting (i.e., burning) the refuse with air or oxygen at temperatures ranging from 1000–1400°C (1800–2600°F) by which it is possible to achieve a reduction in volume which ranges from 85 to 90% for conventional incinerators to some 97% for the high-temperature slagging incinerators.

In assessing the combustion of solid wastes in incinerators it is of interest to compare the heating value and other properties of refuse with those of other fuels. Such a comparison is presented in Table 8.12, where it is seen that compared to coal, refuse is a low-density, low-calorific value fuel, which also has a rather high moisture and ash content. However, refuse appears to have a "respectable" heating value compared with peat or wood.

TABLE 8.12

Characteristic Properties of Refuse and Selected Fuels

Fuel	Moisture (%)	Ash (%)	Density (as fired) (lbm/ft³)	Lower heating value (Btu/lbm)	kcal/kg
Refuse[a]	28.00	20.00	17.00	5000	2800
Wood	46.9	1.5	60.5	3190	1770
Peat	64.3	10.0	25.0	2390	1330
Bituminous coal (high-volatile)	8.6	8.4	50.0	11,735	6300
Fuel oil (No 6)	1.5	0.08	61.5	17,280	9600

[a] The figures for refuse are average values for the U.S. and may vary by ±30%.

It is to be noted here that the calorific value of 2800 kcal/kg is typical for refuse in the U.S. Refuse in Western Europe has a calorific value of about 2000 kcal/kg [98] and the net heating value of refuse in Hong Kong is only about 1000–1200 kcal/kg because of its very high moisture content.

The recent energy shortages have focused on refuse as a potential energy source; indeed many of the larger incinerators in Europe and the more recent installations in the U.S. do have a provision for steam raising, which in turn may be used either for space heating or for power generation. However, this matter needs to be put into perspective. While the energy generated from the combustion of refuse could and should make an important contribution to the energy balance of a given local community, overall the total energy supply from refuse could at best supply only about 3% of the U.S. energy requirements.

In discussing the actual equipment for incineration we should distinguish between the conventional low-temperature systems that represent the vast majority of the installations in current use and some novel systems of which only prototypes or pilot plants exist at present.

Conventional incinerators have a capacity of some 400–1500 tons/day of refuse. In the majority of these units refuse is fed through a hopper onto a moving grate, where it is ignited and brought into contact with an air stream that is passed through the bed in an upward direction. The flue gases thus produced are then used for firing a boiler, or several boilers.

Figure 8.29 shows a typical modern installation in Dusseldorf, Germany and the general layout of a power plant fired by both refuse and pulverized coal is shown in Fig. 8.30.

Further information on the layout and mechanical features of conventional incinerators may be found in the proceedings of the Biennial National Incinerator Conferences, published by the American Society of Mechanical

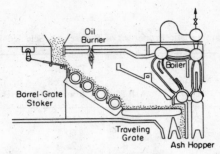

FIG. 8.29. A municipal incinerator in Dusseldorf, Germany. Garbage is first burned on barrel grates and then delivered onto a traveling-grate stoker.

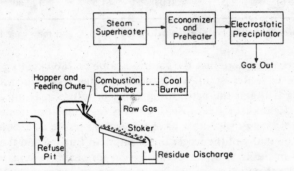

FIG. 8.30. Power plant in Munich using auxiliary burners to combine refuse incineration and pulverized coal burning.

Engineers. These proceedings also contain a wealth of information on the consistency and composition of urban waste and on many peripheral problems associated with incineration.

The newer concepts in incinerator development include high-temperature slagging incinerators, which operate at 1400–1650°C (2500–3000°F) and produce a liquid slag, which may be granulated. These slagging incinerators, examples of which include the ANDCO–Torrax system, the PUROX system using oxygen enrichment, and the FLK/Dravo system, can effect a substantial reduction in the bulk of the refuse—up to 3% of the volume.

Other less well developed but potentially attractive concepts in incineration involve fluidized-bed processing, the combustion of sorted and shredded refuse, and the recovery of high BTU fuels from refuse by pyrolysis.

MATHEMATICAL MODELING AND DESIGN OF INCINERATORS

Ideally one would expect that the modeling and design of incinerators would follow the pattern described in the preceding chapters, viz., kinetic data obtained on single particles would be translated into multiparticle situations as outlined in Chapter 7. However, the heterogeneity, the unavailability of kinetic information, and the ill-defined nature of refuse precludes such a procedure. At present the mathematical modeling of incinerators is at a very preliminary stage and design is usually done on a purely empirical basis. Moreover, the limited amount of work that has been done on the incineration or pyrolysis kinetics of the components of solid waste is not in a form that is immediately applicable for design purposes. In the present discussion we shall confine our attention to a brief review of the modeling work done and of some of the relevant kinetic studies.

MATHEMATICAL MODELING OF INCINERATORS

In concept, the movement of the reaction zone(s) in a moving-grate incinerator, such as depicted in Fig. 8.29 or 8.30, shows some similarity to the behavior of sinter beds described in Chapter 7.

A schematic representation of the various reaction zones is given in Fig. 8.31. Another schematic sketch of these zones showing their location within the actual physical system is given in Fig. 8.32. It is seen that with an underfiring arrangement the solid refuse upon being fed onto the grate undergoes drying, pyrolysis, ignition, and burnout. The solid residue (ash) is then discharged. The gaseous pyrolysis products (hydrogen, methane, aldehydes, tars)

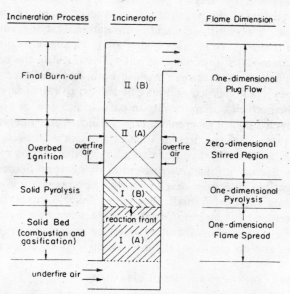

FIG. 8.31. Simplified incinerator model [100].

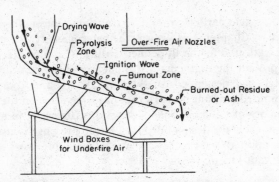

FIG. 8.32. Representation of processes occurring on an incinerator grate.

and some of the elutriated solids are burned above the bed upon being brought into contact with the overfire air.

Relatively little work has been published on the mathematical models describing the temperature and concentration profiles in incinerators. Perhaps the most successful approach to date has been that of Essenhigh and co-workers [99, 100], who through suggesting rather simplified models were able to provide an adequate representation of their pilot-scale measurements as illustrated in Fig. 8.33. A somewhat more basic approach was taken by Flanagan [101] and by Szekely [102] but at this time it is questionable whether the quality of the physical information available on these systems would warrant the sophistication and computational labor involved.

The point that must be stressed is that the modeling of incinerators poses much more complicated problems than those involved in the representation of ordinary packed-bed reactors, because both the solid feed composition and consistency are ill defined and may be subject to rather large variations. Because of the marked nonuniformity in particle size and consistency, channelling and bypassing are inherent features of incinerator operation.

We may summarize the present state of incinerator modeling by stating that there is a good qualitative understanding of the various reaction zones that exist in a conventional moving-grate incinerator. Some preliminary modeling equations have been developed, which do provide a good means for interpreting data obtained in small, pilot-scale units. No comprehensive model has been developed for incinerators that could serve as a basis for design. At present the design of incinerators is based essentially on heat balance considerations and on specifying the mechanical features of the system [103].

KINETIC DATA ON THE PYROLYSIS OF CELLULOSE

Since paper is one of the predominant combustible components of urban solid waste, the kinetics and the mechanism of the combustion of cellulose

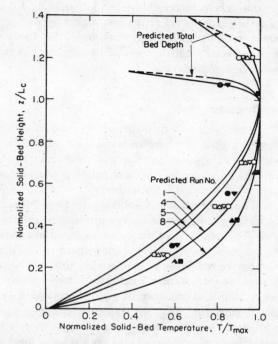

FIG. 8.33. Comparison of predicted and experimental temperature profiles through a packed-bed incinerator [100]. Experimental runs: ○, 1; ▽, 2; △, 3; □, 4; ●, 5; ▼, 6; ▲, 7; ■, 8.

are of particular interest. While some useful work has been done in this field, much of the information available at present is essentially qualitative and not immediately useful for mathematical modeling or design purposes.

Investigators have generally agreed that combustion of cellulose occurs in multiple stages; initially drying and pyrolysis takes place, which involves the generation of the residual structure and the reaction of the vapors formed with the excess air. Useful work in this area has been done by Bamford *et al.* [104], Weatherford and Sheppard [105], Squire [106], and others [107, 108].

It has been suggested [107–109] that the volatilization of moisture and endothermic pyrolysis usually precedes the exothermic decomposition of cellulose; thus the endothermic reactions occur at lower temperatures than the exothermic reactions.

The actual combustion of cellulosic materials has been studied by Andersen [110] and by Essenhigh [111]. Both these authors proposed quantitative kinetic expressions that were shown to be consistent with their measurements.

For a good comprehensive review on the kinetic data for the decomposition of cellulose the reader is referred to an article by Roberts [112].

We feel, however, that much more work could be done in the area of pyrolysis kinetics, especially with a view of developing kinetic models ultimately suitable for design purposes.

8.7 Concluding Remarks

The specific examples that formed the subject matter of this final chapter were presented with a dual objective in mind, namely both to illustrate the economic importance of some of the key gas–solid reaction systems and to provide a list of the key references for these reaction systems for readers who may wish to have more detailed information.

The reader will have seen that gas–solid reactions do indeed play a major role in the materials-processing and energy-producing technologies of most industrial nations. The quality and the quantity of information available on specific systems ranges from extensive studies involving in-depth investigations, exemplified by the reduction of iron oxides, to the very meager, almost nonexistent data on the incineration kinetics of urban wastes.

The obvious economic importance of gas–solid reaction systems and the inevitable pressures stemming from the shortages of energy and of resources for metals production together with pressures associated with environmental problems, make gas–solid reactions an attractive field for further research. A word of caution may, however, be appropriate here. The ultimate objective of engineering research into gas–solid reaction systems must be the development of novel or at least improved processes through the better quantitive understanding of these operations. In this regard some areas may appear to be more fruitful for future work than others.

Iron oxide reduction is reasonably well understood and here the major challenge seems to be the application of the knowledge gained on single-particle behavior to possible process improvement in the large-scale units, such as the iron blast furnace or direct reduction processes.

The thermodynamics of the roasting of "simple sulfides" is well understood and experience has shown that the overall rates of the roasting reactions in industrial-scale equipment tend to be controlled by heat transfer considerations, rather than by chemical kinetics. It would seem therefore that further detailed studies on the roasting kinetics of simple sulfides might not be of a very high priority. However, the selective oxidation of complex sulfide ores, segregation roasting, and the selective reaction of solid mixtures could be a very attractive field for further work. The distributed models, or some derivatives thereof, described in earlier chapters could find very good use for the interpretation of such experimental work.

The absorption of SO_2 by limestone or by some porous alkaline substance that may be regenerated is of paramount technical importance, because SO_2 emissions from power stations using solid fuel constitute a major environmental problem. The development of suitable absorbents together with appropriate contacting arrangements and possible regeneration schemes is a task that requires a level of sophistication similar to that of catalyst development. As indicated by some of the publications cited, a very promising start has been made in this area through the application of distributed (grain) models and the use of sophisticated techniques for the characterization of the solid reactants.

The shortages of oil and natural gas in the United States have stimulated a great deal if interest in coal gasification. While the rates of many coal gasification processes will tend to be mass transfer (and possibly heat transfer) controlled, it is thought that there is much scope for the application of the concepts of gas–solid reactions outlined in this text to many coal-processing operations, such as the hydrosulfurization of coal and coal hydrogenation. Because coal is a very complex substance this is likely to be a difficult task, but a very fruitful area for further research.

Finally, the recovery of heating values and possibly certain chemicals from solid urban wastes is a problem being forced upon us by both environmental and economic considerations. The combustion and pyrolysis kinetics of solid wastes provide one of the major challenges facing research workers in the gas–solid reaction field.

References

1. "Commodity Data Summaries, 1975," p. 83. U.S. Bureau of Mines.
2. J. Strassburger (ed.), "Blast Furnace—Theory and Practice," p. 439. Gordon and Breach, New York, 1969.
3. F. A. Berczynski, In "Blast Furnace Technology," (J. Szekely, ed.). p. 345. Dekker, New York, 1972.
4. R. Wild, Chem. Process Eng. London, 55 (February 1969).
5. J. R. Miller, J. Metals 25, 52 (October, 1973).
6. J. A. Astier, J. Metals 25, 62 (March, 1973).
7. Proc. South Africa Int. Ferroalloys Congr. (April 1974).
8. L. von Bogdandy and H.-J. Engell, "The Reduction of Iron Ores." Springer-Verlag, Berlin and New York, 1971.
9. L. von Bogdandy and W. Janke, Z. Elektrochem. Ber. Bunsenges. Phys. Chem. 61, 1146 (1957).
10. M. Wiberg, Iernkontorets Ann. 124, 172 (1940).
11. H. K. Kohl and H.-J. Engell, Arch. Wiss. 34, 411 (1963).
12. J. M. Quets, M. E. Wadsworth, and J. R. Lewis, Trans. Met. Soc. AIME 218, 545 (1960).
13. W.-K. Lu and G. Bitsianes, Can. Met. Quart. 7, 3 (1968).

14. W. M. McKewan, *Trans. Met. Soc. AIME* **218**, 2 (1960); **221**, 140 (1961); and **224**, 2 (1962).

15. R. H. Spitzer, F. S. Manning, and W. O. Philbrook, *Trans. Met. Soc. AIME* **236**, 726 (1966).

16. R. G. Olsson and W. M. McKewan, *Trans. Met. Soc. AIME* **236**, 531 (1966); *Met. Trans.* **1**, 1507 (1970).

17. B. B. L. Seth and H. V. Ross, *Trans. Met. Soc. AIME* **233**, 180 (1965).

18. T. L. Joseph, *Trans. AIME*, **120**, 72 (1936).

19. A. E. El-Mehairy, *J. Iron Steel Inst. London* **179**, 219 (1955).

20. G. D. McAdam, R. E. A. Dall, and T. Marshall, *N.A. Journal of Sci.* **12**, 649 (1969).

21. E. T. Turkdogan, and J. V. Vinters, *Met. Trans.* **2**, 3175 (1971); **3**, 1329 (1972).

22. E. T. Turkdogan, R. G. Olsson, and J. V. Vinters, *Met Trans.* **2**, 3189 (1971).

23. R. H. Tien and E. T. Turkdogan, *Met. Trans.* **3**, 2039 (1972).

24. P. K. Strangway and H. V. Ross, *Trans. Met. Soc. AIME* **242**, 1981 (1968).

25. M. Ishida and C. Y. Wen, *AIChE J.* **14**, 311 (1968).

26. N. A. Warner, *Trans. Met. Soc. AIME* **230**, 163 (1964).

27. W. A. Edmiston and R. E. Grace, *Trans. Met. Soc. AIME* **236**, 1547 (1966).

28. P. F. J. Landler and K. L. Komarek, *Trans. Met. Soc. AIME* **236**, 138 (1966).

29. R. D. Walker and D. L. Carpenter, *J. Iron Steel Inst. London*, 67 (1970).

30. N. J. Themelis and W. H. Gauvin, *AIChE J.* **8**, 437 (1962).

31. Minerals Yearbook, Vol. II, U.S. Dept. of Interior, U.S. Bur. of Mines Publ.

32. T. R. Ingraham, Sulfate stability and thermodynamic phase diagrams with particular reference to roasting, "Applications of Fundamental Thermodynamics to Metallurgical Processes," p. 187. Gordon and Breach, New York, 1967.

33. H. H. Kellogg and S. K. Basu, *Trans. Met. Soc. AIME* **218**, 70–81 (1960).

34. T. Rosenqvist, "Principles of Extractive Metallurgy." McGraw-Hill, New York, 1974.

35. N. B. Gray, M. R. Harvey, and G. M. Willis, Roasting of sulphides in theory and practice, *Proc. Richardson Conf. Phys. Chem. Process Metallurgy Imperial College*, London (1973).

36. T. R. Ingraham and H. H. Kellogg, Thermodynamic properties of zinc sulfate, zinc basic sulfate and the system Zn–S–O, *Trans. Amer. Inst. Min. Eng.* **227**, 1419–1426 (1963).

37. O. Kubaschewski, The thermodynamic properties of double oxides (A review), *High-Temp.-High Pressures* **4**, 1–12 (1972).

38. R. L. Benner and H. Kenworthy, The thermodynamic properties of the $ZnO–Fe_2O_3–Fe_3O_4$ system at elevated temperatures. 1. The thermodynamic properties as related to spinel structure. U.S. Bureau of Mines, Rep. of Investigations 6754, pp. 1–44 (1966).

39. J. N. Ong, Jr. M. E. Wadsworth, and W. M. Fassell, Jr., Kinetic study of the oxidation of sphalerite, *Trans. Amer. Inst. Min. Eng.* **206**, 257–263 (1956).

40. K. J. Cannon and K. G. Denbigh, Studies on gas–solid reactions, 1. The oxidation rate of zinc sulphide. *Chem. Eng. Sci.* **6**, 135–154 (1957).

41. K. G. Denbigh and G. S. G. Beveridge, The oxidation of zinc sulphide spheres in an air stream, *Trans. Inst. Chem. Eng.* **40**, 23–24 (1962).

42. K. Natesan and W. O. Philbrook, Oxidation kinetic studies of zinc sulfide pellets. *Trans. Amer. Inst. Min. Eng.* **245**, 2243–2250 (1969).

43. J. Gerlach and W. Stichel, Kinetics of the oxidising roasting of zinc sulphide, *Z. Erzbergbau Metallhuetteniwes.* **17**(8), 427–433 (1964); *Chem. Abstr.* **63**, 13930h (1964).

44. E. Mendoza, R. E. Cunningham, and J. J. Ronco, Oxidation of zinc sulfide pellets. Application of a model of diffusion with simultaneous reaction under effective diffusivity and surface area profiles, *J. Catal.* **17**, 277–286 (1970).

45. K. Natesan and W. O. Philbrook. Oxidation kinetic studies of zinc sulphide in a fluidized bed reactor, *Met. Trans.* **1**, 1353–1360 (1970).
46. M. M. Rao and K. P. Abraham. Roasting of zinc sulfide and mixed zinc sulfide–iron sulfide compacts, *Indian J. Technol.* **3** (9), 291–293 (1965); *Chem. Abstr.* **64**, 346d (1966).
47. A. W. Sommer and H. H. Kellogg, Oxidation of sphalerite by sulfur trioxide. *Trans. Amer. Inst. Min. Eng.* **215**, 742–744 (1959).
48. M. Watanabe and T. Yoshida, Roasting sulfide ores. II. Roasting of zinc sulfide in an atmosphere with high partial pressure of sulfur dioxide. *Tohoku Daigaku Senko Seiren Kenkyusho Iho* **18** (2), 131–140 (1962); *Chem. Abstr.* **62**, 8734g (1965).
49. T. R. Ingraham and P. Marier, Kinetics of the thermal decomposition of $ZnSO_4$ and ZnO. $2ZnSO_4$ at low temperatures, *Can. Met. Quart.* **6**, 249–261 (1967).
50. B. K. Dhindaw and J. C. Sircar, Kinetics and mechanism of sulfation of zinc oxide, *Trans. Amer. Inst. Min. Eng.* **242**, 1761 (1968).
51. S. E. Woods and C. F. Harris, Heat transfer in sinter roasting, *Symp. Chem. Eng. Met. Ind. (Inst. Chem. Eng.)* 77–86 (1963).
52. G. S. G. Beveridge, The prediction of reaction zone propagation rates in beds of reacting solids—the effect of moisture content in the solids, *Symp. Chem. Eng. Met. Ind. (Inst. Chem. Eng.)* 87–104 (1963).
53. H. H. Kellogg, Equilibrium considerations in the roasting of metallic sulfides. *Trans. Amer. Inst. Min. Eng.* **206**, 1105–1111 (1956).
54. E. H. Roseboom, Jr., An investigation of the system Cu–S and some natural copper sulfides between 25° and 700°C, *Econ. Geol.* **61**, 641–672 (1966).
55. T. R. Ingraham, Thermodynamics of the thermal decomposition of cupric sulfate and cupric oxysulfate, *Trans. Amer. Inst. Min. Eng.* **233**, 359–363 (1965).
56. A. Espelund, A Literature Survey of the System Fe–Cu–O at Oxidizing Conditions, pp. 1–10 Univ. of Trondheim (1970).
57. J. M. Floyd and G. M. Willis, Phase relations and oxygen dissociation pressures in the system Cu–Fe–O. *Rec. Prog. Res. Chem. Extract. Met. (Aus I.M.M.)* 61–64 (1967).
58. R. F. Blanks, Sulphate–oxide equilibria, Ph.D. thesis, Univ. of Melbourne (1961).
59. C. L. McCabe and J. A. Morgan, Mechanism of sulfate formation during the roasting of cuprous sulfide, *J. Metals* **8**, 800A (1959).
60. M. E. Wadsworth, K. L. Leiter, W. H. Porter, and J. R. Lewis, Sulfating of cuprous sulfide and cuprous oxide. *Trans. Amer. Inst. Min. Eng.* **218**, 519–525 (1960).
61. P. G. Thornhill and L. M. Pidgeon, A microscopic study of sulfide roasting, *J. Metals* **9**, 989–995 (1957).
62. I. D. Shah and S. E. Khalafalla, Chemical reactions in the roasting of copper sulfides, U.S. Bur. Mines. Rep. Invest. 7549, pp. 1–21 (1970).
63. I. D. Shah and S. E. Khalafalla, Kinetics and mechanism of the conversion of covellite (CuS) to digenite ($Cu_{1.8}S$). *Met. Trans.* **2**, 2637–2643 (1971).
64. T. A. Henderson, Oxidation of powder compacts of copper–iron sulfides, *Bull. Inst. Min. Met.* **620**, 497–520 (1958).
65. V. V. Rao and K. P. Abraham, Kinetics of oxidation of copper sulfide, *Met. Trans.* **2**, 2463–2470 (1971).
66. M. G. Hocking and C. B. Alcock, The kinetics and mechanism of formation of sulfates on cuprous oxide, *Trans. Amer. Inst. Min. Eng.* **236**, 635–642 (1966).
67. J. D. Esdaile, Equilibria in the lead–oxygen–sulphur system. *Rec. Progr. Res. Chem. Extract. Met. (Aust. I.M.M.)* 65–80 (1967).
68. E. Rosen and L. Witting, Thermodynamic studies of high temperature equilibria. IV. Experimental study of the equilibrium reaction $5PbO + \frac{1}{2}S_{2(g)} + \frac{3}{2}O_{2(g)} \rightarrow 4PbO + PbSO_{4(s)}$, *Acta Chem. Scand.* **26**, 2427–2432 (1972).

69. J. S. Anderson, The primary reactions in roasting and reduction processes, *Discuss. Faraday Soc.* **4**, 163–173 (1948).

70. J. Doughty, K. Lark-Howovitz, L. M. Roth, and B. Shapiro, The structure of lead sulfide films, *Phys. Rev.* **79**, 203 (1950).

71. H. Wilman, The structure of photosensitive lead sulphide and lead selenide deposits and the effect of sensitization by oxygen, *Proc. Phys. Soc.* **60**(2), 117–132 (1948).

72. N. B. Gray, N. W. Stump, W. S. Boundy, and R. V. Culver, The sulfation of lead sulfide, *Trans. Amer. Inst. Min. Eng.* **239**, 1835–1840 (1967).

73. A. V. Slack, SO_2 Removal from Waste Gases, Noyes Develop. Corp. (1971).

74. G. A. Mills and H. Perry, *Chem. Tech.* **3**, 53 (1973).

75. I. A. Raben, Statue of technology of the commercially offered lime and limestone flue gas desulfurization systems, Presented at the *EPA Flue Gas Desulfurization Symp., New Orleans, Louisana* (May, 1973).

76. J. K. Burchard, G. T. Rochelle, W. R. Schofield, and J. O. Smith, Some general economic considerations of flue gas scrubbing for utilities. Internal Rep., EPA, Contr. Syst. Div.

77. R. W. Bartlett and N. H. Huang, *J. Metals* **25**, 28 (December 1973).

78. W. T. Reid, *Trans. ASME Ser. A* **92**, 11 (1970).

79. R. H. Borgwardt and R. D. Harvey, *Environm. Sci. Tech.* **6**, 350 (1972).

80. R. L. Pigford and G. Sliger, *Ind. Chem. Eng. Process Design Develop.* **12**, 85 (1973).

81. R. H. Borgwardt, *Environ. Sci. Tech.* **4**, 59–63 (1970).

82. R. W. Coutant *et al.*, Investigation of the reactivity of limestone and dolomite for capturing SO_2 from the flue gas, Battelle Memorial Inst., Columbus, Ohio (November 20, 1970).

83. J. H. Russel *et al.*, Bureau of Mines, R.I., 7582 (1971).

84. N. G. Krishnan and R. W. Bartlett, *Environ. Sci. Tech.* **7**, 923 (1973).

85. R. W. Bartlett *et al.*, *Chem. Eng. Sci.* **28**, 2179 (1973).

86. N. G. Krishnan and R. W. Bartlett, *Atmos. Environ.* **7**, 575 (1973).

87. M. C. Van Hecke and R. W. Bartlett, *Met. Trans.* **4**, 941 (1973).

88. J. L. Reed, *Chem. Eng. Progr.* (November 72, 1973).

89. J. S. S. Brame and J. G. King, "Fuel," 6th ed., Chapter 5. St. Martin's Press, New York.

90. "The 1970 National Power Survey, The Federal Power Commission," Part 1. U.S. Gov. Printing Office, Washington, D.C., 1971.

91. "Steam, Its Generation and Use." Babcock and Wilcox, New York, 1963.

92. W. H. Hadley, Effect of gas turbine efficiency and fuel cost on the cost of producing electric power, Contract 68-02-1320. U.S. Environ. Protection Agency to Monstanto Res. Corp., Dayton Lab., Dayton Ohio, 1972.

93. Evaluation of Coal Gasification Technology, Part I, Pipeline Quality Gas. Ad Hoc Panel on the Evaluation of Coal Gasification Technol. of Nat. Acad. Eng., Washington D.C. (1972). •

94. Evaluation of Coal Gasification Technology, Part II, Low Btu Gas. Ad Hoc Panel on the Evaluation of Coal Gasification Technol. of NAE, Washington, D.C., 1974.

95. J. W. Farnsworth *et al.*, K-T: Koppers commercially proven cool gasifier. Presented to the Ass. of Iron and Steel Eng. Convention, Philadelphia, Pennsylvania (1974).

96. Municipal Incineration Review of the Literature, U.S. Environ. Protection Agency (June 1971).

97. D. D. Wilson (ed.), "Urban Solid Waste," p. 100. Technomic Publ., Westport, Connecticut, 1972.

98. P. J. Scott and J. R. Holmes, *Proc. Incinerator Conf.*, p. 115. Amer. Soc. Mech. Eng., New York (1974).

99. R. H. Essenhigh and T. J. Kuo, *Proc. Nat. Incinerator Conf.*, p. 261. Amer. Soc. Mech. Eng., New York (1970).

100. R. H. Essenhigh, *Proc. Nat. Incinerator Conf.* p. 87. Amer. Soc. Mech., Eng., New York (1968).

101. J. E. Flanagan and G. A. Hosack, *Proc. Nat. Incinerator Conf.* p. 65. Amer. Soc. Mech. Eng., New York (1970).

102. J. Szekely and J. H. Chen, *Proc. Int. Symp. Dechema, Frankfurt* (1973).

103. R. C. Corey (ed.), "Principles and Practice of Incineration." Wiley, New York, 1969.

104. C. H. Bamford, J. Crank, and P. H. Malan, *Proc. Cambridge Phil. Soc.* **42**, 166 (1946).

105. W. D. Weatherford and D. M. Sheppard, *Int. Symp. Combust. 10th*, 897 (1965).

106. W. Squire, *Combust. Flame* **7**, 1 (1963).

107. S. Martin, *Int. Symp. Combust. 10th*, 877 (1965).

108. P. L. Blackshear and K. A. Murty, *Symp. Int. Combust. 10th*, 911 (1965).

109. E. R. Tinney, *Int. Symp. Combust. 10th*, 925 (1965).

110. W. H. Andersen, *Combust. Sci. Tech.* **2**, 213 (1970).

111. R. H. Essenhigh, M. Kuwate, and J. P. Stumbar, *Int. Symp. Combust. 12th*, 663 (1969).

112. A. F. Roberts, *Combust. Flame* **14**, 261 (1970).

113. R. W. Bartlett, N. G. Krishnan, and M. C. Van Hecke, *Chem. Eng. Sci.* **28**, 2179 (1973).

114. G. J. Vogel *et al.*, NTIS number PB-277 058 (1973).

115. D. H. Archer *et al.*, NTIS number PB-212 806 (1971).

116. D. L. Kearns *et al.*, NTIS number PB-231 162 (1972).

117. S. Ehrlich and W. A. McCurdy, *Intersoc. Energy Conversion Conf.*, *9th, San Francisco, California* (1974).

118. E. T. Turkdogan *et al.*, *Trans. Soc. Min. Eng. AIME* **254**, 9 (1974).

119. M. Hartman and R. W. Coughlin, *Ind. Eng. Chem. Process Design Develop.* **13**, 248 (1974).

120. M. F. R. Mulcahy and I. W. Smith, *Rev. Pure Appl. Chem.* **19**, 81 (1969).

121. P. L. Walker *et al.*, *Advan. Catal.* **11**, 133 (1959).

Suggested Reading

Transport Phenomena

"Transport Phenomena," R. B. Bird, W. E. Stewart, and E. N. Lightfoot, Wiley, New York, 1960.

"Rate Phenomena in Process Metallurgy," J. Szekely and N. J. Themelis, Wiley, New York, 1971.

"Transport Phenomena in Metallurgy," G. H. Geiger and D. R. Poirier, Addison-Wesley, Reading, Massachusetts, 1973.

"Mass Transfer," T. K. Sherwood, R. L. Pigford, C. R. Wilke, McGraw-Hill, New York, 1975.

"Molecular Theory of Gases and Liquids," J. O. Hirschfelder, C. F. Curtiss and R. B. Bird, Wiley, New York, 1956.

"Mass Transfer in Heterogeneous Catalysis," C. Satterfield, M.I.T. Press, Cambridge, Massachusetts, 1970.

Experimental Techniques

"Property Measurements at High Temperatures," W. D. Kingery, Wiley, New York, 1959.

"Physicochemical Measurements at High Temperatures," J. O.'M. Brockris, J. L. White, and J. D. Mackenzie (eds.), Butterworths, London and Washington, D.C., 1959.

"Particle Size Measurement," 2nd ed., T. Allen, Halsted Press (Chapman and Hall), London, 1975.

Numerical Methods

"Numerical Methods for Partial Differential Equations," W. F. Ames, Barnes and Noble, New York, 1969.

"Quasilinearization and Invariant Imbedding," E. S. Lee, Academic Press, New York, 1968.

"Digital Computation for Chemical Engineers," L. Lapidus, McGraw-Hill, New York, 1962.

"Numerical Methods for Digital Computers," Vols. I and II, A. Ralston and H. F. Wilf (eds.), Wiley, New York, 1960.

"Methods for the Numerical Solution of Partial Differential Equations," D. V. von Rosenberg, Amer. Elsevier, New York, 1969.

General

"Blast Furnace Theory and Practice," Vols. I and II, J. H. Strassburger (ed.), Gordon and Breach, New York, 1969.

"High Temperature Chemical Reaction Engineering," F. Roberts, R. F. Taylor, and T. R. Jenkins (eds.), Inst. Chem. Eng., London, 1971.

"Fluidization Engineering," D. Kunii and O. Levenspiel, Wiley, New York, 1969.

"Fluidization," J. F. Davidson and D. Harrison (eds.), Academic Press, New York, 1971.

Reaction Kinetics

"Heterogeneous Kinetics at Elevated Temperatures," G. R. Belton and W. L. Worrell (eds.), Plenum Press, New York, 1970.

"Kinetics of Chemical Processes," M. Boudart, Prentice-Hall, Englewood Cliffs, New Jersey, 1968.

Proc. Int. Symp. Reactivity of Solids, 6th, J. W. Mitchell (ed.), Wiley, New York, 1969.

"Introduction to the Principles of Heterogeneous Catalysis," J. M. Thomas and W. J. Thomas, Academic Press, New York, 1967.

"Chemical Reaction Analysis," E. E. Peterson, Prentice-Hall, Englewood Cliffs, New Jersey, 1965.

"Chemical Reaction Engineering," 2nd ed., O. Levenspiel, Wiley, New York, 1972.

Index

395